IFIP Advances in Information and Communication Technology 411

IFIP – The International Federation for Information Processing

IFIP was founded in 1960 under the auspices of UNESCO, following the First World Computer Congress held in Paris the previous year. An umbrella organization for societies working in information processing, IFIP's aim is two-fold: to support information processing within its member countries and to encourage technology transfer to developing nations. As its mission statement clearly states,

> IFIP's mission is to be the leading, truly international, apolitical organization which encourages and assists in the development, exploitation and application of information technology for the benefit of all people.

IFIP is a non-profitmaking organization, run almost solely by 2500 volunteers. It operates through a number of technical committees, which organize events and publications. IFIP's events range from an international congress to local seminars, but the most important are:

- The IFIP World Computer Congress, held every second year;
- Open conferences;
- Working conferences.

The flagship event is the IFIP World Computer Congress, at which both invited and contributed papers are presented. Contributed papers are rigorously refereed and the rejection rate is high.

As with the Congress, participation in the open conferences is open to all and papers may be invited or submitted. Again, submitted papers are stringently refereed.

The working conferences are structured differently. They are usually run by a working group and attendance is small and by invitation only. Their purpose is to create an atmosphere conducive to innovation and development. Refereeing is also rigorous and papers are subjected to extensive group discussion.

Publications arising from IFIP events vary. The papers presented at the IFIP World Computer Congress and at open conferences are published as conference proceedings, while the results of the working conferences are often published as collections of selected and edited papers.

Any national society whose primary activity is about information processing may apply to become a full member of IFIP, although full membership is restricted to one society per country. Full members are entitled to vote at the annual General Assembly, National societies preferring a less committed involvement may apply for associate or corresponding membership. Associate members enjoy the same benefits as full members, but without voting rights. Corresponding members are not represented in IFIP bodies. Affiliated membership is open to non-national societies, and individual and honorary membership schemes are also offered.

George L. Kovács   Detlef Kochan (Eds.)

Digital Product and Process Development Systems

IFIP TC 5 International Conference, NEW PROLAMAT 2013
Dresden, Germany, October 10-11, 2013
Proceedings

Springer

Volume Editors

George L. Kovács
Technical University Budapest
and Hungarian Academy of Sciences
Computer and Automation Institute
Kende u. 13-17, 1111 Budapest, Hungary
E-mail: gkovacs@sztaki.hu

Detlef Kochan
HTW Dresden
Zentrum für Angewandte Forschung und Technologie e.V.
Friedrich-List-Platz 1, 01069 Dresden, Germany
E-mail: detlef.kochan@zaft.htw-dresden.de

ISSN 1868-4238
e-ISSN 1868-422X
ISBN 978-3-642-41328-5
e-ISBN 978-3-642-41329-2
DOI 10.1007/978-3-642-41329-2
Springer Heidelberg New York Dordrecht London

Library of Congress Control Number: 2013949272

CR Subject Classification (1998): J.6, J.1, J.7, H.4, I.6, K.4, I.2

Typesetting: Camera-ready by author, data conversion by Scientific Publishing Services, Chennai, India

Printed on acid-free paper

Springer is part of Springer Science+Business Media (www.springer.com)

Preface

The triennial PROLAMAT conferences focus on computer applications in manufacturing and already have a respectable history. The first conference was in Rome, Italy in 1969 and the last one in Shanghai, China, in 2006. They addressed the issue of programming languages for machine tools, and created the acronym PROLAMAT. While the acronym has not changed, the topic has shifted considerably, due to the fast paradigm changes in the past years. The last conferences have addressed advances in CAM, software for discrete manufacturing, human aspects of CIM and globalization and virtual enterprises; etc. Finally TC5 of IFIP, which has traditionally been the main organizer of PROLAMAT, as flagship conference of TC5, however most WGs organize their regular meetings as well, decided to change the name a little and agreed to use the name NEW PROLAMAT to show and emphasize the changes and the continuity at the same time. It was already defined for the Shanghai PROLAMAT that it should cover all possible aspects of computerized design and production and operation of products and production systems from old style traditional to sophisticated great real ones through virtual and mixed ones. Today we can add that all aspects are taken into account during the total life span of the product and production system life-cycle of real, virtual or mixed products as well. As there are distinctions between systems and virtual and extended systems in the production level, we make distinctions between simple and extended products and between tangible and intangible features of products. In our recent sense everything can be understood as a product, including a production system, a car or any part of a clock, a leaflet, a service station or an advertisement, etc.

In order to demonstrate this new orientation towards IT-Innovations the label PROLAMAT is now interpreted as "Project Research On Leading-edge Applications and Methods for Applied Information Technology", closely related to industrial process chains. Knowledge, knowledge acquisition, management, distribution and applications are main issues of all up-to-date tools and means, which are taken into account in the research and applications of computers and networks for manufacturing automation and robotics are included in the topics of New PROLAMAT.

One of the best attended PROLAMAT conferences took place in 1988 in the city of Dresden, well known as a center of mechanical engineering, mechatronics, and computer sciences. It is an honor for the organizers to welcome again the PROLAMAT conference participants to the NEW PROLAMAT 2013 in Dresden.

The effective mastering and further developments of industrial process chains require continuous improvements by means of product and process innovations.

The scope of the NEW PROLAMAT covers engineering decision making and automated process realization, which are closely related to powerful and sophisticated software support. At the beginning of the development of computer application for automated manufacturing, tasks as major pioneer work were realized mostly by university or academic research institutions together with large industrial companies.

Nowadays the situation has totally changed. Over the last decades an industry for software and manufacturing automation has established itself in the market. This actual situation will be taken into consideration at the NEW PROLAMAT 2013.

Because of the entire dynamic development in all engineering and research-fields, knowledge and experience exchange related to highly developed software is of utmost importance.

International competition requires innovations along the entire process chain. Product development, design and manufacturing of high-quality products has to be constantly improved in order to allow for realization in shortest times at reasonable costs as well as reduced energy and resources consumption.

The essential applications are realized in the automotive and aerospace industry; machine and plant engineering; consumer goods and others.

The automotive industry in Germany and throughout Europe is noticed to be the pathfinder in designing and implementing innovative and integrated process chains. This particular market constantly sets new standards and represents the reference points for both established and cutting-edge technology. Under the general framework of complete PLM strategies some of the most essential working fields will be focused on at the NEW PROLAMAT 2013

This book contains about 40 peer reviewed and carefully selected papers, which were presented at the meeting and were sent to the volume editors in due time. The authors represent several European and some overseas countries, and industry and academia, SMEs and multinational firms both sent their experts to distribute their knowledge and opinion, and to discuss differences and suggested ways to go.

The main topics of the papers are:

- Introduction, industrial challenges
- Digital Product- and Process Development
- Challenges for Digital Product and Process Design Systems
- IT support for Product and Process Development and Future Perspective
- Additive Manufacturing – Fast Prototyping
- Quality Management
- Standardization and Knowledge Development - Knowledge Management
- Modeling and Simulation of Procedures and Processes
- Industrial success stories, case studies,

On behalf of the International Program Committee and of the National Organizing Committee we express our gratitude to everybody, who assisted in preparing and running such an excellent international conference in Dresden. We underline only a few of them: IFIP TC5 people, who supported our efforts, people, who assisted in local management, or in editing the papers (most editing was done by Adam Kisari of SZTAKI), and finally editors of Springer Publishers, who prepared this beautiful book.

August 2013

Detlef Kochan
George Kovács

Organization

Organized and sponsored by

IFIP (International Federation of Information Processing)

Netzwerk Automobilzulieferer Sachsen (AMZ)
AMZ is for 13 years now the central network of the automotive industry, its suppliers and companies in research, development and engineering in Saxony. The central idea of the AMZ network is to initiate and support product and technology development for sustainable growth of Saxon companies in the automotive supply industry.

Main focus of AMZ are powertrain, body in white, interior, vehicle safety and future technologies like electric and hydrogen power.

Different research groups and innovative events of the AMZ network such as NEW PROLAMAT are focused in the individual support of Saxon automobile supplying companies, for example ergonomic work place, process management, financing the growths and insurance, innovative human management as well as training and education.

The network AMZ is managed by the "RKW Sachsen GmbH Service and Consulting" that offers individual service and professional development for economic enterprises and public establishments. RKW is partner of the nationwide network of RKW Association in Germany.

RKW Sachsen
Wege für die Wirtschaft

MTA SZTAKI (Computer and Automation Research Institute, Hungarian Academy of Sciences, Budapest)

International Program Committee

Chair

George L. Kovács, Hungary

Co-chairs

Alain Bernard, France
Chris McMahon, UK
Detlef Kochan, Germany

Members

H. Afsarmanesh, The Netherlands
C. Avram, Australia
A. Bernard, France
P. Bernus, Australia
P. Bertok, Australia
U. Bititci, Scotland
X. Boucher, France
A. Bouras, France
L.M. Camarinha Matos, Portugal
G. Cascini, Italy
D. Cavallucci, France
S. Cho, USA
B.K. Choi, Korea
G.O. Codina, Spain
T. Czachorski, Poland
P. Dallemagne, Switzerland
Rattan K. Datta, India
R. Denzer, Germany
G. Doumeingt, France
D. Dutta, USA
R. A. Earnshaw, United Kingdom
P. Falster, Denmark
Man-Wook Han, Austria
S.P. Frysinger, USA
R. Gan, China
B. Gurumoorthy, India
G. Haidegger, Hungary
Man-Wook Han, Austria
G. Halevi, Israel
G. Henning, Argentina
J. Hřebíček, Czech Republic
P. Johnson, Sweden
S. Kassel, Germany
E.F. Kimura, Japan
D. Kiritsis, Switzerland
T. Kjellberg, Sweden
D. Kochan, Germany
G.L. Kovacs, Hungary
L. Kutvonen, Finland
L. Camarinha. Matos, Portugal
M.B. McGrath, USA
I. Mljadenovic, Serbia
K. Mertins, Germany
J. Muhlbacher, Austria
E. Neuhold, Austria
G.J. Olling, USA
K.R. Pardasani, India
H.J. Pels, The Netherlands
R.J. Rabelo, Brazil
N.L. Rovira, Mexico
T. Runhue, China
G. Schimak, Austria
R. Smeds, Finland
V. Stich, Germany
D. Swayne, Canada
M. Taisch, Italy
S. Umeda, Japan
D. Velev, Bulgaria
J. Vlietstra, USA
K. Wang, Norway
R. Waxman, USA
R.H. Weston, UK
M. Wozny, USA
X. Xu, China
N. Yussupova, Russia

National Program Committee

Chair

Detlef Kochan, Germany

Members

Prof. R. Anderl, TU Darmstadt
Prof. J. Bliedtner, FH Jena
Prof. E. Beyer, TU Dresden
Prof. J. Dietrich, HTW Dresden
Dr. R. Dumitrescu, IPT Paderborn
Prof. G. Eckert, HTW Dresden
Prof. K. Grote, TU Magdeburg, VDI
Prof. K. Kabitzsch, TU Dresden
Prof. A. Katzenbach, Daimler AG Stuttgart
Prof. F. Flocke, RWTH Aachen / Fraunhofer IPT
Prof. W. Krug, Dualis, VDI Dresden
Dr. B. Machner, BMW Group
PD. Dr. A. Nestler, TU Dresden
Prof. S. Nowotny, FhG IWS
Prof. C. Otto, HTW Dresden
Dr. A. Richter, EXAPT Aachen
Dr. J. Rohde, SAP AG
Dr. C. Schöne, TU Dresden
Prof. C. Schumann, FH Zwickau
Dr. R. Seidler Schwindt, CAD/CAM Technologie
Prof. R. Stark, IPK Fraunhofer Berlin
Prof. K.D. Thoben, Uni Bremen
Prof. S. Vajna, Uni Magdeburg
Prof. G. Witt, TU Duisburg/Essen, VDI
Prof. T. Wiedemann, VDI, HTW Dresden
Prof. M. Zäh, TU München
Prof. D. Zühlke, DFKI Kaiserslautern

Table of Contents

Keynote and Plenary Papers - State of the Art, Challenges, Future

Manufacturing Innovation and Horizon 2020 1
Erastos Filos

IT Support for Product and Process Development in Japan and Future Perspective 11
Fumihiko Kimura

3D Experiences – Dassault Systèmes 3DS Strategy to Support New Processes in Product Development and Early Customer Involvement ... 24
Andreas Barth

Challenges for Digital Product and Process Design Systems at BMW ... 31
Dietmar Trippner, Stefan Rude, and Andreas Schreiber

Industry 4.0: A Best Practice Project of the Automotive Industry 36
Franz E. Gruber

JT Format (ISO 14306) and AP 242 (ISO 10303): The Step to the Next Generation Collaborative Product Creation 41
Alfred Katzenbach, Sebastian Handschuh, and Steven Vettermann

Digital Product- and Process- Development

Economic Advantages by CAD/CAM Use in Compound with Production Data Organization 53
Arndt Richter

A 'Lean' Fuzzy Rule to Speed-Up a Taylor-Made Warehouse Management Process 61
Lapo Chirici and Kesheng Wang

Improving the Usability of Collaboration Methods and Technologies in Engineering 73
Lars Wolter, Haygazun Hayka, and Rainer Stark

Towards Product Avatars Representing Middle-of-Life Information for Improving Design, Development and Manufacturing Processes 85
Karl Hribernik, Thorsten Wuest, and Klaus-Dieter Thoben

Challenges and Solutions in Customer Process Based on Collaborative Engineering 97
Sergej Bondar, Leo Potjewijd, and Josip Stjepandic

Additive Manufacturing

Additive Manufacturing as Integral Part of the Digital Solution Process - An Industrial Short Note 108
Klaus Müller-Lohmeier

Additive Manufacturing – A Growing Possibility to Lighten the Burden of Spare Parts Supply 112
Steve Rommel and Andreas Fischer

Laser Beam Melting for Tooling Applications – New Perspectives for Resource-Efficient Metal Forming and Die Casting Processes 124
Bernhard Müller, Ralf Hund, Roland Malek, and Nancy Gerth

Design-Opportunities and Limitations on Additive Manufacturing Determined by a Suitable Test-Specimen 138
Karlheinz P.J. Hoeren and Gerd Witt

Mechanical Properties of Laser Beam Melting Components Depending on Various Process Errors 153
Stefan Kleszczynski, Joschka zur Jacobsmühlen, Jan T. Sehrt, and Gerd Witt

New Fiber Matrix Process with 3D Fiber Printer – A Strategic In-process Integration of Endless Fibers Using Fused Deposition Modeling (FDM) 167
Andreas Fischer, Steve Rommel, and Thomas Bauernhansl

From Rapid Prototyping to eManufacturing 176
Nikolai Zaepernick

Energy Turnaround: Printing of Thermoelectric Generators 181
Ines Dani, Aljoscha Roch, Lukas Stepien, Christoph Leyens, Moritz Greifzu, and Marian von Lukowicz

Generative Manufacturing and Repair of Metal Parts through Direct Laser Deposition Using Wire Material 185
Steffen Nowotny, Sebastian Thieme, David Albert, Frank Kubisch, Robert Kager, and Christoph Leyens

Quality Management

Capp Model for Prismatic Parts in Digital Manufacturing 190
Vidosav D. Majstorović, Tatjana V. Šibalija, Bojan M. Erčević, and Marko M. Erčević

Product Quality Inspection Combining with Structure Light System, Data Mining and RFID Technology 205
Kesheng Wang and Quan Yu

Additional Methods to Analyze Computer Tomography Data for Medical Purposes and Generatively Produced Technical Components 221
Philipp Sembdner, Stefan Holtzhausen, Christine Schöne, and Ralph Stelzer

Correction of Highlight Line Structures 230
György Gyurecz and Gábor Renner

Re-make of Sheet Metal Parts of End of Life Vehicles - Research on Product Life-Cycle Management 239
George L. Kovács and Imre Paniti

Standardization and Knowledge Management/Developments

Enabling the Crowd Sourcing of Very Large Product Models 254
Martin Hardwick, David Loffredo, Joe Fritz, and Mikael Hedlind

Knowledge Retrieval in Complex Information Landscapes Related to Products and Production 273
Marcus Schichtel

Advanced Engineering Visualization with Standardized 3D Formats 286
Christian Emmer, Arnulf Fröhlich, and Josip Stjepandic

Scientific Automation Rises the Productivity of Production Facilities 296
U. Frank, H. Anacker, and Dirk Bielawny

Prospective Evaluation of Assembly Work Content and Costs in Series Production 310
Ralf Kretschmer, Stefan Rulhoff, and Josip Stjepandic

Model-Based Approach for Self-correcting Strategy Design for Manufacturing of Small Metal Parts 320
Mikhail Borzykh, U. Damerow, Christian Henke, Ansgar Trächtler, and W. Homberg

Digital Eco-factory as an IT Support Tool for Sustainable Manufacturing 330
Michiko Matsuda and Fumihiko Kimura

Enterprises Monitoring for Crisis Preventing Based on Knowledge Engineering 343
Nafisa I. Yusupova, Gyuzel R. Shakhmametova, and Elmira Kh. Dusalina

The Knowledge on the Basis of Fact Analysis in Business Intelligence 354
Alexander Vokhmintsev and Andrey Melnikov

Simulation of Procedures and Processes

Composite Simulation as Example of Industry Experience 364
Joachim Bauer

Simulation and Optimization Based Flexible Job Scheduling of Powder Coating Lines 369
Heike Wilson and Reik Zettl

Event Based Identification and Prediction of Congestions in Manufacturing Plants 376
Clemens Schwenke, Thomas Wagner, and Klaus Kabitzsch

Intelligent Manufacturing Operations Planning, Scheduling and Dispatching on the Basis of Virtual Machine Tools 391
Gerald Rehage, Frank Bauer, Jürgen Gausemeier, Benjamin Jurke, and Peter Pruschek

Methodological Issues in Support of Selected Tasks of the Virtual Manufacturing Planning 401
Marius Eßers, Martin Erler, and Andreas Nestler

A Comprehensive Framework for the Computer-Aided Planning and Optimisation of Manufacturing Processes for Functional Graded Components 413
Marcus Petersen and Jürgen Gausemeier

As-Is Analysis and Efficient 3D Layout Planning during the Conversion of Old and Design of New Production Facilities by Means of 3D-CAD 423
Björn Schuster

Author Index 427

Manufacturing Innovation and Horizon 2020

Erastos Filos

European Commission, DG Research and Innovation, Brussels, Belgium
erastos.filos@ec.europa.eu

Abstract. A current trend in crisis-laden Europe is de-industrialisation. The European Union has launched ambitious policies and proposed a new research and innovation framework programme, Horizon 2020, to counter this trend and to foster innovation in manufacturing. The paper highlights the key issues related to this upcoming programme. R&D and innovation efforts, however, need to be underpinned by smart policies that in particular link up relevant R&D and innovation strategies at regional, national and European level and offer incentives for growth and differentiation as well as for leveraging private investment.

Keywords: crisis, innovation, R&D, investment.

1 Introduction

Manufacturing is the activity to make goods, usually on a large scale, through processes involving raw materials, components, or assemblies with different operations divided among different workers. Manufacturing encompasses equipment for materials handling and quality control and typically includes extensive engineering activity such as product and system design, modelling and simulation, as well as tools for planning, monitoring, control, automation and simulation of processes and factories. It is increasingly seen as a priority area of economic activity especially for economies that have been hit by the recent financial and economic crisis.

The paper aims to identify issues pertaining to manufacturing innovation in the framework of Horizon 2020. Smart policies will be required to link up R&D and innovation strategies at regional, national and European levels and to offer needed incentives for growth and differentiation as well as to leverage investments towards the ambitious goal of countering Europe's de-industrialisation.

2 Addressing the Challenges of Europe's Manufacturing

The challenges for manufacturing in Europe can be summarised as in fig. 1:

G.L. Kovács and D. Kochan (Eds.): NEW PROLAMAT 2013, IFIP AICT 411, pp. 1–10, 2013.

Strengths

- European mechanical engineering (machines, equipment & systems industry)
- Industrial software (embedded software including factory automation + robotics, ERP + analytics, PLM software)
- High & flexible automation equipment enabling mass customisation
- Sustainability as a driver

Weaknesses

- Despite strengths in industrial ICT, Europe's ICT industry is globally insignificant & fragmented
- Lack of large integrators (e.g. OEMs offering system-level products rather than components)
- Incoherent trade/competition/industrial policies

Opportunities

- Concerns over climate change & nuclear risks drive demand for sustainability-conscious European technology
- Growing wealth of global consumers drives demand for customized quality products
- Manufacturing increasingly popular with policymakers
- Local capabilities enhanced by 3D printing, photonics-based equipment & methods, micro-nano- & bio-based materials, personal robotics, ICT infrastructures (e.g. Clouds)

Threats

- De-industrialisation: Manufacturing/engineering unpopular with large parts of society, outsourcing trends, low investment in R&D&I
- Lack of support for exporting SMEs
- Lack of incentives to skills development in STEM

Fig. 1. SWOT analysis of European manufacturing

Manufacturing accounts for around 16% of Europe's GDP. It remains a key driver of R&D[1], innovation, productivity growth[2], job creation and exports. 80% of innovations are made by industry. 75% of EU exports are in manufactured products. Each job in manufacturing generates two jobs in services. Since the beginning of the financial and economic crisis, however, EU employment in manufacturing has fallen by 11%. The recent moderation of the recovery has kept employment levels in manufacturing low and the perspectives of a fast rebound remain rather weak. Over the last decade the

[1] 80% of private investment.

[2] 3% on average vis-à-vis 1% of the economy between 2000-2007.

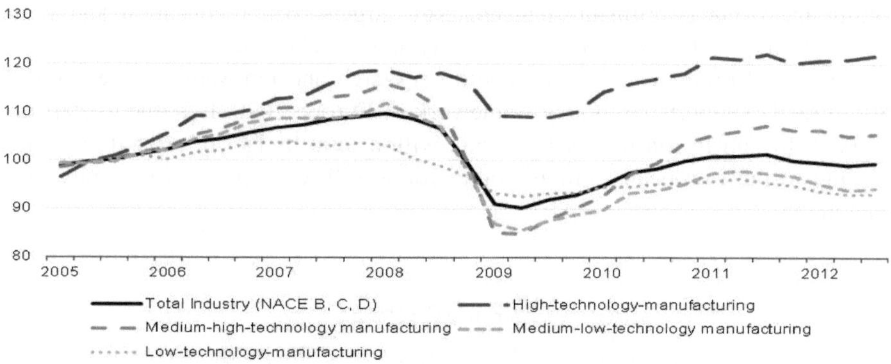

Fig. 2. Index of production for total industry and main technology groups in manufacturing, EU27, 2005-2012, seasonally adjusted (2005=100) - Source: Eurostat

share of manufacturing in EU employment and value-added has been declining. In 2011 it accounted for over 14% of employment and over 15% of total value added, representing 23% and 22% respectively, in the non-financial business economy.

As shown in fig. 2, the crisis has had severe consequences over manufacturing in Europe. However, high-technology manufacturing[3] has performed better than the rest of the industry.

Recent econometric research[4] points to the fact that a country's or a region's capacity to export products only few others can make is based on an accumulation of manufacturing knowledge and capability leading to its competitive advantage over others.

A recent World Economic Forum report on "The Future of Manufacturing"[5] highlights the key role of enabling technologies and infrastructures in enabling "manufacturing to flourish and contribute to job growth". As these are growing in importance and sophistication it becomes challenging to develop and maintain them.

At the beginning of this decade the European Union put forward Europe 2020, a ten-year growth strategy that aims at more than just overcoming the crisis that continues to afflict many of Europe's economies. It aims to address the shortcomings of the current growth model by supporting "smart, sustainable and inclusive growth" along five objectives[6]: more employment, more investment in R&D and innovation,

---

3 High-technology manufacturing comprises aerospace, pharmaceuticals, computers, office machinery, electronics/communications and scientific instruments sectors with relative high R&D expenditure on value added.

4 See "Complexity and the Wealth of Nations" in: `http://harvardmagazine.com/print/26610?page=all` and "The Art of Economic Complexity" in: `http://www.nytimes.com/interactive/2011/05/15/magazine/art-of-economic-complexity.html?_r=0`

5 "The Future of Manufacturing. Opportunities to Drive Economic Growth", World Economic Forum Report, April 2012; "Special Report on Manufacturing and Innovation", The Economist, 19 April 2012; "Why Manufacturing Still Matters", New York Times, 10 February 2012; "Future Factories", Financial Times, 11 June 2012.

6 Europe 2020 strategy, see relevant documents under: `http://ec.europa.eu/europe2020/index_en.htm`

attaining the triple-20 climate change/energy targets, better education and higher skills levels and finally by eliminating poverty and social exclusion.

The next EU Framework Programme for research and innovation, called Horizon 2020[7], directly supports the policy framework of Europe 2020. It is due to start on 1 January 2014 and to run over seven years with a total budget of around 80 billion, according to the European Commission's proposal. The proposal is currently under debate in the

Fig. 3. The Horizon 2020 Framework Programme

European Parliament and the Council of the European Union. The proposed structure consists of basic priorities related to: (1) excellent science; (2) industrial leadership; (3) societal challenges (see fig. 3).

Compared to previous EU Framework Programmes Horizon 2020 will bring together research and innovation in a single programme, it will focus on multidisciplinary societal challenges European citizens face and it will aim to simplify the participation of companies, universities, and other institutes in all EU countries and beyond.

The programme's activities related to manufacturing are mostly concentrated in the activity area called "Leadership in enabling and industrial technologies". Key activities will focus on roadmap-based research and innovation involving relevant industry and academic research stakeholders.

In 2012 the Commission put forward an update of its Industrial Policy Communication[8] which supports a further 2020 objective which aims to reverse Europe's de-industrialisation: Industry to account for 20% of the EU GDP by 2020.

---

[7] Horizon 2020: The EU Framework Programme for research and innovation, see relevant documents under: `http://ec.europa.eu/research/horizon2020`

[8] "A stronger European industry for growth and economic recovery", COM(2012) 582 final.

3 Towards Manufacturing Innovation

3.1 The Factories of the Future PPP

The Factories of the Future initiative was launched as a public-private partnership (PPP) in 2009, constituting a EUR 1.2 bn part of the European Economic Recovery Plan[9].

In total 98 projects were launched after three calls for proposals[10] and 52 projects in the final call of 2012, currently under grant agreement negotiations. The projects represent research and technology development and innovation related activities that cover the full spectrum of manufacturing, from the processing of raw materials to the delivery of manufactured products to customers, across many sectors, covering both large-volume and small-scale production, dealing with matters such as supply chain configurations, virtual factories, material processing and handling, programming and planning, customer-driven design, energy efficiency, emissions reductions, new processing technologies, new materials, upgrading of existing machines and technologies, including many projects involving elements of, or wholly focused on, various uses of Information and Communication Technologies (ICT).

3.2 High Performance Manufacturing

There are also a diverse range of projects that are clustered under the theme of High Performance Manufacturing. Projects here focus on: production of high precision plastic parts; development of photonic device production capabilities; development of precision glass moulding processes; enhancing tool-making technologies for high-precision micro-forming; development of new manufacturing routes for micro- and nano-scale feature manufacturing; zero defect manufacturing; control of milling processes for thin-walled work-pieces; production of grapheme, etc. There are also projects investigating additive manufacturing and robotics for maintenance tasks applied to large scale structures. Consideration is also being given in some projects to modular reconfigurable production systems and the development of capabilities to customise products and to put in place the necessary manufacturing capabilities needed for customisation.

Impacts for this theme consist of improved dynamics for cutting processes, higher precision, and improved reliability during changing process conditions. But also evident are impacts in terms of lowering commissioning and ramp-up times, and

---

9 A European Economic Recovery Plan, COM(2008) 800 final, 26 November 2008, electronically available under: `http://eur-lex.europa.eu/LexUriServ/LexUriServ.do?uri=COM:2008:0800:FIN:en:PDF`

10 Interim Assessment of the Research PPPs in the European Economic Recovery Plan, Brussels, 2011, electronically available: `http://ec.europa.eu/research/industrial_technologies/pdf/research-ppps-interim-assessment_en.pdf`

avoiding investments in new production machinery through enhancing re-usability of manufacturing technologies and systems, and extending the life of machinery through add-ons and upgrades. Some developments are evidently also enabling, in the sense that once the processes and associated technologies are developed they will enable those who purchase them to engage in their own innovation (additive manufacturing is one example of this).

3.3 Sustainable Manufacturing

Under the theme Sustainable Manufacturing, projects, also diverse in character and focus, are dealing with matters that have relevance to creating a more environmentally responsible approach to manufacturing. Here there are projects that are also addressing customisation issues, with one example being customised clothing that allows the customer to choose more environmentally benign materials. There are also other projects dealing with customisation with an eco-dimension, delivering environmental assessment tools while also providing web-based access to support the customisation process. Work is also being undertaken on the development of the supporting manufacturing and supply chain infrastructure necessary for a shift away from large volume standardised products, to small scale production of customised products. Other projects are working on topics such as: environmental footprint reduction for metal formed products; resource efficient manufacturing systems; waste energy recovery; eco-efficient firing processes; use of predictive maintenance to achieve optimal energy use; condition and energy consumption monitoring; harmonisation of product, process and factory lifecycle management, etc.

For projects with a sustainability focus, impacts tend to be with respect to reductions in costs, energy consumption, emissions, and material wastage, but also in terms of increased availability of processes and machines and improved plant efficiencies. Also evident are impacts such as better understanding of customer needs, reduced time to market, reductions in delivery times, reduced transportation costs, and more time and cost effective customisation. Also, improved decision-making as a result of more in-depth understanding of environmental impacts is an impact. Potentially, the capability to produce customised products cost-effectively on a small scale could result in a shift away from large expensive centralised production facilities (with their associated significant transportation-driven carbon emissions) to more local production closer to the point of sale.

3.4 ICT for Manufacturing

In the domain of ICT there are projects that focus on platforms, which can be considered as the hardware, system architectures and software, necessary to undertake a range of related tasks. Typically these platform projects are focused on specific interests such as: manufacturing information and knowledge management; supply chain configuration; creation of virtual factories assembled from multiple independent factories; and collaborative engineering. But there are also ICT projects addressing simulation, game-based training, support for end-of-life material recovery and re-manufacturing, advanced robotics, energy efficiency monitoring, laser welding, etc.

Impacts by these projects include reduced costs, accelerated product and process engineering, greater flexibility, improved quality, better equipment availability, reduced use of consumables, lower energy consumption, etc. ICT is also in many of these projects more than just a technology, but also an enabling technology in that it has the potential to bring about changes in method, procedures, or to open up new possibilities. Some of the targeted impacts in this respect include developing specific software targeted at SMEs to take account of their specific constraints (for example, limited time and expertise). Another enabling impact is greater flexibility, both in terms of capabilities of specific machines and processes, but also in the capability to redeploy these in new configurations (factory layouts), to meet changing demands. In some of the ICT projects it is also possible to see consideration being given to add-ons and upgrades to existing machines and systems, which will help to improve the performance of existing equipment and to extend its useful life.

3.5 Lessons from Four Years of Operation of the PPP

In a report from a recent workshop[11] involving all Factories of the Future projects launched under the first three years of the PPP's operation, the following observations were made.

Overall the Factories of the Future PPP has been a good initiative to generate a family of industry-related projects. Industry participation has increased to over 50 % (30 % SME participation). The PPP also helps to create a broader spectrum of research projects, addressing higher levels of technology readiness than basic research projects. For SMEs, who need solutions designed to fit their constraints (time, money, expertise) the PPP proves to be very valuable if sufficient consideration is given to projects that address these constraints.

Academia and research institutions participate in these projects, and for them there is an added value of being able to provide students with research projects driven by a clear industrial need, and also to provide contacts for these students with industry personnel and facilities. However, greater emphasis needs to be given to the research training potential of those PPP projects, to maximise the benefits for young researchers that come from working with industry. Development of manufacturing oriented Living Labs may have a role to play here in enhancing industry-academia interactions.

Projects address a wide range of technologies and issues relevant to enhancing the competitiveness of European manufacturing, and many potential impacts are evident in these projects. The challenge is how to turn impacts into tangible business benefits in the marketplace, which requires more focus on business and market issues. Projects

---

[11] Impact of the Factories of the Future Public Private Partnership, Workshop Report, 22 April 2013, available electronically:
`http://ec.europa.eu/research/industrial_technologies/pdf/fof-workshop-report-11-12-032013_en.pdf`

are beginning to realise that a much stronger business and market focus is needed within the PPP, and they have identified activities in this area that could form the basis for clustering.

One of the great benefits of clustering is bringing different perspectives and expertise together, which points towards clustering that is based on diversity, rather than similarity. However, some balance between the two does need to be achieved. Some clustering activities are relatively easy to do, such as joint dissemination, or collective contributions to standardisation. Some activities however are more challenging, such as addressing business and market related issues, and also sharing results, which raises confidentiality and IPR related issues. Yet in terms of that which will have most impact for projects and the PPP as a whole, is the hard-to-do clustering that is likely to be most effective. Projects have identified that clustering which addresses business and market related matters can be of great importance with regard to helping to take results to the market. Self-evidently, difficulties such as coping with IPR issues and guidance on how best to undertake challenging clustering activities will be needed, but none of the problems that people raise to undertaking this type of clustering are insurmountable.

4 Conclusions

R&D and innovation activities under a public-private partnership scheme are a way forward for Europe's industry to address effectively the challenges of manufacturing innovation.

It will be essential to address new and promising technologies such as 3D-printing, model-based controls, composites and nano-based materials as well as industrial ICT to stimulate innovation in manufacturing and offer a competitive advantage to European enterprises.

4.1 Develop and Implement 'New Manufacturing'

Manufacturing is likely to evolve along three paths[12]:

- **On-demand manufacturing:** Fast-changing demand from internet-based customers requires mass-customised products. The increasing trend to last-minute purchases and online deals requires from European manufacturers to be able to deliver products rapidly and on-demand to customers. This will only be achievable through flexible automation and effective collaboration between suppliers and customers.
- **Optimal (and sustainable) manufacturing:** Producing products with superior quality, environmental consciousness, high security and durability, competitively priced. Envisaging product lifecycle management for optimal and interoperable product design, including value added after-sales services and take-back models.

[12] http://www.actionplant-project.eu/images/stories/vision.pdf

- **Human-centric manufacturing:** Moving away from a production-centric towards a human-centric activity with greater emphasis on generating core value for humans and better integration with life, e.g. production in cities. Future factories have to be more accommodating towards the needs of the European workforce and facilitate real-time manufacturing based on machine data and simulation. 'Assisted working' should aid an aging workforce to leverage skills and knowledge effectively for the creation of innovative products.

4.2 Continue the Structuring of Europe's Industrial Landscape

The introduction of the Recovery Plan PPPs has contributed to this structuring effect by encouraging strategic thinking in terms of roadmapping, financial commitment as well as 'impact-driven' thinking among European industrial players. The formation of industrial research associations such as EFFRA[13], in the case of the Factories of the Future PPP, alongside the four annual calls for proposals in this domain has had a significant structuring effect in synergy with the European Technology Platforms - which in the case of MANUFUTURE has also led to the creation of national Technology Platforms in almost every EU Member State and beyond.

4.3 Leverage the Benefits of Advanced Manufacturing across the EU

Smart specialisation in advanced manufacturing is a way forward through identification and strengthening of competitive advantages existing in EU regions (e.g. skills, R&D capability, industrial output, ICT infrastructure and complementarities to neighbouring regions). Whilst national and European R&D programmes in the past have been focusing on developing new and powerful technologies, they lacked however the incentives and the institutional setup to encourage their dissemination and take-up across sectors, across countries and regions. Roadmap-driven R&D and innovation partnerships (PPPs) could play a significant role in bringing together the relevant stakeholders (industrial firms, academia, standardisation bodies, funding bodies (including VC investors) and public administrations at all levels) to generate market impact and thus strengthen Europe's industrial capability.

In Europe, it seems that entrepreneurial skills are required to drive results exploitation forward, from EU funded projects, via national/regional level demonstrators (pilot plants), to cross-fertilisation and uptake in different industrial sectors[14].

---

[13] The European Factories of the Future Research Association (EFFRA), `http://www.effra.eu`

[14] José Carlos Caldeira, From Research to Commercial Exploitation – The Challenges Covering the Innovation Cycle, presentation at Advanced Manufacturing Workshop, Brussels, 27 May 2013, electronically available here: `http://ec.europa.eu/enterprise/policies/industrial-competitiveness/amt/public-hearings-workshops/index_en.htm#27.05.2013`

5 Disclaimer

The views outlined in this publication are the views of the author alone and do not necessarily reflect the official position of the European Commission on this matter.

IT Support for Product and Process Development in Japan and Future Perspective

Fumihiko Kimura

Hosei University, Tokyo, Japan
fumihiko.kimura@hosei.ac.jp

Abstract. Due to the globalization of market and manufacturing activity, manufacturing industry in industrially advanced countries are facing difficult problems, such as severe competition with the low-cost production in developing countries and radical changes of customer requirements in industrially mature countries, etc. For coping with these problems, it is important to identify hidden or potential customer expectation, and to develop systematized design and manufacturing technology to augment human expertise for innovative product development. It is known that the strength of Japanese manufacturing industry comes from the intimate integration of sophisticated human expertise and highly efficient production capability. For keeping the competitiveness of the Japanese industry, it is strongly required to systematize the product and process development throughout the total product life cycle, and to introduce IT methods and tools for supporting creative and intelligent human activities and for automating well understood engineering processes. In this paper, current issues in manufacturing industry are generally reviewed. Future directions of manufacturing industry are described, and important technological issues and their IT support solutions are discussed. Finally future perspective for advanced IT support is investigated.

Keywords: Product development, Process development, IT support, CAD, CAM.

1 Introduction

Manufacturing is a basic discipline for sustaining economy for advanced industrialized countries, such as Japan, USA and Europe. However due to the recent trends of globalization in manufacturing activities, difficult issues are arising, such as severe competition with very low cost production of developing countries, risk management for global distribution of product development and production activities, environmental problems for achieving sustainable manufacturing, etc. In this paper, the recent advances of product and process development technology, current issues and future perspectives are reviewed from the standpoint of IT support, and particularly Japanese activities are discussed for keeping the competitiveness in the future.

G.L. Kovács and D. Kochan (Eds.): NEW PROLAMAT 2013, IFIP AICT 411, pp. 11–23, 2013.

It is well known that the strength of Japanese manufacturing industry comes from the intimate integration of sophisticated human expertise and highly efficient production technology. For keeping the competitiveness of the Japanese industry, it is essential to maintain the quality and volume of the expert human work force, but it is predicted that the population of Japanese working age people will decease about half in the year of 2050. Therefore it is strongly required to systematize the product and process development technology throughout the total product life cycle, and to introduce IT methods and tools for supporting creative and intelligent human activities and for automating well understood engineering processes. This new IT supported way of product and process development will rationalize the current human-dependent processes, and achieve efficient global collaboration among industrially developed countries and developing countries. The underlying research and development activities are discussed in this paper.

In the next section, current issues in manufacturing industry are generally reviewed. Future directions of manufacturing industry are described in section 3. Then important technological issues and their IT support solutions are discussed in sections 4 to 7. Finally future perspective for advanced IT support is shown in section 8.

2 Current Issues in Manufacturing

Technological, economical, and social situations for manufacturing are changing rapidly in recent years. There are many issues manufacturing industry is facing today, especially in industrially advanced countries. Major issues are reviewed, and specific problems with Japanese industry are discussed.

- Energy and resource constraints

It is needless to say that industrial technology in advanced countries cannot be fully diffused to developing countries due to over waste of energy and resources. For example, today's standard automobiles cannot be directly spread, and small light-weighted energy-efficient vehicles should be developed for mass use in the developing countries. Technology innovation is required for such new developments.

Japan and other advanced countries import large amount of energy sources and other resources. There are always risks of disruption of those supplies due to natural disaster and political problems.

- Competition in global market

Big consumer market is emerging in developing countries, and completely new categories of very cheap products are required for adapting to the global market needs. It is fairly hard for Japanese industry to change their technology from high-end products to commodity products. This change requires not only new product strategy but also new technology.

- Radical changes of world market

In industrially advanced countries, most of the products have been already spread among the consumers, and, generally speaking, consumer's products are not sold well. It is a big problem how to inspire hidden potential demands for new products.

- Service design for product production

Fundamentally it is very important for manufacturing industry to capture potential social expectation for the future, and to propose vision and scenario to approach to it. For example, it is an urgent issue to realize society of higher resource efficiency. Maintenance of social infrastructure, such as roads and bridges, is a good example, and new technology is urgently desired for efficient life cycle management.

- Population problem

In Japan, and similarly in other advanced countries, labour force in manufacturing industry will decrease rapidly in coming 50 years. For sustaining the industrial level, it is very important to amplify human intellectual ability by IT support, and to automate production processes as much as possible.

3 Future Direction for Manufacturing Technology Innovation

For analyzing the issues explained in the previous section, a role of design and manufacturing engineering is investigated, and the issues are classified according to the customer expectation and technology systematization.

Figure 1 shows a key role of design and manufacturing engineering. For enhancing the quality of life (QOL), social expectation for technology development and new products is expressed in the form of explicit demands, social scenario or general social vision, through market mechanism and other social /political mechanisms. Based on the level of industrialization of the society, the expectation is very clearly expressed, or is not expressed explicitly. Depending on such expectation or independently, design and manufacturing engineering tries to develop and propose various attractive technology options in the form of new systems and products to the society, based on the contribution from basic science and engineering.

Traditionally what customers want to get was clear, and known to customers themselves and to engineers. In such cases, according to the social expectation, appropriate technology options can be identified, and the developed systems and products are accepted by customers for better QOL. This case corresponds to the Class I and Class II problems in Figure 2, and is further explained below.

Today especially in industrially advanced society, customer demand or social expectation is not explicitly expressed, but only potentially noticed. Then, even though advanced technology and products are available, they may not be accepted by the society, and the customers are frustrated.There seems to be a big discrepancy between customer wish and producer awareness. This case corresponds to the Class III problem in Figure 2, and is further explained below.

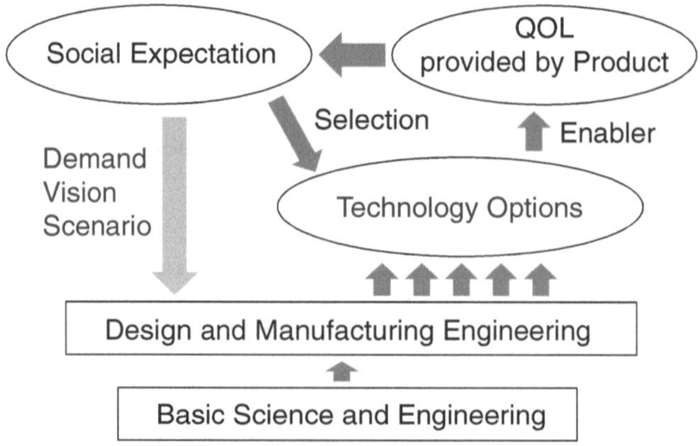

Fig. 1. Importance of design and manufacturing engineering

For clarifying the current issues in manufacturing, it is effective to classify the problems into three classes described in Figure 2 [1]. The classification is based on the level of recognition of the following two factors: social expectation and adaptability to various conditions surrounding the manufacturing.

Class I Problem: Customer expectation is clearly expressed, and surrounding conditions for manufacturing activity is well known to producers. Therefore the problems are well understood, and systematized technology can be effectively applied.

Class II Problem: Customer expectation is clearly expresed, but surrounding conditions are not well captured or not known. In this case, product and production technology must be adapted to the changing and unknown situations. Integration of human expertise is required for problem solving.

Class III Problem: Customer expectation is not explicitly expressed or is not known. In this case, the problems are not clearly described, and tight coupling exists between identification of customer expectation and correponding technology for problem solving. Co-creative approach between customers and producers is mandatory.

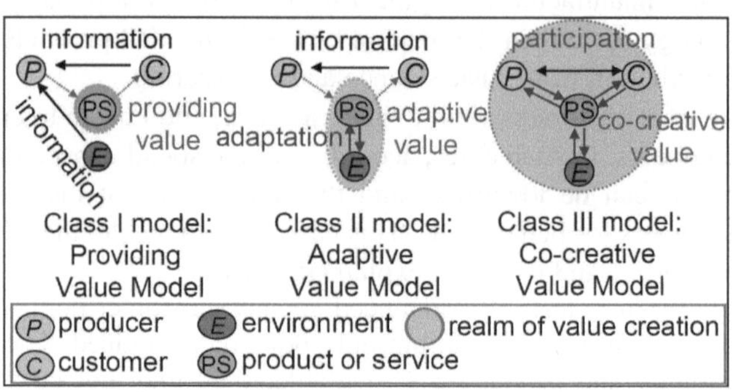

Fig. 2. Classification of design and manufacturing problems [1]

The current situation of Japanese manufacturing industry is explained by use of the above classification. Figure 3 shows the manufacturing problem classification with two coordinates: customer expectation and adaptability to surrounding conditions. Here adaptability to surrounding conditions is considered to depend on technology systematization, and it can be characterized by technology innovativeness and maturity.

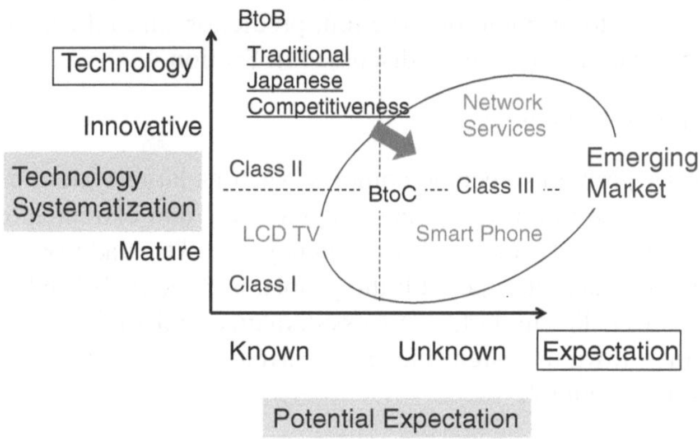

Fig. 3. Current trend of product development

Under the clear target setting situation, traditionally Japanese industry is very strong, even with varying conditions and unknown technology issues, to adapt to the difficult technology problems by applying front-edge technology and very sophisticated human involvement, and to produce very high-quality attractive products. This is a Class II problem, and Japanese approach is called as "Suriawase" in Japanese, which means sophisticated integration of human expertise with problem solving.

If the problem is well understood, and technology to solve the problem is well matured, the whole product development and production process can be systematized, and possibly automated. This is a Class I problem. Products which belong to this Class tend to be mass production products, and their competitiveness mainly depends on cheap price. The current difficulty of Japanese manufacturing industry is the technological and organizational inflexibility to adapt to this problem. By simply applying the sophisticated Class II problem solving method to Class I problems, it results in very expensive products with excessive quality.

If we look at the world market situation, two expanding or emerging market areas are recognized, as shown in Figure 3. One area is a mass production commodity product area, where products belong to the Class I problem, and the price is the most critical competitive factor. Another area is an innovative product area, where products belong to the Class III problem. The most important factor for Class III products is to identify customers' potential or hidden expectation, and to incorporate appropriate advanced knowledge from basic science and engineering toward product innovation.

Based on the above discussion, important future directions for manufacturing technology development are considered.

- Identification of customer expectation

As the product technology is advancing so rapidly, the customers are normally uable to capture the vision of future society, and tend to be frustrated with the proposed products and systems from the producers. It is important to develop a methodology to search for the hidden and potential social expectation, and to identify various requirements of customers for daily life products and social infrastructure. It is effective to utilize IT methods to promote observation, prediction and information sharing for mass population society. This issue is discussed in Section 4.

- Systematization of technology

The manufacturing problems become complex, such as large system problems, complexity problems due to multi-disciplinary engineering, and extremely difficult requiremts toward product safety and energy efficiency, etc. The traditional human dependent approaches only cannot cope with the problems effectively, and it is mandatory to introduce advanced IT support, and to systematize and to integrate various engineering disciplines toward design and engineering methods. These issues are discussed in Sections 5, 6 and 7.

4 Potential Customer Expectation

It is often argued that recent innovative products, such as a smart phone or a hybrid vehicle, could not be commercialized by the conventional market research activity, because impacts of those innovative products are difficult to imagine by normal consumers due to their technological difficulty. Many customers have vague expectation for new products, but they cannot express their wish correctly, therefore their expectation cannot be satisfied. Manufacturers can offer new products based on their revolutional technology, but it is not easy to match their design intention with customers' real wish. It is very important to develop a methodology to capture hidden or potential customer expectation.

In recent years, Japanese industry and research community have heavily disussed this issue, and proposed various practical methods for observing the potential social expectation [2]. It is still pre-mature to develop systematic methods, but several useful existing methods are discussed:

- systematic survey of existing literature,
- multi-disciplinary observation,
- observation of social dilemma and trade-off,
- collection of people's intuitive concern,
- deep analysis of already known social concern,
- re-examiation of past experiences.

Modelling, simulation, and social experiments are useful tools for prediction and information sharing. IT support is very effective for data mining and bibliometrics. Combination of information network and sensor capability has a big potential for extracting unconsciously hidden social wish. Promising approaches are advocated as a Cyber-Physical System [3]. A huge number of sensors are distributed into the society, and various kinds of information are collected and analysed. There are many interesting trials in Japan, such as energy consumption trends of supermarkets, combination of contents and mobility information, zero-emission agriculture, healthcare information, etc. Important aspects are collection of demand-side information and combination of different kinds of industrial activities. It is expected that, by capturing latent social or customer wish, social infrastructure and individual QOL are better servicified by the manufacturing industry.

5 Large System and Complexity Problems

By the diversity and vagueness of customer requirements, industrial products and systems tend to become large in scale and complicated. Large system problems typically occur for designing social infrastructure or complicated products like a space vehicle, etc. Complexity problems are often related with multi-disciplinary engineering, such as designing mechatronics products.

For coping with large system problems, various system engineering methods are already well developed, but these methods are not fully exploited for product design and manufacturing practices. Those methods include the following system synthesis steps [4]:

- common understanding via system modelling,
- subsystem decomposition and structuring,
- quantitative analysis of subsystem behaviour,
- scenario setting and system validation.

There are important viewpoints for system design, such as harmonization of local optimization and global optimization, multi-scale consideration, structural approach to sensitivity analysis, etc. Standardization and modular concept are essential for effective system decomposition. The V-Model approach in system engineering is valid, but decomposition, verification and validation processes become very complicated with multi-disciplinary engineering activities in product design and manufacturing.

Various kinds of model-based approaches are proposed, and standardization for model description scheme and languages is progressing. For coping with large scale and complexity problems, it is important to take into account of the following aspects:

- modelling and federation at various abstraction levels and granularity,
- modular and platform approach,
- multi-disciplinary modelling.

For system decomposition and modularization, it is effective to utilize a concept of function modelling as a basis, instead of physical building blocks. Product development processes are modelled, starting from requirement modelling, via function modelling and structure modelling, to product modelling. Those modelling should be performed in multi-disciplinary domains, and appropriately federated. Many research works are being performed, but industrial implementation are not yet fully realized.

6 Upstream Design Problems

It is argued that inappropriate product functionality and product defects are often caused at the upstream of product development processes. It is very expensive and time-consuming to remedy such problems at the later stages of product development, because many aspects of products have been already fixed. It is very effective to spend more time and effort at the stages of product requirement analysis, concept design and function design.

In Japan, it is currently a big problem that products tend to have excessive functions, and become the so-called "Galapagos" products, as shown in Figure 4. "Galapagos" products mean products designed to incorporate available leading technology as much as possible for product differentiation, and it results in very expensive products. Now the big market is expanding into developing countries. As the "Galapagos" products cannot be sold well in such market, it is required to eliminate excessive functions, and to make the products cheaper. But, it is difficult to compete with the cheap products designed especially for such market from scratch.

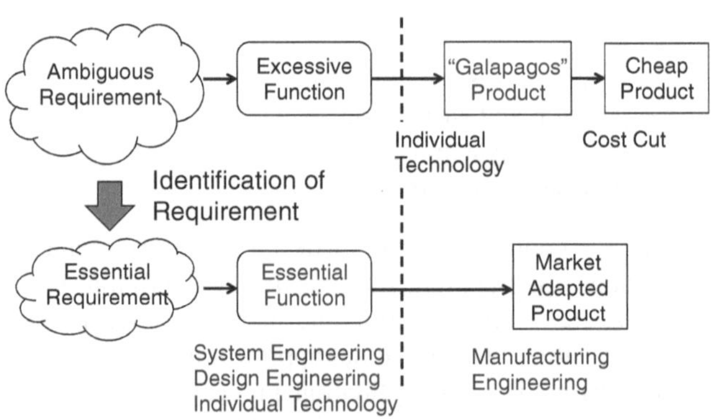

Fig. 4. Identification of essential requirement

This problem happens from the ambiguity of product requirement identification. The following approach is important for coping with this problem:

- identification of essential product requirements,
- realization of essential functions with science-based methods,
- rationalization and simplification of traditional functions and processes,
- minimization of required resources.

The above approach cannot be implemented by conventional technology only, but requires dedicated advanced technology specifically tailored for the target products, such as extremely light-weighted materials, highly energy-efficient processes, etc. This is a way that Japanese industry can take for competitiveness.

A systematic approach to upstream design is a very keen research issue in Japan. An interesting approach is advocated by the name 1DCAE [5]. There are many IT tools available today for doing precise engineering simulation. However it is very cumbersome to use those tools for upstream conceptual design activity. Also those tools are inconvenient for thinking and understanding product functional behaviour in an intuitive way. 1DCAE tries to establish a methodology to systematically utilize any available methods and tools to enhance the true engineering understanding of product characteristics to be designed, and to support the conceptualization of the products. 1DCAE is to exhaustively represent and to analyse product functionality, performance and possible risks at an early stage of design, and to provide a methodology for visualizing the design process and promoting the designer's awareness of possible design problems.

Figure 5 shows a 1DCAE approach for mechatronics product design. For conceptual design, various existing IT tools are used based on mathematical analysis and physics, such as dynamics, electronics and control engineering. Through good understanding of product concept and functional behaviour, detailed product models are developed.

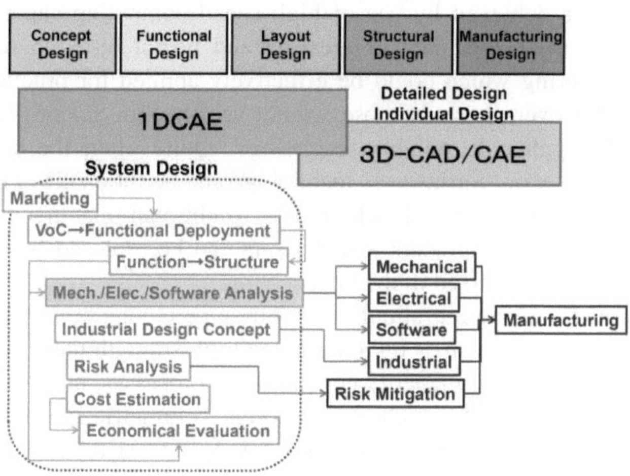

Fig. 5. 1DCAE approach for mechatronics product design [5]

Figure 6 represents a 1DCAE approach for large scale system design, such as a spacecraft. In this case, system optimization and risk mitigation are the very important design target. Design process is optimized by using DSM (Design Structure Matrix) method, and every necessary technological aspect of the products is modelled at appropriate granularity for simulation.

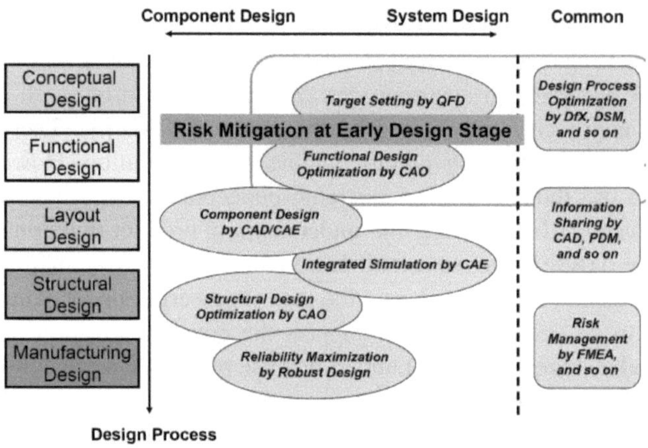

Fig. 6. 1DCAE approach for large system design [5]

7 Importance of Basic Technology

In addition to the various system technologies, the basic individual technology is also important. In recent years, remarkable progress of product and process development technology has been achieved by use of high speed computing capability. As indicated in Figure 1, there are many interesting and useful research results in basic science and engineering which could be effectively applied for practical design and manufacturing. However, many of those are not yet utilized. Science-based approach enables generally applicable reliable technology. Quite often the so-called expert engineering know-how or "Suriawase" technology can be rationalized by the science-based developments. By these developments, traditional manufacturing practices relying on veteran expert engineers and workers can be replaced by comprehensive automated systems, as discussed in the next section.

By sophisticated engineering simulation with supercomputers, extreme technologies have been developed, such as light-weighted high-strength materials, low-friction surfaces, nano-machining, low-energy consumption processes, etc. Advanced modelling technology is developed, which can represent volumetric information including graded material properties and various kinds of defects in the materials. Powerful measurement methods, such as neutron imaging, are being available to visualize internal structure of components and assemblies. By such precise modelling, accuracy of computer simulation is very much enhanced, and delicate engineering phenomena can be captured, which is difficult for physical experiments.

Ergonomic modelling and robotics technology have evolved, and behaviour of human-robot interaction can be simulated precisely by computer. This is a basis for designing comprehensively automated production systems, as discussed in the next section.

8 Future Perspective for Advanced It Support

Various kinds of CAD/CAM systems are effectively utilized in industry today, and they have already become indispensable tools for daily product and process development works. However their functionality is not satisfactory from the future rquirements for IT support discussed in the previous sections. Two important aspects for advanced IT support for product and process developments are identified. One is comprehensive support of intelligent human engineers for creative product design, and the other is systematic rationalization and automation for well developed engineering processes.

The possible configuration of advanced IT support for product and process development is shown in Figure 7. "Virtual Product Creation" deals with intelligent human support for product design, and "Error-Free Manufacturing Preparation" performs comprehensive support and automation of well developed engineering processes. A core part of the system is "Integrated Life Cycle Modelling", which represents all the necessary product and process information for intelligent support and automation. Technologies discussed in Sections 5 to 7 are somehow integrated in these system modules.

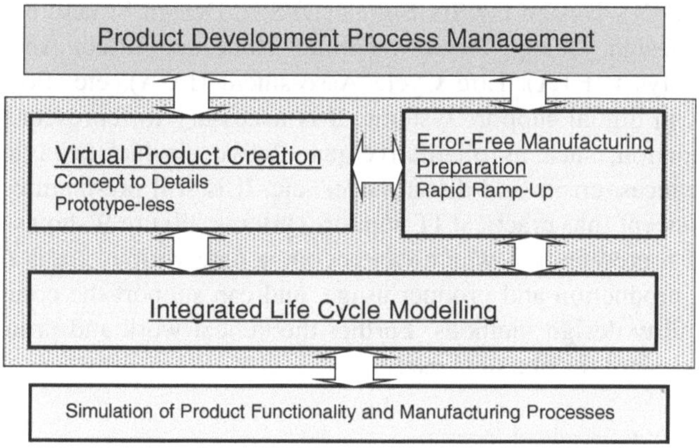

Fig. 7. IT support for product and process development

In Japanese industry, some part of those system functionalities are implemented individually as in-house applications, and some are realized as commercially available IT support systems. Figure 8 shows an example of digital pre-validation of production lines for electronics components. Recently Japanese companies operate such factories in foreign countries. By using the digital pre-validation, most of the required engineering works can be done in Japan before actual implementation of the factory equipments in foreign contries. Line design work load of human expertise can be radically reduced by this support system. This system incorporates many sophiscticated modelling and evaluation engineering know-hows, and exhibits differentiated characteristics from the conventional factory simulation systems.

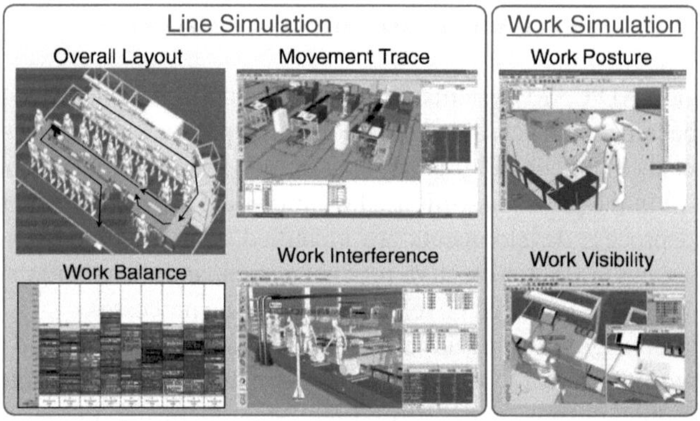

Fig. 8. Digital Pre-Validation of Production Lines [6]

There are several critical issues for realizing future-oriented IT support systems as shown in Figure 7. One of the important problems is to integrate well developed practical design methods into IT support systems. There are many such methods as Quality Function Deployment (QFD), Functional Modelling, First Order Analysis (FOA), Design for X (DfX), Design for Six Sigma (DFSS), Design Structure Matrix (DSM), Optimization Design, Design Review, Failure Mode and Effect Analysis (FMEA), Fault Tree Analysis (FTA), Life Cycle Assessment (LCA), etc. For implementing those methods in digital support systems, it is necessary to represent pertinent engineering information, such as qualitative/quantitative product behaviour, functional structure, tolerances, errors and disturbances, etc. It is still pre-mature to install such engineering concept into practical IT support systems. Figure 9 shows an example of product model representation which can accommodate various kinds of disturbances arising during production and product usage, and can support the computerization of practical reliability design methods. Further theoretical work and prototyping implementation are desired for practical use.

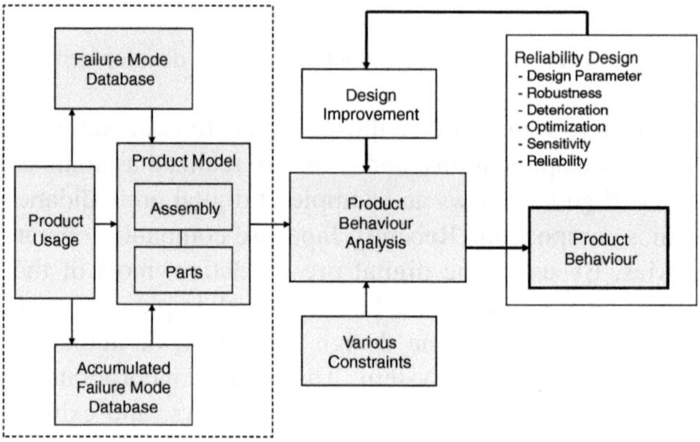

Fig. 9. Model based engineering with disturbances [7]

9 Summary

With the globalization of market and manufacturing activity, manufacturing industry in industrially advanced countries are facing difficult problems, such as severe competition with the low-cost production in developing countries and radical changes of customer requirements in industrially mature countries, etc. For coping with these problems, it is important to identify hidden or potential customer expectation, and to develop systematized design and manufacturing technology to augment human expertise for innovative product development. As Japan expects the radical decrease of population in coming 50 years, it is very important to systematize the product and process development technology throughout the total product life cycle, and to introduce IT methods and tools for supporting creative and intelligent human activities, and for automating well understood engineering processes. In this paper, current issues in manufacturing industry are generally reviewed. Future directions of manufacturing industry are described, and important technological issues and their IT support solutions are discussed from the viewpoints of potential customer expectation identification, large system and complexity problems, upstream design problems and important basic technology. Finally future perspective for advanced IT support is investigated.

References

1. Ueda, K., Takenaka, T., Vancza, J., Monostori, L.: Value Creation and Decision-Making in Sustainable Society. CIRP Annals 58(2), 681–700 (2009)
2. Discovery of Social Wish through Panoramic Observation, CRDS-FY2010-SP-08, Japan Science and Technology Agency (2010) (in Japanese)
3. Lee, E.A.: Computing Foundations and Practice for Cyber-Physical Systems: A Preliminary Report, TR UCB/EECS-2007-72, University of California at Berkeley (2007)
4. Towards Solving Important Social Issues by System-Building Through System Science and Technology, CRDS-FY2010-SP-04, Japan Science and Technology Agency (2010) (in Japanese)
5. Ohtomi, K.: The Challenges of CAE in Japan; Now and Then, Keynote Speech, NAFEMS World Congress, Salzburg (2013)
6. VPS/GP4, Fujitsu Ltd. (2013)
7. Kimura, F.: Product Quality Evaluation Based on Product Life Cycle Modelling with Disturbances. In: Frey, M. (ed.) Closed-Loop Product Life Cycle Management - Using Smart Embedded Systems, pp. 215–232. ISA (International Society of Automation) (2011)

3D Experiences – Dassault Systèmes 3DS Strategy to Support New Processes in Product Development and Early Customer Involvement

Andreas Barth

VP and Managing Director, Dassault Systèmes Deutschland GmbH
andreas.barth@3ds.com

Abstract. In many industries the market sees a lot of change happening these days; new trends and challenges obviously need to be addressed as part of the product creation process. Dassault Systèmes (3DS) is introducing a 3DEXPERIENCE PLATFORM to expand the usability of the digital smart product development in an accurate virtual universe beyond Product Lifecycle Management (PLM). The target is to provide the ability to place the customer at the heart of a system, integrating both company business processes and product development processes. It is a combination of science, art and technology to bring value to our customers by helping them to respond to the needs of their customers and creating 'magnetic' products with strong market appeals. 3DExperiences can be a catalyst for innovation, enabling any enterprise stakeholder to participate in the innovation process, contributing to drive value for the end consumer and create smart products from design to recycling. This presentation highlights some aspects of the new Dassault Systèmes strategy; shows some of the solution experiences and how customers respond to them. 3DS is a scientific company positioned among the top 10 software companies worldwide and for more than 30 years has been helping companies to transform the way they design and produce their products.

Keywords: 3DExperience, CATIA, Innovation, DMU, Smart Product.

1 Introduction

Innovation is driven by change and the consumer markets do change rapidly these days. Mobility requirements for instance do have strong impacts on the Tele-Communication Industry (mobile devices such as smartphones, tablets …) and on the Transportation & Mobility Industry (Electro mobility, personal mobility, sustainability …). It's the first time in industrial history that innovation is more driven by end consumer needs rather than by engineering and manufacturing companies.

2 Why Is Change Necessary?

It is a complex challenge to move from traditional engineering to Smart Product development. Forward-thinking executives are, quite rightly, demanding more and

G.L. Kovács and D. Kochan (Eds.): NEW PROLAMAT 2013, IFIP AICT 411, pp. 24–30, 2013.

better consumer insight and smarter research to measure impacts of potential new innovations. Companies can gain an understanding of their customers which is better, faster, and less expensive by doing four things called the "digital research revolution":

1. Leverage the digital society power to rapidly create and deliver detailed information about what would be products customer acceptance information.
2. Recognize the limitations of the measurement systems based on projected past or isolated focus groups.
3. Seek to understand what customers value the most and how technologies and innovation can deliver and enhance the product value.
4. Link investigations into consumer insights based on virtual product candidates that are sorting value very close to their customer relationships management systems and create criteria such as "Cool Feature", "Must Have", etc.

2.1 Leverage the Power of the Digital Society

The Internet is fast. It enables any type of digital exchange and can leverage data in many forms plus has the ability to create many more iterations with a turnaround in a few days, rather than weeks required for traditional methods. Marketing strategy segmentations for innovation that used to take months can be completed in weeks. Working digitally is also less expensive than prototyping/ face-to-face studies and can be conducted with dramatically larger sample sizes. Finally, customers' online comments, search behavior, and other revelations offer a prove of data.

2.2 Decrease Reliance on Traditional Prototypes and Focus Groups

Focus groups were once the default qualitative research approach. Companies gathered small numbers of consumers in a facility around real prototypes for some hours, promised them a small payment or free products, and asked them to give simple written or oral answers to questions that may or may not have provided real insight. Despite the emergence of better alternatives, little has changed yet.

Full usage of Virtual Product candidates now enables leading marketers to create deep immersion experiences with a large number of users organized through communities to generate consumer insights and new products; consider values and positioning; and redefine a product's competitive set. These ideas can then be tested quantitatively, either through further research or a market pilot.

2.3 Understand How Successful Products Are Created

Companies have long focused on what and why people buy. Research data from consulting companies like Oliver Wyman show that this is not sufficient.
They clearly show that only few high-quality products are magnetic[1] , defining "magnetic" by a simple equation: M = F×E. Magnetic equals best functionality times

[1] Magnetic = having a great power of attraction over people.

most powerful emotional connection with customers. Very good functionality is obviously not enough anymore.

2.4 Connect What You Know about Consumer Value

Why do consumers disproportionately demand one product over a seemingly similar one, often by a factor of four or five to one? Functionally and technically, products might be close; emotionally, they can be worlds apart. Why do seemingly similar products produce radically different demand curves?

Demand creators, people who design products that truly excite consumers, are obsessed with understanding customer values, and connecting the dots from multiple value chains to fix it. They don't assume that buying = wanting. They use the value map to recognize the huge gaps between what people buy and what they really value including emotionally — and use those value map as a springboard to see differently. Demand creators crack the mystery of the demand equation by doing a radically better job answering a small set of critical value and differentiation questions very well.

3 The Concept of 3dexperiences By Dassault Systèmes

How can a Software tool vendor support demand creators?

3DS has initialized a large change to apply the power of the Digital Society to help product creators with the new challenges described above: Introduce the ability to create Smart Products from an unified digital representation that are all interacting together to create a "Virtual" Smart Product.

To reach customers but also to ensure that fast development cycles based on value of a product that can actually be produced can be achieved digitally, a "Virtual" Smart Product needs to be digitally accurate. This product accuracy is not sufficient to help creators if it is not also possible to benefit from similar accuracy in the digital modeling of the environment in which the products will be used.

And of course measuring value and making product creators able to deliver the magnetic products described above means that ability to interface with large communities and finally partners including suppliers, channels and consumers are requesting interacting socially in fast and efficient digital loops.

3DS has started to create a Product Creator Digital Platform that has the ambition to complete the digital landscape compared to other platform technologies

- Facebook as a social platform
- Google as a search platform
- Wikipedia as a knowledge platform
- App stores as a services and applications platform
- Android as a development platform
- YouTube, iTunes and others as a content platform

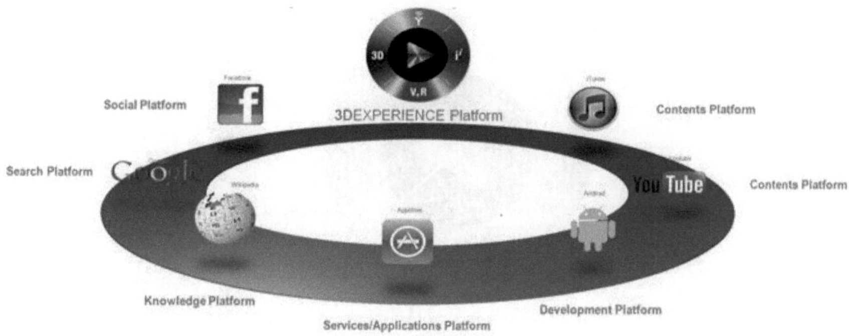

Fig. 1. 3DExperience Platform compared to other platform technologies

This Product Creator Digital Platform is named 3DExperience platform and is designed to socially link the large communities of contributors in product developments required to create Smart Products.

The integration is done around real accurate Digital Virtual Smart Products that are leveraging the 3DS 20 + years of experience in creating accurate engineering around 3D based digital engineering.

The target is to provide the same accuracy that was obtained on geometry to the other engineering domains. The target is not to provide game platforms but to enable the magnetic product creation to be done socially based on the real physics and manufacturability of the products - this can be seen as a natural extension to the Product Lifecycle Management market (PLM).

The representation of this strategy is indicated by a user interface to access the software and data platform in the form of a compass as a synonym for a highly integrated application environment and a platform concept.

4 User Experiences By Industry

The 3DS new 3DExperience platform is to be delivered for 12 industries – that produce Smart Products including the more traditional customers of the company such as Automotive, Aerospace and Industrial Equipment to new customers such as high volume Consumer Packaged Goods, Retail, Energy, Process and Utilities, Financial and Business Services, Life Science and more.

Fig. 2. Dassault Systèmes supports 12 Industries

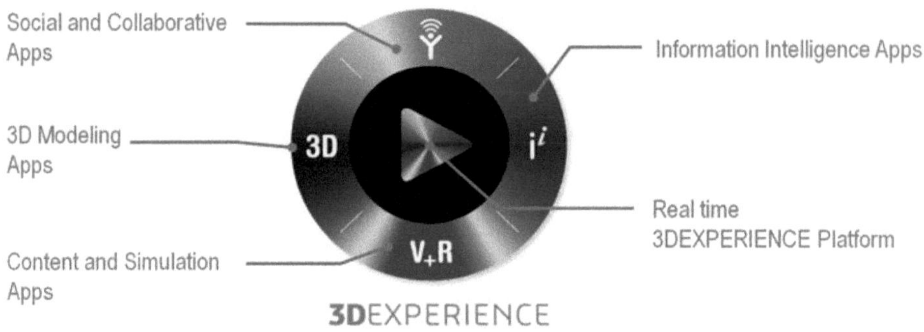

Fig. 3. 3DExperience Platform Compass

4.1 PLM Innovation By Engineered Business Experience

Overall the changes, whether it is the market, the consumer buying behavior, new products, changed mobility etc. have a significant impact on the way engineering will work in the future.

From	**Discipline Collaboration**	to	**Social Industry World**
From	**Product**	to	**Business Modeling**
From	**Document Management**	to	**Experience Management**
From	**Search**	to	**Dashboard Intelligence**
From	**Product attributes**	to	**Consumer Experience**

Fig. 4. PLM Innovation

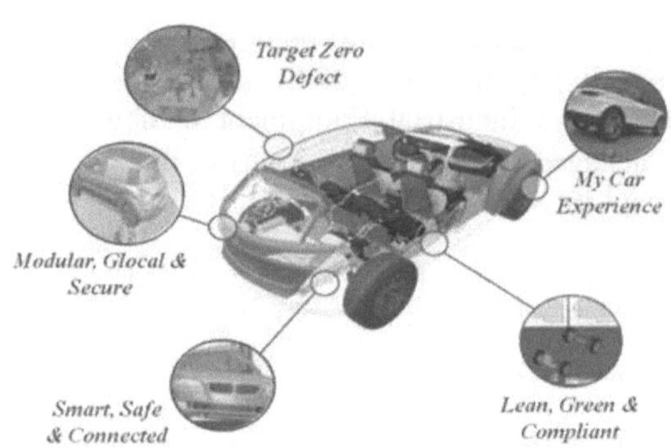

Fig. 5. Transportation and Mobility Industry Solution Experience

5.1.1 My Car Experience

The purpose of this experience is to understand how the customers are expecting the next generation of the mobility services by providing some innovative ways to collect their needs and identify the market trends, to enable crowd-sourcing and co-creation around new concepts and social innovative design. Powerful 3D design tools to boost the industrial innovation process and efficient tools to perform virtual car clinics and showroom, with augmented reality and immersive reality, will be introduced. A virtual in-car feeling by enabling new in-car experience, including driving experience will lead to new design and engineering front loading experience, thus reducing the time to market and reducing physical prototypes.

5.1.2 Lean, Green and Compliant

The purpose of this experience is to improve vehicle energy efficiency upfront with light weight and green propulsion and to explore new ways of progress by optimizing the energy balance and the weight. Providing an end-to-end solution for EV and HEV propulsion concepts development accompany companies in their will to be more environmentally responsible. An integrated platform for vehicle validation and certification to define, manage and monitor the validation and verification process (physical and virtual test) insures requirements validation. The compliance towards global regulations and certifications (material compliance, FMVSS, CAVA, passive safety, etc.) help to optimize the vehicle performance globally and by providing a multi-physics simulation platform to evaluate and optimize all the performance of the vehicle, increased product quality and satisfying sustainability results can be harvested.

5.1.3 Smart, Safe and Connected

The market today is challenged by a re-invention of the mobility experience as 80% of vehicle innovations are coming from embedded systems and development moves from vehicle attributes to vehicle experience. A huge complexity in the number of ECUs, lines of code of software and an increasing number of functional requirements makes car design error prone, resulting in 20 to 40% of vehicle development cost spent in testing and diagnostics. The market demands to improve quality and safety and decreasing the number of warranty cost and cars recalls. To demonstrate functional safety against class actions lawsuits and to standardize on industry initiatives such as Software re-use (AUTOSAR), Strengthen safety (ISO26262) and early validation are mandatory to support next generation car developments. The concept of Functional Mock up that integrates mechanical, electronics, software and simulation, is key to design and simulate the car of the Future: smarter, safer and always connected.

5.1.4 Modular, Glocal[2] and Secure

This experience allows defining the technical requirements from a marketing analysis and ideation process done on unstructured data, to define the vehicle portfolio and

[2] Glocal = Global and local.

monitor the program: project dashboards, program 360° and quality plans (incl. suppliers). Platforms and modules, interfaces as well as the vehicle architecture (volumes, technological choices, etc.) are managed along the defined configurations. The management of the EBOM is based on a modular approach and support the Digital Mock-up review processes (Design, Manufacturing and Simulation) in a distributed environment (Suppliers, R&D centers,). The MBOM manages the assembly processes per manufacturing plant and the supply chain, from engineering to after sales, including sourcing, RFQ/Supplier choice process, and the contract and cost management, etc. are integrated as well.

5.1.5 Target Zero Defect

To ensure a zero defect process End-to-End on all the domains, extend the zero geometry defect to the End-to-End, and taking full benefit of new system engineering and collaboration environments is the purpose of this experience.

Capitalizing and reusing knowledge globally for design, manufacturing and simulation will promote innovative tools and methodologies in Architecture and Conceptual phases of the processes. Augmented simulation contribution and key differentiators in all steps of the processes will improve productivity and efficiency to reduce cycles by providing Best-of-Class solutions.

5 Conclusions

With the 3DEXPERIENCE platform companies create "social enterprises" that involve their customers in the innovation process. With its online architecture the 3DEXPERIENCE environment enables businesses to test and evaluate – anywhere in the development lifecycle of a product or service – the eventual experience they will deliver to their customers.

References

1. Charles, B., Menghini, M.: 3DS Corporate Strategy presentation 2012
2. 3DS 3DExperience Universe for Sustainable Innovation, `http://www.3ds.com/company/about-dassault-systemes/our-vision/`
3. Sappin, O., Belloncle, G.: 3DS Transportation and Mobility Industry Referential presentation (2012)
4. 3DS Transportation and Mobility 3DExperience Platform, `http://www.3ds.com/solutions/transportation-mobility/overview/`
5. Transportation and Mobility Smart, Safe and Connected Solution Experience, `http://www.3ds.com/solutions/transportation-mobility/industry-experiences/smart-safe-and-connected/`
6. 3DS Annual Report 2012

Challenges for Digital Product and Process Design Systems at BMW

Dietmar Trippner, Stefan Rude, and Andreas Schreiber

{dietmar.trippner,stefan.rude,andreas.as.schreiber}@bmw.de

Abstract. The methods of model based product development are today well understood by industry across branches as in automotive and aerospace including their supply chain. Actual challenges of these industries however let question the capacity of the concept. Such major changes in Automotive are light weight design, electro-mobility as well as modern mobility concepts, all increasing the product complexity. Increasingly an overall approach is requested that is able to consistently integrate Requirements, Functions, Logic and Physics (RFLP approach). Not only with regards to mechanical aspects, but also considering demands of electric/electronic as well as software development.

The Systems Engineering approach is addressing the consistent availability and linkage of product information. Well established in aerospace and defence since decades the concept arrives in Automotives just right now, driven for example by the introduction of integrated product development environments. However the Systems Engineering introduction to an automotive OEM still might be a challenge. Typical obstacles are different structured and detailed design requirements (customer requirements versus vehicle requirements and component characteristics) as well as missing consideration of configuration, effectivity and maturity. Additionally the two dimensions of automotive consistency (Vehicle to component – vertical integration – versus integration of early concept stages via development, verification, approval until the final start of production – horizontal integration) contribute to the challenge.

Finally also an improvement of interdisciplinary interaction is needed (mechanics including simulation, electric/electronics and software). The introduction of Systems Engineering is not only challenging the coordination of process IT infrastructure (Authoring systems, TDM and PDM) but also needs to consider organizational aspects (process and appropriate organizational units, integration of design partners and suppliers). Frequent acquisitions among IT system vendors – especially in the CAD/PLM/CAE market – as well as the need to select a system considering functional as well as economic aspects drive the demand for open interfaces.

The presentation highlights the success and payoff of the application of Systems Engineering for Automotive processes. Implications for the process IT infrastructure coordination will be introduced.

Necessary process IT adaptations will be addressed using target templates for processes, design system infrastructure, integration and mapping.

Finally BMW's concept to enforce openness from all OEM, Supplier and System Vendor based on the Codex of PLM Openness (CPO) will be discussed. Success stories as well as lessons learned will be shared.

Keywords: Systems Engineering, PDM, TDM, CAE, Simulation, RFLP, Codex of PLM Openness (CPO)

G.L. Kovács and D. Kochan (Eds.): NEW PROLAMAT 2013, IFIP AICT 411, pp. 31–35, 2013.

1 Introduction

Digital Product- and Process development are getting more and more important at BMW and all related brands, as Mini or Rolls-Royce, for example. See Fig. 1. and Fig.2.

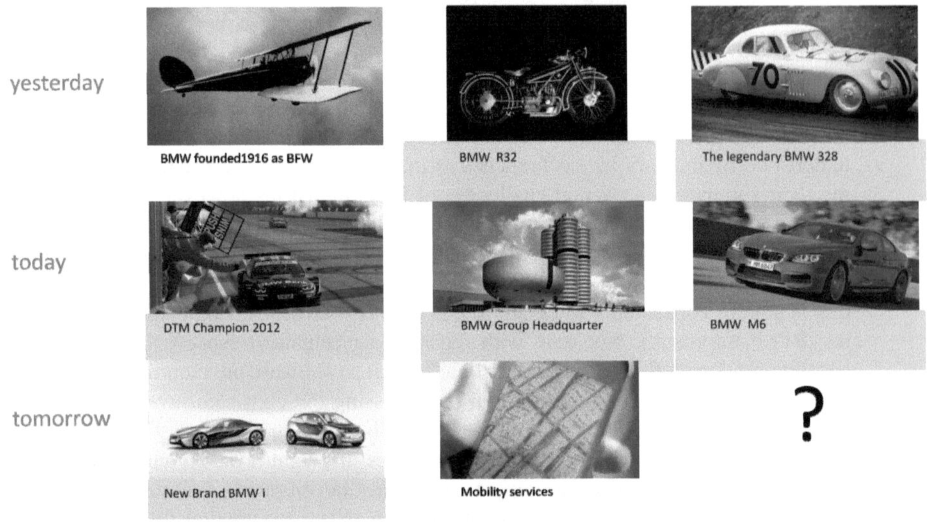

Fig. 1. BMW Group – Company Portrait

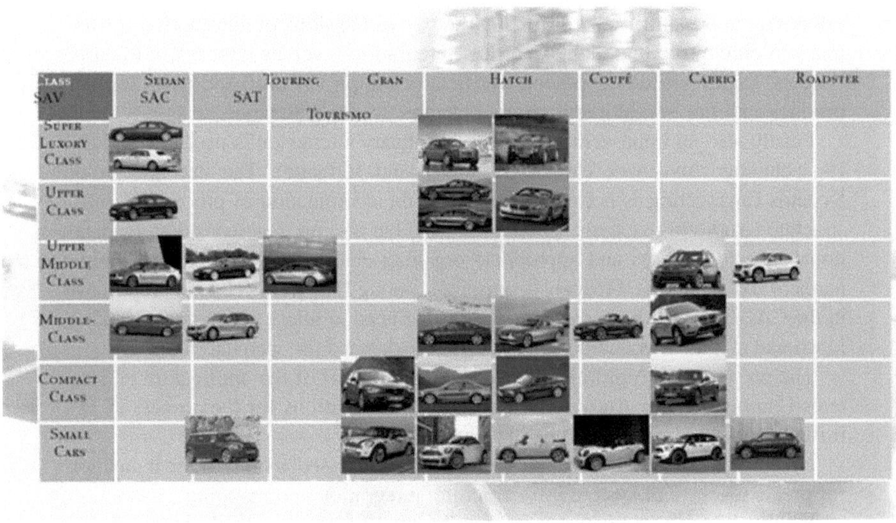

Fig. 2. The BMW Group Product Portfolio

Major challenges in Automotive are light weight design, electro-mobility as well as modern mobility concepts, all increasing the product complexity (see Fig. 3).

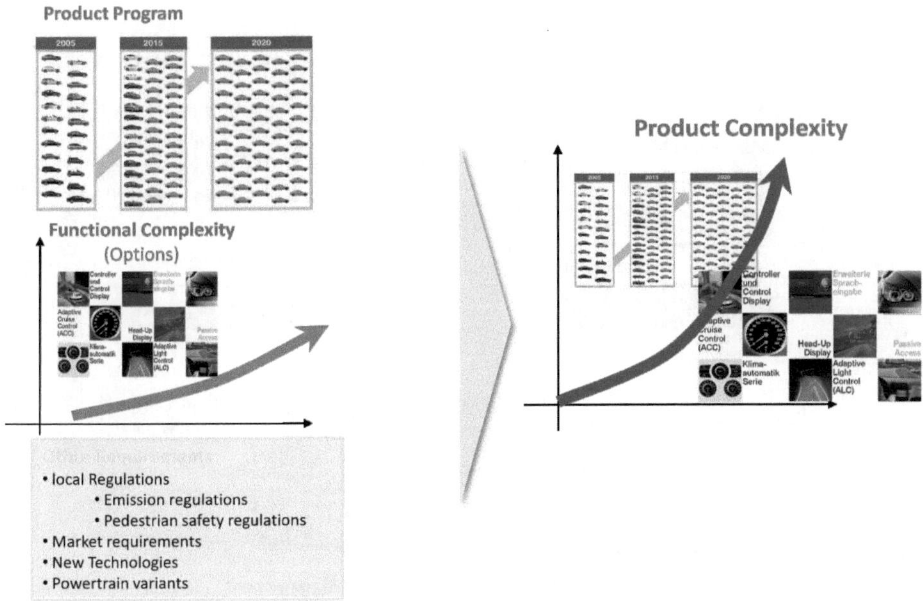

Fig. 3. Increased product complexity in product design

2 Requirements, Coming from the Development Process

To meet these different requirements for the development of a complex product program a highly optimized design methodology is needed (see Fig. 4).

The Systems Engineering V-Model is a good base for such a design process (see Fig. 5).

The use of virtualization dramatically increased over the years. But its success was created by introducing many IT applications covering the various aspects of simulation. The next level of efficiency therefore needs to address the topic of consistent integration of all these applications (see Fig. 6).

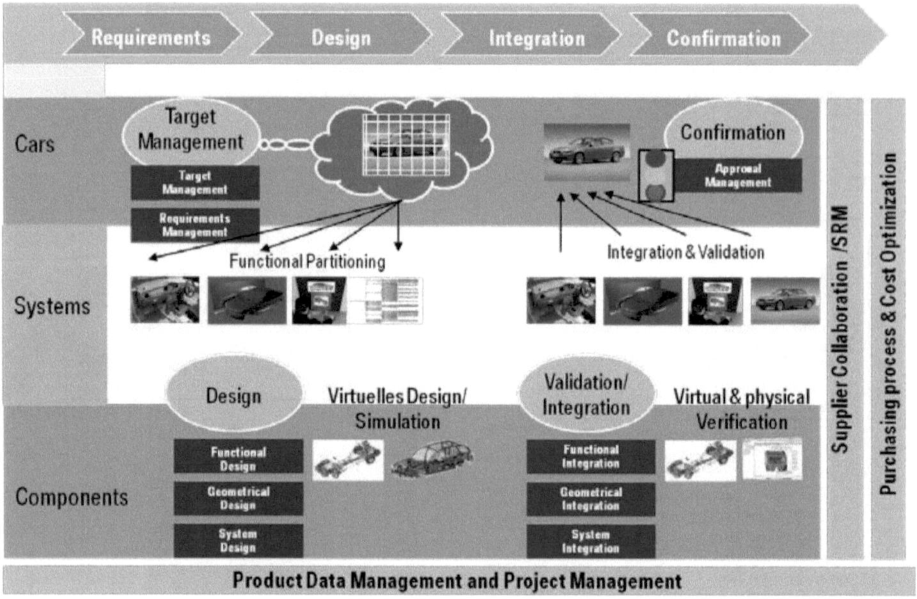

Fig. 4. Design process principles

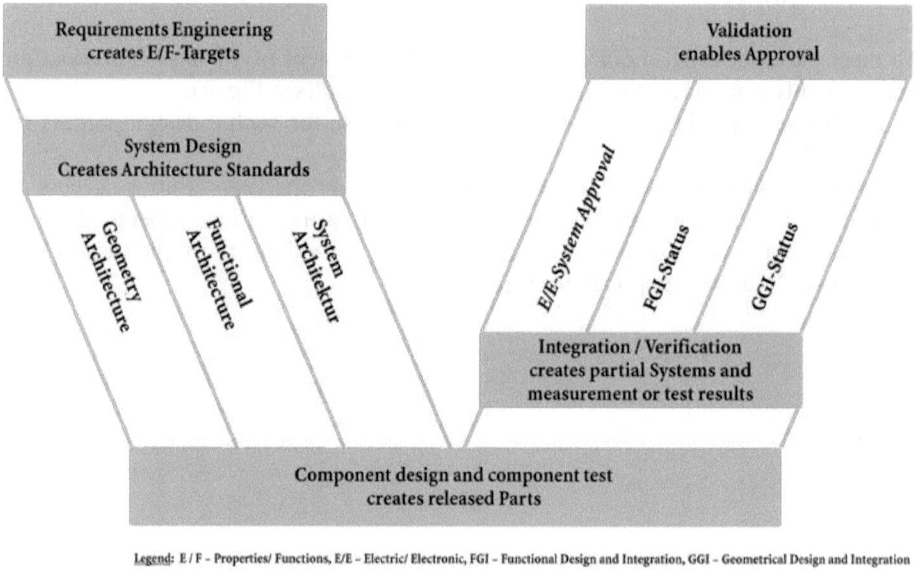

Fig. 5. The development process in the V-Model

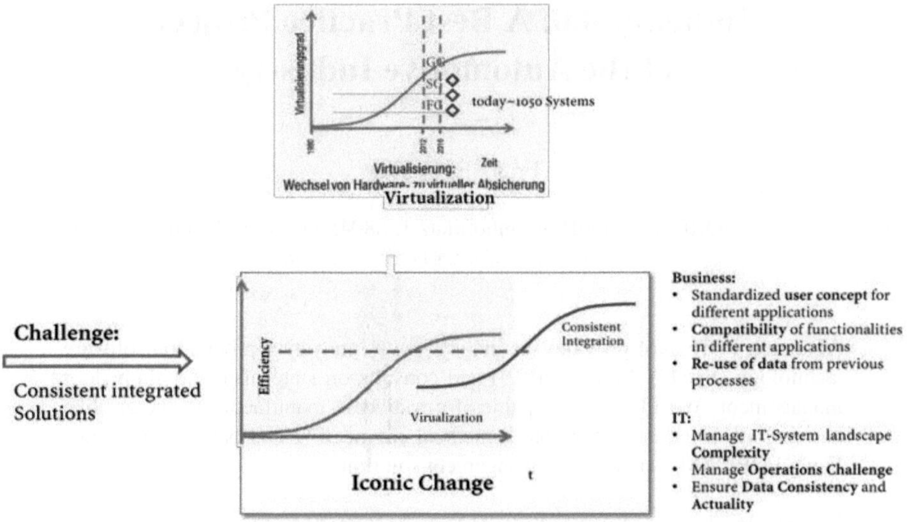

Fig. 6. Product design challenges

Industry 4.0: A Best Practice Project of the Automotive Industry

Franz E. Gruber

FORCAM GmbH, Bahnhofplatz 1, 88045 Friedrichshafen
franz.gruber@forcam.com

Abstract. FORCAM provides the industry with innovative production software technology (Factory Framework™) and consults on state-of-the-art shop floor management. World class companies from diverse manufacturing sectors like automotive, machinery, aerospace, medical engineering and packaging are utilizing FORCAM's shop floor management solution.

Keywords: shop floor, automotive, sensors, best practice, FORCAM.

1 Introduction

It is worldwide known and accepted that FORCAM provides the industry with innovative production software technology (Factory Framework™) and consults on state-of-the-art shop floor management. World class companies from diverse manufacturing sectors like automotive, machinery, aerospace, medical engineering and packaging are utilizing FORCAM's shop floor management solution. It enables companies to measure performance of their machines and plants in real-time to quickly recognize and correct errors and waste, and to continuously optimize their productivity.

2 FORCAM Advantages and Results

With FORCAM technology, elite companies have increased productivity by up to 25% within the first year, measured by TRUE Overall Equipment Effectiveness (OEE).

- Daimler (Mercedes-Benz)
  - Factory of the Year Award (by A.T. Kearney)
  - Delivering monitoring and optimization of more than 14,000 machine tools
- Audi
  - Audi Stamping Plants – 20% increase in TRUE OEE within 12 months
  - Audi Components Plants – 10% increase in TRUE OEE within 12 months
- BorgWarner
  - Exceeding expectations with unique shop floor management strategy
  - 4.3% increase in TRUE OEE within 6 months
- Weir Minerals
  - 12% increase in TRUE OEE within 6 months

G.L. Kovács and D. Kochan (Eds.): NEW PROLAMAT 2013, IFIP AICT 411, pp. 36–40, 2013.

FORCAM offers you a unique solution, based on the Lean Manufacturing Philosophy of how to successfully manage and operate manufacturing facilities.

"FORCAM will bring True OEE to your company."

3 At FORCAM, We Value TRUE OEE

With the help of FORCAM, elite companies have been able to increase productivity measured with TRUE OEE by up to 25% within the first year.

The Recipe for Success - holistic Shop Floor Management that effectively combines the most advanced Shop Floor Management System with the Lean Management philosophy.

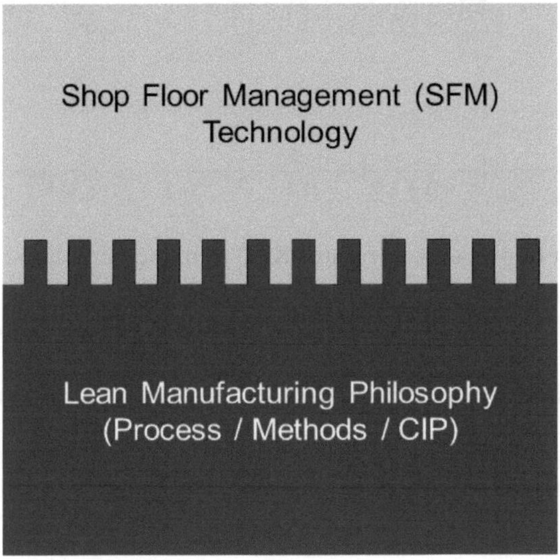

A Recipe for Success in Shop Floor Management

- Measuring REAL OEE - what you cannot not measure you cannot improve
- Adopting Lean Management principles and lead by OEE
- Creating a Continuous Improvement Process (CIP) organization with closed-loop communication
- Delivering the right information to right people at the right time
- Delivering planned vs. actual information to shop floor operators
- Establishing a PDCA (Plan-Do-Check-Act) cycle with actions and measures
- Setting regular TRUE OEE goals for continuous improvement

4 Overall Equipment Effectiveness Is Optimized

OEE - Overall Equipment Effectiveness

Shop Floor Management (SFM) is a strategy designed for the continuous optimization of the plant availability by eliminating errors and waste. The most important key figure to measure the performance is the OEE – Overall Equipment Effectiveness.

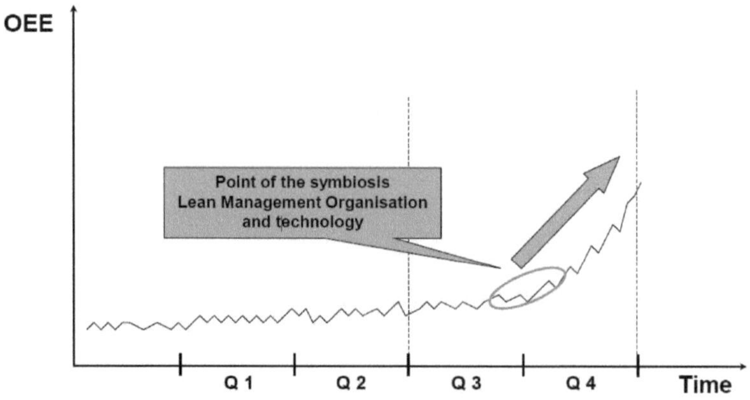

- OEE is the most important key figure (KPI) to implement the SFM strategy.
- OEE is the measure for the added value of a machine.
- OEE visualizes exactly which machine and process losses can be minimized.

OEE - Three Crucial Performance Metrics: Availability, Productivity, Quality

As key performance indicator, OEE results from three performance data which FORCAM monitors and visualizes in real-time, and therefore available anywhere and at any time:

- How long has it been producing?
- How fast has it produced?
- How well has it produced?

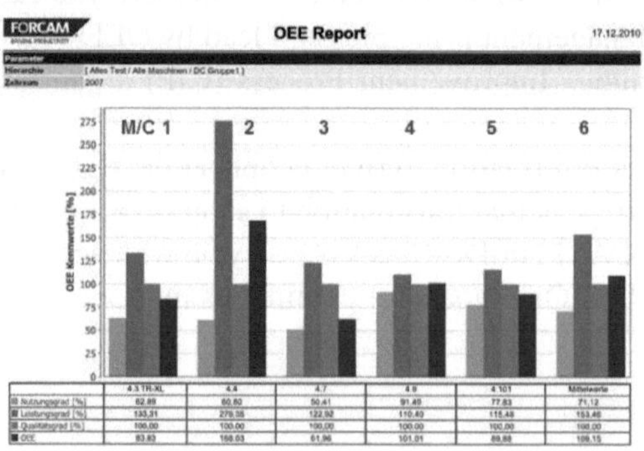

5 Advanced Shop Floor Management Creates the Transparent Factory

Measuring and Optimizing the Performance of Manufacturing Plants in Real-Time

FORCAM provides the industry with innovative production software technology Factory Framework™ and consults on modern shop floor management. Leading companies from diverse manufacturing sectors like automotive, machinery, aerospace, medical engineering and packaging are utilizing FORCAM shop floor management solution. It enables companies to measure performance of their machines and plants in real-time to quickly recognize and correct errors and waste, and to continuously optimize their productivity.

Planning and Production are Now Synchronized

"Nowadays, companies need resource efficiency in processes and in the manufacturing."

FORCAM's Chairman, Franz Gruber, states. Everything has to intertwine in a way that errors and waste can be recognized immediately and eliminated sustainably. The FORCAM technology complements standard ERP-software on the planning level (top floor) using objective performance data coming directly from all factory assets (shop floor) – from one single machine to multiple plants worldwide. The efficiency factor of standard ERP-software is enhanced due to the real-time data capture by FORCAM. Concurrently, FORCAM enables customers to apply the technology themselves in their company following a pilot phase.

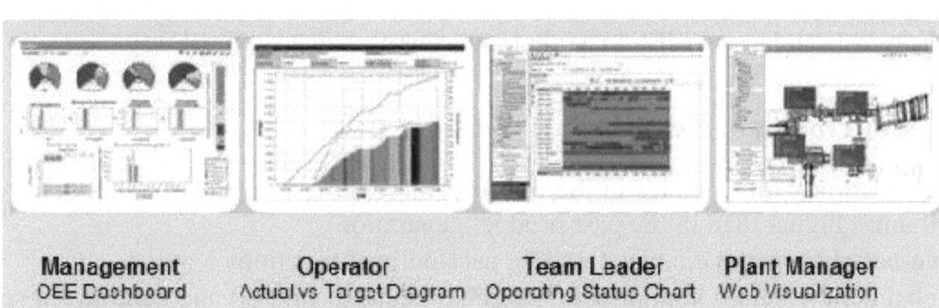

6 State-of-the-Art Shop Floor Management

Flexible Web-Based Connectivity of a Diverse Number of Machine Controls

FORCAM's Factory Framework™ solution is technologically leading Manufacturing Execution System (MES). It enables improvement processes on all levels of the production. Factory Framework™ connects, collects, and standardizes real-time data from most different machine controls. For every production level, the signals provide comparable web-based operational states of machines and plants (production, malfunctions, and downtimes). Result: Transparent and reliable performance data – runtime, speed, quality. Additionally, Factory Framework™ connects production data with ERP data from corporate planning such as targets for orders and resources.

From the Operator to the Manager

All roles involved are able to continuously optimize productivity within their role.

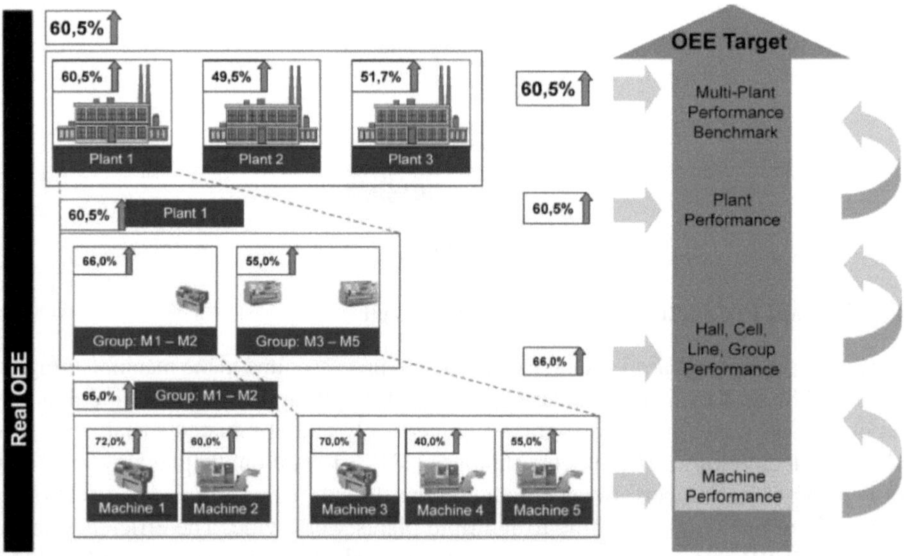

Leading-edge Technology: Complex Event Processing (CEP) with Live Cache FORCAM uses Complex Event Processing (CEP) with Live Cache, a technology used for real-time analysis of event stream from the financial sector. FORCAM combines CEP with the proven and tested MES in a rule-based approach in an optimal way.

7 Conclusions, Results

- Fast reaction with latencies in the millisecond range and high throughput (factor is 100 times higher than in the past product generation)
- Rule-based approach enables flexible, user-defined functions
- Global deployment potential with Unicode language support and centralized server architecture

JT Format (ISO 14306) and AP 242 (ISO 10303): The Step to the Next Generation Collaborative Product Creation

Alfred Katzenbach[1], Sebastian Handschuh[2], and Steven Vettermann[3]

[1] Executive Consultant
alfred.katzenbach@daimler.com
[2] Head of JT Workflow Forum
sebastian.handschuh@daimler.com
[3] General Manager of ProSTEP Association
steven.vettermann@prostep.com

Abstract. JT format became ISO Standard ISO14306 in December 2012 [1]. It is the beginning of new effective CAD collaboration in Multi-CAD environments. In parallel it gives innovative IT providers the chance to contribute with advanced solutions to challenge the established IT vendors. Through smart combination with the upcoming STEP AP 242 XML for the first time a process-oriented solution for supporting automotive data exchange requirements is available.

Keywords: JT Format, ISO 14306, STEP AP 242, ISO 10303, Engineering Collaboration, Product Creation.

1 Introduction

Digital Engineering Visualization is an approach which gives a broad number of individuals, being part of the product creation processes, the access to 3D data without using a traditional CAD system. As depicted in Figure 1, the number of people who only need to view information is 10 times higher than of its creators.

Over the last 5 to 8 years, JT Format (at this time owned by Siemens PLM) became more and more a de facto standard for 3D visualization in the automotive and aerospace industry. Hence, the Global Automotive Advisory Group, a forum of the international heads of automotive engineering IT, recommended Siemens PLM to handover the JT file format description to ISO for international standardization. To make this happen, ProSTEP iViP Association, the well-known organization for standardization in the field of engineering collaboration took over the operational lead to provide all documents and arrangements to support the ISO organization in the formal process to get ISO 14306 approved.

The JT format contents not only a definition of a tessellated representation with dedicated level of detail. It defines also an exact geometry description based on BREP and e.g. product and manufacturing information.

G.L. Kovács and D. Kochan (Eds.): NEW PROLAMAT 2013, IFIP AICT 411, pp. 41–52, 2013.

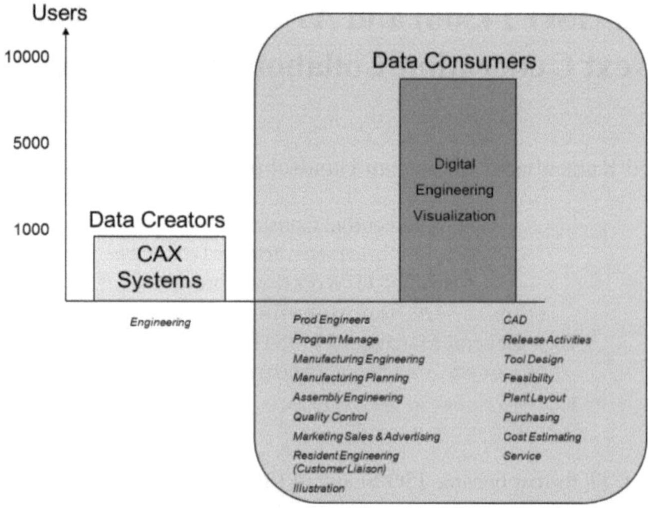

Fig. 1. Figure 1 - Hugh need for viewing information [1]

But for supporting all industrial visualization requirements JT itself could reach its full potential only by applying it in combined application with the smart XML-functionalities of the new STEP AP 242. Thus, in 2010, the decision was taken to establish STEP AP 242 XML and JT as the standard-combination for enabling lightweight data exchange. STEP AP 242 XML, as the process backbone, for assembly, meta, kinematic data etc. And JT as lightweight visualization format for 3D industrial data. By this, for the first time a process-oriented solution for supporting automotive data exchange requirements was agreed.

2 JT Based Processes

2.1 Process Overview

Only to provide an international standard without defining, how this standard can be used gains only limited value. So it is crucial to define processes first and underline these processes by concrete use cases.

Figure 2 gives an overview of the possible work share of native CAD and JT in an international automotive enterprise. Here we have to distinguish between internal company value adding processes which still require native CAD representation and integration processes between the OEM and the various numbers of suppliers. For this integration tasks, JT can contribute to be more flexible, faster and especially cost efficient.

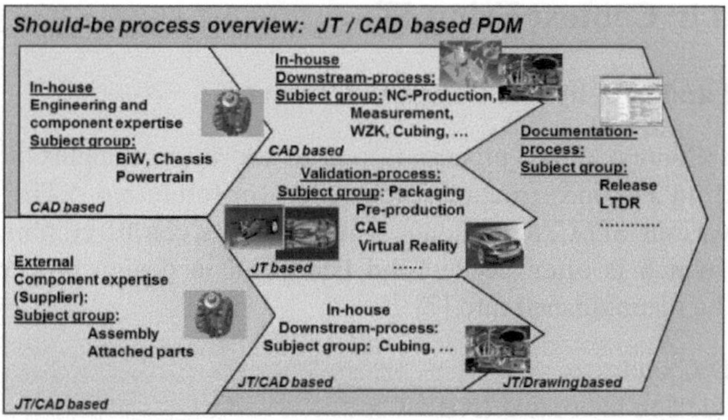

Fig. 2. Figure 2 -Future work share between CAD and JT (Source: Daimler, cf. [3])

2.2 JT Use Cases

To go into more detail it is recommended to detail the usage of JT in dedicated workflow descriptions.

The JT Workflow Forum, a working group under the umbrella of Pro STEP iViP Association defined such scenarios and published it in a whitepaper (cf. [3]).

The different use cases were classified in three categories, related to the complexity of usage

- Viewing
- Digital Mock Up
- Design in Context

2.2.1 Viewing

Viewing was the foundation of using JT. This includes also high end visualization, bidding and inquiry processes as well as material specification.

2.2.2 Digital Mock Up

Beside DMU analysis, the use cases in this cluster define processes for tolerance analysis, drawingless manufacturing, tolerance studies, supplier integration and al processes of digital factory.

2.2.3 Design in Context

Here the more advanced processes are defined. This includes hybrid design in context, all simulation activities as long term data retention.

3 Desgin in Context Using JT

3.1 Hybrid and Non-hybrid Design in Context

The Design in Context (DiC) process is one of the most complex processes which could be based on JT- and structure information (in future e.g. STEP AP 242XML) coming from various 3D-CAD-systems. These formats can be combined with native 3D-CAD-data which is often needed and beneficial in design processes (no media break). It is to be distinguished into [7]:

- Hybrid- (HDiC)and
- Non-hybrid design in context (NHDiC).

The hybrid design in context process directly uses JT-/ structure data as an input format without data translation into native 3D-CAD-formats (cf. Figure3).

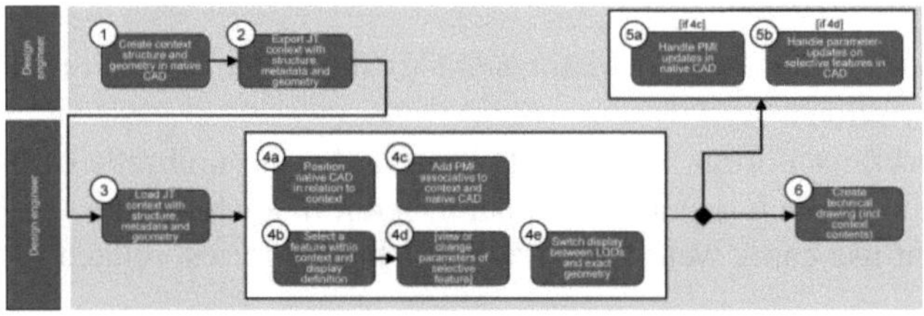

Fig. 3. Hybrid design in context use case [3]

In today's design in context processes often the exchange of geometrical change information is needed. The design in context process often refers to interface regions of (e.g. externally created) parts. Information located in these regions (e.g. geometrical elements) can be used for DiC-processes. The main categories in which this information can be divided are:

- Geometrical- and
- Technological information

Geoemtrical information is often used to define geometrical elements in surrounding parts on customer/OEM side. Therefore information such as diameters, curves or supplemental geometry is used. In case technological information is needed a broad range of information is thinkable in interface regions of parts e.g. mechanical-, electrical-information [7]. These interface related feature information can be transfered out of the sourcing 3D-CAD-system into JT-data using e.g. Product Manufacturing Information (PMI) or in future using a combination of JT and XML-based accompanying dataformats such as STEP AP242XML for the transfer of additional information. Work on this topic was done e.g. at the University of Kaiserslautern in context of value extracting use of open lightweight data formats (such as JT) in automotive collaboration and development process chains [7], [8]. Due to this fact the next chapter gives an overview of the value added chain in automotive industry and describes (as an example) a JT-based data exchange pilot project.

4 Supplier Integration

4.1 JT Based Supply Chains in Automotive Industry

The value added in many industries is decreasing on OEM side and increasing on supplier side. Several studies record this (cf.Figure 4) [5],[6]

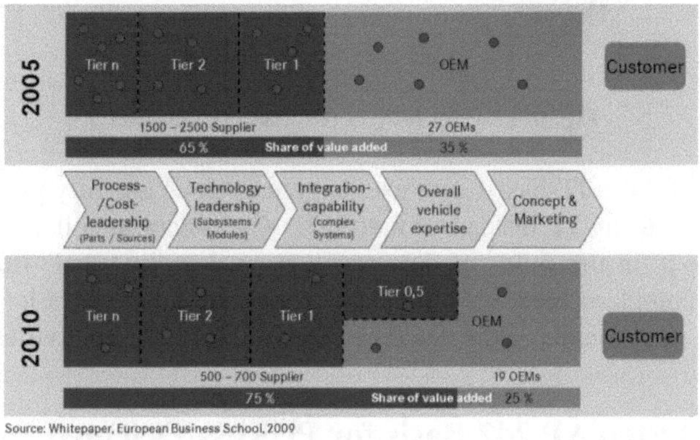

Fig. 4. Value added on OEM/Supplier side [5]

In the automotive industry often suppliers deliver complete car moduls such as door panels or dash boards.

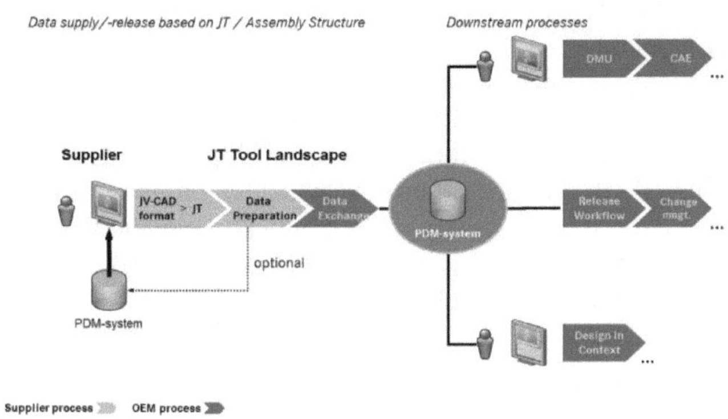

Fig. 5. Big Picture JT-based process land-scape

While system landscapes in context of global sourcing processes are typically heterogeneous the usage of ISO standardized data formats (JT /STEP APA242) for data exchange is getting more and more important. Figure 5 gives an overview of the processlandscape in such supplier integration scenarios. These processes are validated in automotive industry in several pilot projects e.g. with global automotive suppliers (cf. Figure 6).

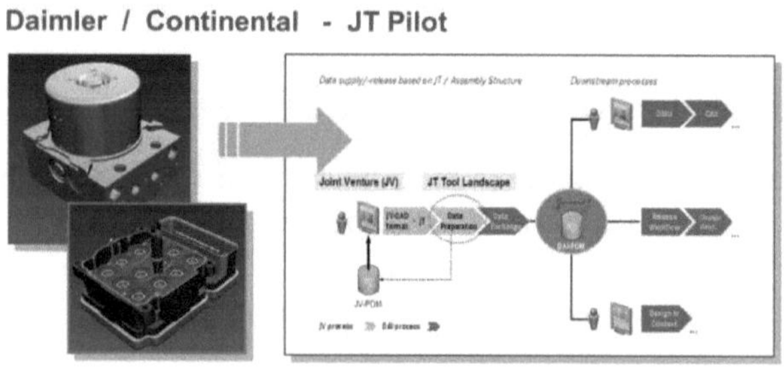

Fig. 6. JT-Pilot project (exchange of an ECU)

The results of these JT-based data exchange processes are very promising and further potentials in using JT instead of native 3D-CAD-data in collaboration processes are identified e.g. in the fields of data-preparation,-handling and -storage. To reach such a duo of complemented and ISO standardized data formats a lot of work has to be done. In the next chapter some detail of the standardization activities are shown.

5 JT and Step AP 242 Basis for Process-Oriented Data Exchange

5.1 Following Modern Standardization Concepts

Modern IT standards must be flexible and fulfil the industrial requirements. The challenge is to figure out the 20% that are needed to cover the 80% of practical applications (Pareto principle).

Standards can only provide industry with the necessary value added if there is a stable basis on which additional functionality can be developed to respond to context-specific needs. This may take the form of new developments or, preferably, developments that are already in widespread use but which can now be simply transferred to other areas. It should not be necessary to continuously reinvent the wheel. Within the context of Multi-CAD environments this has been seriously addressed by starting the standardization JT and STEP AP 242.

5.2 STEP Forms the Backbone

In 2010 representatives of leading aerospace and automotive companies as well as standardization organization came together in order to discuss measures for strengthen the future position of STEP. By this time, with regard to 3D engineering the two major standards "ISO 10303-203 - Configuration controlled 3D design of mechanical parts and assemblies" and "ISO 10303-214 - Core data for automotive mechanical design processes" reached an overlap of functionality by 75% over the years (cf. Figure 2). This was regarded as costly and a dissemination hindrance. Thus, the merger of these two standards was decided: STEP AP 242 (ISO 10303-24 - Managed model based 3d engineering).

AP 242 is designed to address the common aerospace and automotive use cases and to provide a single data backbone. Its publication as ISO standard is planned for beginning of 2014.

AP 242 offers more and/or enhanced functionalities compared with its parents AP's, AP 203 and AP 214. But it also offers additionally a new concept: the Business Object Model, the so-called BO-Model.

Major enhancements are provided in AP 242 by adding functionalities for (cf. Figure 3):

- Composites
- Kinematics
- Product Manufacturing Information
- Rules and Requirements
- Shape Data Quality

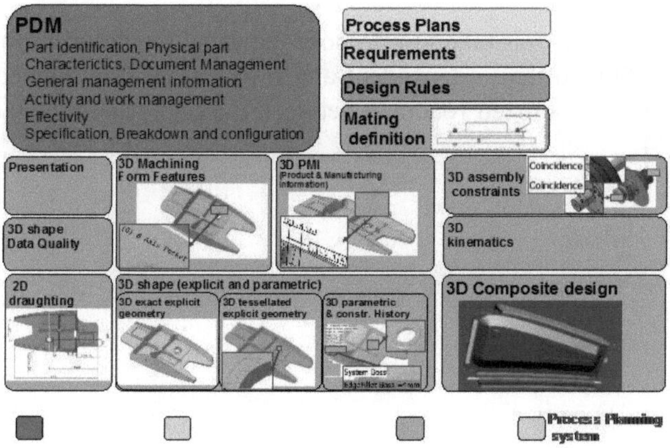

Fig. 7. Figure 3: AP 242 information model overview

Additionally, the new External Element Reference Mechanism is integral part of AP 242. Unlike in the past, by this mechanism it is not only possible to reference a file (e.g. from an assembly tree to parts), it is now possible to reference elements within a (STEP)file (e.g. a curve, an edge, an axis). This is extremely important for e.g. kinematics, where one needs to reference certain constraints.

The BO-Model is a subset of the AP 242 functionalities. It covers all data relevant for data management, e.g. meta data. This also means, that it does not cover e.g. 3D geometry itself, it only links to it. Beside the tradition representation of a STEP-File via a STEP Physical File (SPF), the BO-Model also provides a representation expressing the equivalent information in XML. Thus, the intent behind the BO-Model is to provide a standard-based representation of product data management data, as SPF and in XML.

Especially the new capabilities to express PDM-relevant data in XML and the described External Element Reference Mechanism make AP 242 so attractive to use it together with JT in the field of visualization. And this was one of the main motivations to go for this.

5.3 JT as the Leafs of a (STEP)tree

In the past, STEP became the standard for the exchange of product model data. There is no better. But in one field it failed: visualization.

One could now state: So, let's invest resources and in some years we could close this gap. But in 2010 industry required a standard 'right now'.

In the same time, STEP matured to what it is today, JT matured to the leading visualization solution. Thus, it was obvious to start the standardization of JT via ISO and to take care, that the two standards could by applied separately, but if they are used together, that they form a process-supporting solution which is more powerful than its pieces.

In 2010, the decision was taken to propose the standardization of the complete JT Version 9.5 via ISO, so that independent and royalty-free solutions can be realized. Figure 4 show the harmonized roadmap between the two standardization projects JT and STEP AP 242.

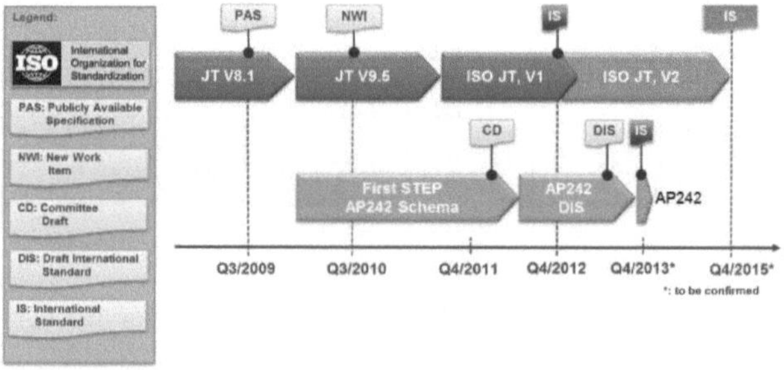

Fig. 8. Figure 4: ISO Roadmap for JT and STEP AP242

The agreed aim was to establish a visualization standard which could use the enhanced XML-functionalities of STEP AP 242. By this, JT should be kept lightweight. Frankly said, JT provides the 3D-relevant data within a STEP-tree.

5.4 JT Data Model

The data model for JT includes more than just geometry models (cf. Figure 5). On the one hand, it offers the option of displaying the models in various degrees of precision and, on the other hand, other information that is important for product development and downstream processes can be stored together with the model.

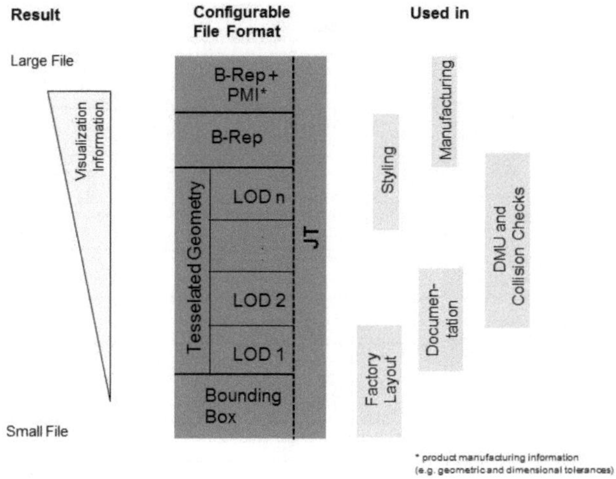

Fig. 9. JT Data Model

In addition to the ability to display geometry, it was the fact that product manufacturing information (PMI) such as tolerances, toleranced dimensions and information about surface properties could be displayed together with the 3D models and processed further that contributed to its fast proliferation.

Geometry

- At one of the lowest levels, simple regular geometry such as cuboids, cylinders and pyramids – for the definition of designed spaces for example – are located in what is referred to as the bounding box.
- Tessellated or faceted geometry allows the representation of solids and surfaces. JT offers the option of defining different levels of detail (LOD), i.e. specifying how precisely something is to be represented. A very low LOD means less precision and a smaller volume of data, a very high LOD means a level of precision approaching the exact geometry and a larger volume of data.
- The latest form of compression is JT ULP, which was released in 2008 and allows the volume of data to be reduced to almost one hundredth the size of the original data.
- BREP (boundary representation) geometry offers the highest level of precision. They are an exact representation of the solids and surfaces that were generated in the CAD system, for example. The exact JT file is also only a fraction of the size of the original file.

Through certain compression algorithms it's possible to reduce the size of a representing the same content tremendously (cf. Figure 6).

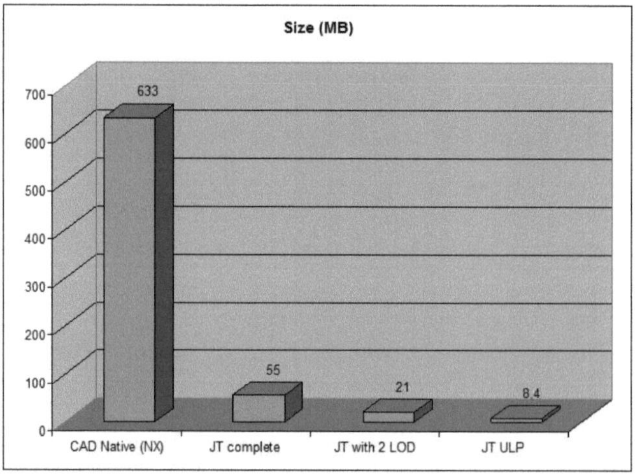

Fig. 10. Figure 6: JT file size comparison (Source: Siemens PLM)

6 Opportunities for Innovative It Provider

The various application scenarios give rise to numerous requirements that need to be met by the JT and STEP data. Questions also rise with regard to the quality of the JT data supplied by the systems involved. In other words, on the one hand they relate to the authoring systems and on the other hand to the translators that generate JT from a native format or can convert JT into a native format.

In order to provide a reliable answer to these questions, systematic benchmarking of the tools involved is needed. Until now, the ProSTEP iViP Association, together with the VDA PLM Working Group, conducted three neutral Benchmarks. The latest was published beginning of 2013 (cf. [4]). In this benchmark the focus was on LOD- und PMI-related functionalities as well as the performance of available JT viewers.

Based on the derived use cases and the related requirements that the JT and STEP need to satisfy the JT Workflow Forum, which comprises a number of companies from the automotive industry, and the JT Implementor Forum, which comprises leading IT vendors and providers of translators and system integrations, set up metrics and test models for conducting state-of-the-art benchmarks. A participants gathered rich information out of the fruitful discussions and took the chance for process and product enhancements.

The next benchmark is already in preparation and the complexity as well as the expectations with regard to the performance of the tested products is rising year by year (cf. Figure 7).

The upcoming benchmark will focus on the two building blocks illustrated in Fig. 8.

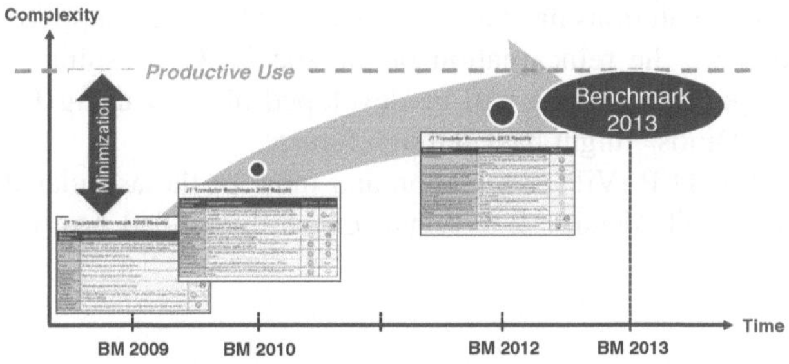

Fig. 11. Complexity of the test scenarios is raising

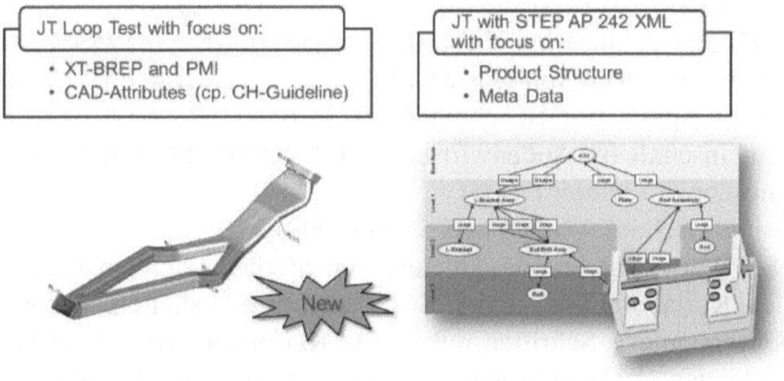

Fig. 12. Building blocks of the JT Benchmark 2013

Speaking about Multi-CAD environments, within these benchmark the involved CAD Systems are CATIA V5, Creo 2, NX 8.5 and Solidworks 2012. And the tested translators are from Core Technologie, Geometrics, Elysium, ITI Transcen-Data, Kronion, Siemens PLM, SpaceClaim, Theorem and T-Systems.

The already gathered information underline that industry took the right step when deciding to address the challenge of Multi-CAD environments with a combination of the standards JT and STEP.

7 Conclusions

JT and STEP AP 242 are on the way to change CAD based development and gain much more benefits out of 3D data than possible in the past.

The ISO Standards provide a reliable and sustainable definition for product describing data including product structure and behaviour models.

Powerful low cost and or freeware solutions will appear for a more intensive use of product data. Special solutions for ultra-light weight descriptions under JT definitions are on the way to be established.

For DMU and other integration processes international and multi partner collaboration can be defined in a new way. Here the specific solutions of each partner can be respected as long as he is committed to the two ISO standards.

Design in context solutions have a better chance when they support JT and AP 242. This will based force the reincarnation of the special CAD solutions for dedicated tasks. Secondly special solutions will be developed directly using JT definition. The one fits all CAD "Dinos" might have a tough future.

In this spirit ProSTEP iViP Association and most of the associated companies are engaged to work on JT Version 2 to enhance to value added position of JT and AP 242.

References

1. ISO: Industrial automation systems and integration – JT file format specification for 3D visualization, ISO 14306:2012 (2012), `http://www.iso.org`
2. SASIG: D-21 Digital Engineering Visualization (2005), `http://www.sasig.org`
3. ProSTEP iViP: White Paper Applying JT (2010), `http://www.prostep.org`
4. ProSTEP iViP: JT Application Benchmark (2013), `http://www.prostep.org`
5. Forschungsgemeinschaft für Außenwirtschaft (FAST), Studie: Innovationsstrategien für erfolgreiche Automobilzulieferer (2008)
6. European Business School, White Paper 2009 (2009)
7. Handschuh, S.: Wertextrahierende Nutzung von offenen leichtgewichtigen Datenformaten in automobilen Kollaborations- und Entwicklungs-prozessketten. Dissertation Technische Universität Kaiserslautern, Schriften-reihe VPE, Kaiserslautern, vol. 11 (2011)
8. Roth, M., Sprickmann, A.: Prozessüber-greifende Nutzung von VR-Technologien im Schiffbau (Tagungsband Go3D). Fraunhofer Institut für Grafische Datenverarbeitung, Rostock (2010)

Economic Advantages by CAD/CAM Use in Compound with Production Data Organization

Arndt Richter

EXAPT Systemtechnik GmbH, Aachen
info@exapt.de

Abstract. Current developments concerning the manufacturing industry cause also new and additional requirements for the use of CAD/CAM systems. It is represented that beside the technological further development especially the virtualization of the planning processes has significant influence. Furthermore the local technical consideration of the CAD/CAM process is not sufficient. The general organization in the factory has to be considered. The integrated consideration of the physical specification of the production means with the organization of the production resources ensures a long-term economic success for the CAD/CAM use. To realize this target the possibilities and advantages are described in this article.

Keywords: CAD/CAM, multi-task machine tool, automatic feature recognition, 3-5-axis-milling, virtualization, simulation, tool management, production data management, blank part actualization, tool sequence optimization, 3D tool models, production resources, communication platform.

1 Introduction

In the manufacturing industry a highly interesting upheaval takes place presently. Not only new production means with new technical possibilities, modified production structures with out- or insourcing or whole production transfer to other production areas, but especially also new approaches for the networking of information about the complete production operations is meant. These new approaches, also summarized under the slogan Industry 4.0, will allow among others

- the further growing together or the parallelism of product development and product design processes
- the better linkage of virtual and real world of the production as well as
- the penetration of the communication in all areas of planning, preparation and execution of the production.

Thereby the digitalization of the production sequences turned out as one of the greatest driver of innovation. The constantly improving junction of the virtual and real production will increasingly be important for an efficient production. Therewith also for classical processes of production planning, e.g. CAD/CAM planning or production

G.L. Kovács and D. Kochan (Eds.): NEW PROLAMAT 2013, IFIP AICT 411, pp. 53–60, 2013.

data and tool management new tasks arise, which only lead to a better economic representation in interaction of all components.

CNC-controlled machine tools always require digital data, that means digitized geometry, technology and organizational data as input for the machine tool control. In the course of the development powerful CAD/CAM systems originated from the simple NC programming, which include under the aspect of the overall system machine tool - clamping device - tool - workpiece extensive planning functions. But for the efficient use of these systems more and more digital data of the process control, production means and production resources are required to create safety and reliable results. Systems for the production organization as PLM-, tool management- or MES-systems have been established parallel to the geometric and technical oriented CAD/CAM systems. All systems require similar, partially equal data concerning the content, to generate the demanded planning results. The required data are managed locally and in general not compatible because of different development approaches and development targets. From the point of view of an economic production the great challenge is to obtain the continuity and consolidation of the organizational and technical oriented systems.

2 Challenges for CAD/CAM Solutions

The requirements of the manufacturing industry concerning the CAD/CAM systems are very versatile and characterized by different factors.

In the machine tool sector currently dominate especially multi-task machines, which enable a mix of processes in many different ways. Therewith it is possible to machine workpieces completely on one machine. However, this causes a more and more complex kinematics of the machine tool and in the large machine tool construction the increasingly dominating influence of change-over units like angle heads, facing heads etc.

Directly associated with the development of the machine tool sector are the innovations in the field of tool technology. New production engineering and production strategies can be realized with according modern tools. Examples are High Speed Cutting or High Performance Cutting, but also adaptive machining strategies and combined procedures, e. g. of classical cutting methods in combination with laser-use.

On the other hand new production organizational strategies have been established in the production. An unbroken change from the batch production "on stock" to a client-specific production as well regarding the order number as also the constructive specification can be recorded. Therewith associated are always smaller batch sizes or piece number 1, whereby the efficient single production gets the primacy. Under this aspect preformed workpieces can only be used in a limited way, whereas the machining "from one block", e.g. also the production of prismatic parts from bar material, is becoming increasingly important.

From the mentioned strategic developments in the production and organization area permanently new changing and more extensive challenges result for the CAD/CAM planning. So e. g. the rate of material removal increases permanently and as a result the expenditure for planning, preparation and program generation for the CNC machine clearly increases. The complexity of programming tasks is also growing. The expenditure per program and the number of necessary NC programs per time unit continue to rise drastically and therewith also the CAD/CAM planning is asked to gain more efficiency.

In addition to the pure technical program generation also the requirements concerning the reliability and safety of the created results and the documentation of the planning results increase continuously. Only with clearly improved and more efficient planning methods the requirements from complex production systems can be realized with the requested quality.

Because of the reduction of batch sizes the start-up time increases also drastically in relation to the entire production time. It is required that additionally to a fast and flexible program supply for the machines all necessary production-relevant documents and information are provided reliably. Only this way it is possible to reduce the start-up time drastically. The requirement to minimize the start-up time is emphasized because the low batch sizes require an enhanced flexibility in the production. This has also an impact on re-planning in the production engineering which has to be realized CAD/CAM system supported. A corresponding system support enables to realize the requirements regarding a flexible production with low batch sizes.

Current and future production facilities cannot be seen without a look to the overall organizational process in the factory. Essential requirements to CAD/CAM systems arise because of these integrated considerations from the product development to the CNC machine including the data, resource and quality management. The integrated consideration of all areas is decisive for a successful economic production.

3 Technical Aspects

The NC preparation has to react fast and flexibly to the often changing requirements from the manufacturing. The results have to be provided with high safety so that the following process steps can be created reliably. For that current CAD/CAM systems offer an lot of possibilities so work fast and efficient. 3D-CAD workpiece models, designed regarding the production technical aspects, are the starting point. This statement is important because for an integrated CAD/CAM planning the responsibility for the measure is based on the CAD model. Therefore pure 3D visualization models of workpieces are not suitable for production engineering. Using an automatic, in modern CAD/CAM systems integrated, feature recognition the required parameters for the production-oriented processes can be determined quickly. Accordingly adapted systems provide very quick and flexible nearly fully automatic results for an efficient production this way (fig. 1).

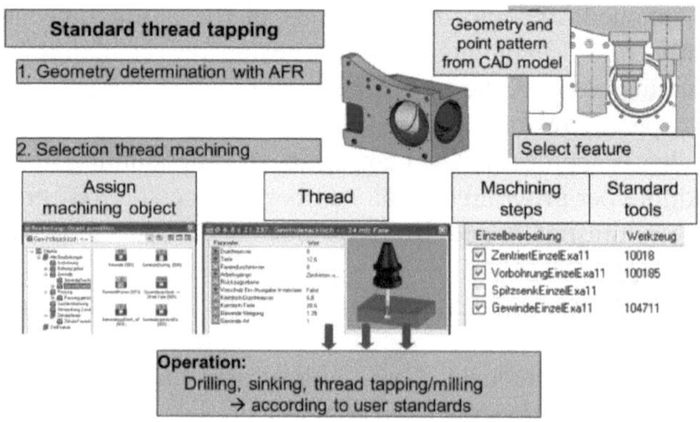

Fig. 1. Automatic feature recognition with automatically generated and client-specific extended machining

Thereby it is important that in addition to standard sequences respective custom-designed machining strategies can be stored to guarantee a know-how protection and also individual company-specific sequences.

For the complete machining of a workpiece with several machining steps in terms of a mix of process with turning, drilling and milling of 2 - 5 axes a dynamic blank part actualization is indispensable. For each machining step the current blank part has to be provided to plan the following machining step with the current geometry data (fig. 2)

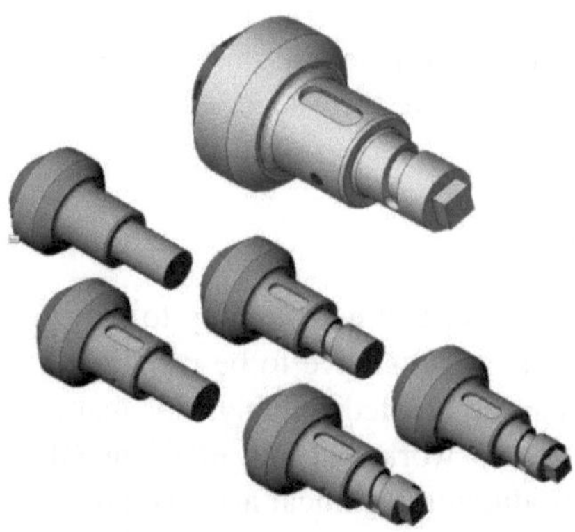

Fig. 2. CAD/CAM planning with blank part actualization in all processes steps

Because of the complexity of the workpiece geometry and the production requirements resulting therefrom it is more and more necessary to use additional sets on machine tools . Only this makes it possible to realize the geometric and qualitative machining requirements. In the large part manufacturing sector such change-over units are absolutely essential, because a relocating of the workpiece is extremely complex, so that the total workpiece has to be machined with one or two settings as possible. The use of different angle heads, facing heads etc. is the rule. Large machine tools have additionally different parallel axes so that the tool can approach to the machining point efficiently. For the CAD/CAM planning such machine kinematics are special challenges. Not only the pure machining has to be regarded and the tool paths to be generated but especially the entire question of the use of aggregates have to be considered. This is the consideration of the tool and attachments handling here, especially the head positioning and the head change. A further emphasis is in this context the tool sequence optimization over several machining sides including the use of change-over units.

Beneath the organizational handling of change-over units including the handling of perhaps changing coordinate systems for the several machining sides of different aggregate applications the visualization of the complete process has the highest priority. For the simulation of the machining sequence it is differentiated basically the tool path-related simulation and the machine simulation based on the NC codes. EXAPT has the target to provide a respective detailed simulation already in the planning of the several process steps. This tool path-related simulation including the display of the kinematics of the machine is completed by a removal simulation with integrated collision control, so that the user gets a largely realistic display. The advantage of the immediate visualization of each machining step during the planning stage is that subsequently no corrections have to be made. Basically both simulation strategies complement one another and lead, at correct application, to high efficiency in the production. All aspects of change-over units have to be regarded and visualized in the case of tool path-oriented simulation, naturally. Especially the Placing of the attachments to the workpiece for the processing and the collision check of the attachments with the workpiece, the clamping devices and other machine elements are important (fig. 3).

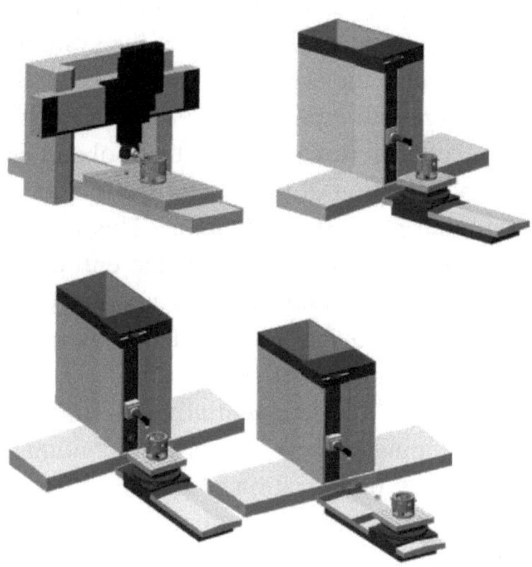

Fig. 3. Simulation of the machine environment based on tool paths

The specific challenge to CAM systems is for this simulation strategy in the planning phase not only to simulate machine motions but also to display transfer motions of the machine tool e. g. concerning the tool or aggregate change, where the tool is not used. The current EXAPT developments have set broad standards in this area. The entire kinematics of the machine tool can be displayed because of the integration of all movements, also the passive machine movements in the processor, and the range of functions of the postprocessor can be reduced explicitly (fig. 4).

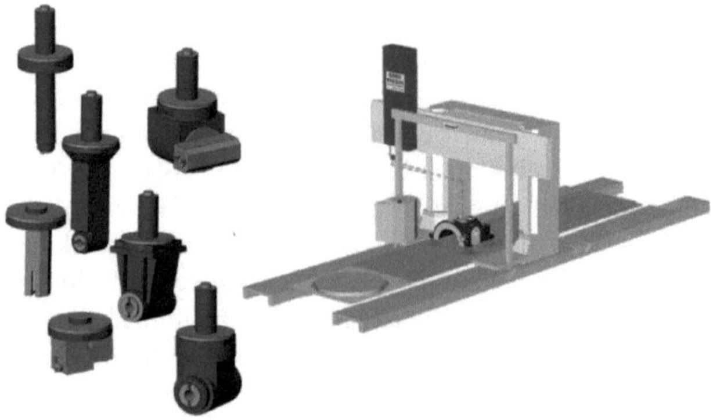

Fig. 4. Large machine tool with tool and attachment change

4 Organizational Aspects

Regarding the described complex manufacturing procedures in detail it quickly becomes clear that beside the physical specification of the system machine tool - clamping device - tool - workpiece also the organizational states of the production resources are eminently important. Therewith especially the tools move into the focus.

In addition to the pure technical description of the tool and especially of the cutting edge also the geometric specification of the assembled tool as a 3D model is required for a safe technology-oriented production planning. This is the condition for a safely collision control and an objective visualization of the entire process. There is a need to distinguish in general that concerning the tool geometry has to be differentiated between a physical and a technological tool model. This quickly becomes clear for asymmetric rotatory tools as boring bars (fig. 5).

The technical description of the tool essentially determines the accuracy and therewith the quality of the planning process. Important economic effects are developed additionally by a corresponding organization of the classic tool department. This are not simply questions e. g. of the storage of tool components or the pre-reservation of special tools for the machining but especially the tool logistic and the job oriented scheduling of tool deployment on the shop floor.

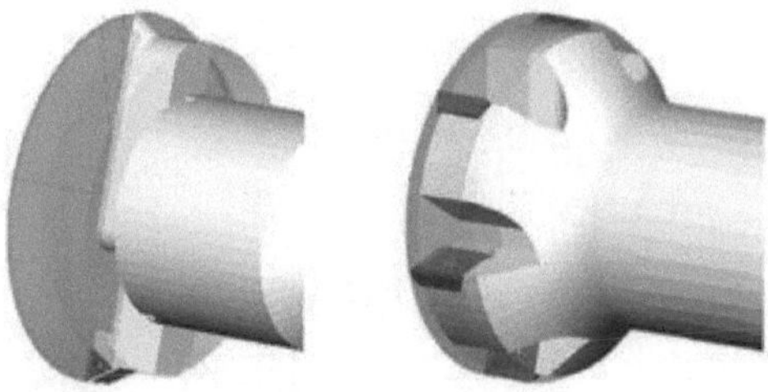

Geometry model
➔ real tool model (stationary)
Technology model
➔ application-oriented tool model (rotatory)

Fig. 5. Appropriate tool data for the NC production

Scheduling of tool deployment thereby comprises the whole process from the tool selection in the CAM process via the generation of orders and the provision of tool data for tool presetting machines up to the transfer of the necessary data to the machine tool. A significant reduction of set-up time can be reached by combining several tool setting orders with an according sequence optimization of the production orders. Thereby the savings are not only in the machine area but also in the tool presetting area, because fewer tools have to be set. Savings up to 30 % of logistic time for tools in relation to normal processes can be verified with this solution. In the comparison between the real set-ups state on the machine with the planned set-up orders are further effects which can be realized with an online communication to the machine tool.

5 Integrated Data Concept between Technique and Organization

The above-mentioned statements clarify that at every point of the process planning safely and reliable information of the state of the production means and production resources have to be available. Considering only one side of the planning tasks the positive results are ruined by other negative influences. Therefore it is necessary to create a corresponding communication to enable the provision of the relevant data in every process step. This is the only way to ensure that the planning results really correspond to reality. Beside reliability and completeness of the required technical characteristics an according communication platform is extremely important (fig. 6).

This communication platform does not only have the function e. g. to exchange data between a CAD/CAM and a tool management system but has to consider the entire process.

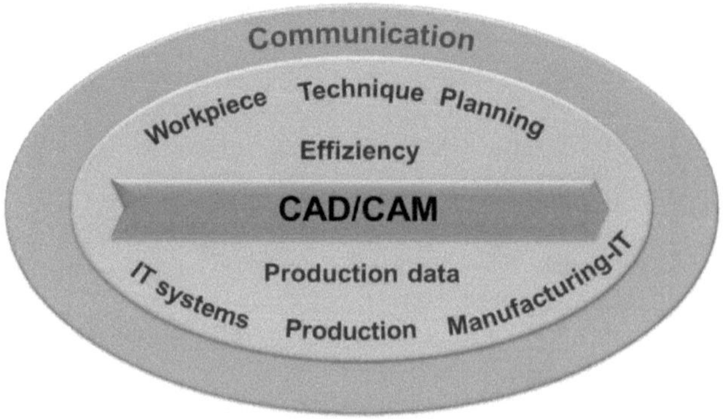

Fig. 6. Technique, organization and communication in the area of the CAD/CAM system

All partners participating in the entire process have to be addressed by suitable communication techniques, e. g. WEB-based solutions, and to be integrated into the general communication. In the area of tool logistic this refers to the questions of communication with different storage systems, the mobile tool identification, the tool presetting or the communication with NC controls. Only this complex logistic consideration makes it possible to obtain the economic effects resulting from this solution. But it is also clear that a corresponding dovetailing of the information is absolutely essential today. With the EXAPT production communication manager EXAPT provides a system which manages a browser-oriented uncomplicated communication with the several stations in the shopfloor. Additionally it is possible provide all production relevant information via respective techniques at every place. Therewith according process sequences can be created more transparent, information can be exchanged immediately and bottlenecks can be displayed transparently more quickly, if necessary.

6 Conclusions

It can be seen that an economic CAD/CAM-use requires a high technological performance of the system combined with according simulation techniques for the visualization and control of the results, but additionally the compound to the organization, especially to the tool management, are indispensable. Due to the close connection of CAD/CAM and organization the economic success can be realized, especially in terms of small run and client-oriented production. Thereby it is not a matter of point to point connections but by building a communication platform the matter is a strategic orientation of the whole communication technique inside a CAD/CAM compound.

A 'Lean' Fuzzy Rule to Speed-Up a Taylor-Made Warehouse Management Process

Lapo Chirici[1] and Kesheng Wang[2]

[1] Department of Business Informatics – University of Pisa (Italy)
lapochirici@gmail.com
[2] Department of Production and Quality Engineering – Norwegian University of Science and Technology (Norway)
kesheng.wang@ntnu.no

Abstract. The minimization of the inventory storage cost and - as a consequence - optimize the storage capacity based on the Stock Keeping Unit (SKU) features is a challenging problem in operations management. In order to accomplish this objective, experienced managers make usually effective decisions based on the common sense and practical reasoning models. An approach based on fuzzy logic can be considered as a good alternative to the classical inventory control models. The purpose of this paper is to present a methodology which assigns incoming products to storage locations in storage departments/zones in order to reduce material handling cost and improve space utilization. The iterative Process Mining algorithm based on the concept of Fuzzy Logic systems set and association rules is proposed, which extracts interesting patterns in terms of fuzzy rules, from the centralized process datasets stored as quantitative values.

Keywords: Logistics, Warehouse management, Putaway process, Fuzzy rules, Data Mining.

1 Introduction

In this era of such drastic and extemporaneous changes, manufacturers with global view pay strong efforts in striving to achieve a lean production, outsource their components, and manage the complexity of supply chain [2].

Warehouse management plays a vital role to be a central actor in any kinds of industry which put-away process is a key activity that brings significant influence and challenges to warehouse performance.

In this dynamic operating environment, reducing the operation mistakes and providing accurate real time inventory information to stakeholder become the basic requirements to be an order qualifier. Here, an OLAP based intelligent system called Fuzzy Storage Assignment System (FSAS) is proposed to easy manipulate the decision support data and rationalize the production in terms of storage location assignment problem (SLAP).

In condition of information's uncertainty, fuzzy logic systems can provide methodologies for carrying out approximate reasoning processes when available.

G.L. Kovács and D. Kochan (Eds.): NEW PROLAMAT 2013, IFIP AICT 411, pp. 61–72, 2013.

Identifying an approach that can bring out the peculiarities of the key operations of warehouse is focal to track the priorities for the storage in terms of Stock Keeping Units (SKUs) [3]. Hence the need to develop a Put-away Decision Tree in order to automates the analysis of possible hidden rules useful to discover the most appropriate storage location assignment decisions. Some examples of SKU features are their dimensions, weights, loading values and popularity. All these are important in order to find out the relationship between the SKU properties and the relative assigned storage location. The aim of the paper is to create an algorithm that would be able to provide the best allocation position for SKUs in a just-in-time manner and with a lean and intelligent stock rotation. This approach provides strategic decisions to optimize the functionality and minimize the costs in a full automated warehouse.

2 The "PUTAWAY'S DILEMMA"

2.1 Manage the SLAP

Warehouse storage decisions influence the main key performance indicators of a warehouse, such as order picking time and cost, productivity, shipping (and inventory) accuracy and storage density (Frazelle, 2002). The customers are looking always to obtain more comprehensive services and shorter response time. The storage location assignment problem (SLAP) results essential to assign incoming products to storage locations in well-defined departments/zones in order to reduce material handling cost and improve space utilization (Gu et al. 2007).

Handling the storage location process is an activity that requires the supervision of several relevant factors. Up to now, some warehouse management systems (WMS) have been developed to acquire "simple data" by the warehouse operators and let recorded to computer support in intelligent slotting (storage location selection) in such way to ensure a constant quality of information available [1].

Besides that, both the lack of relevant data and WMS low customization capability for supporting the put-away process, highlight a common problem the warehouse manager has to deal with. Thus, the put-away decisions are often based on human knowledge, sprinkled unavoidably by a high gradient of inaccuracy (and consequently long order time), which can bring to a negative impacts on customer satisfaction [9].

2.2 Previous Theories on SLAP

Warehouse is used to store inventories during all phases of the logistics process (James et al., 2001). The five key operations in warehouse are receiving, put-away, storage, order picking as well as utilizing and shipping (Frazelle, 2002). Hausman in 1976 suggested that warehouse storage planning involves decisions on storage policy and specific location assignment. In general, there are a wide variety of storage policies such as random storage, zoning, shortest/closest driveway, open location, etc

(Michael et al., 2006). As each of the storage strategy with its own characteristics, there are different ways to solve the storage location assignment problem (SLAP). Brynzer and Johansson (1996) treated SLAP improving a strategy for pre-aggregate components and information for the picking work in storehouses. And this latter leveraging on the product's structure/shape in order to reduce order picking times. Pan and Wu (2009) developed an analytical model for the pick-and-pass system [5], [6]. His theory was founded on three algorithms that optimally allocated items in the storage, analyzing apriori the cases of a single picking zone, a picking line with unequal-sized zones, and a picking line with equal-sized zones in a pick-and-pass system. A nonlinear integer programming model built on a branch-and-bound algorithm was developed to enlighten class-based storage implementation decisions, considering the storage space, handling costs and area reduction (Muppani and Adil, 2008).

2.3 Introducing Fuzzy Logic

Fuzzy logic has already proven its worth to be used as tool to deal with real life problems that are full of ambiguity, imprecision and vagueness [3]. Fuzzy logic is a derivative from classical Boolean logic and implements soft linguistic variables on a continuous range of truth values to be defined between conventional binary. It can often be considered a suspect of conventional set theory. Since fuzzy logic handles approximate information in a systematic way, it is ideal for controlling non-linear systems and for modeling complex systems where an inexact model exists or systems where ambiguity or vagueness is common. A typical fuzzy system consists of a rule base, membership functions and an inference procedure. Fuzzy logic is a super set of conventional Boolean logic that has been extended to handle the concept of partial truth-truth-values between "completely true" and "completely false". In classical set theory, a subset U of asset S can be defined as a mapping from the elements of S to the elements the subset [0, 1], U: S -> {0, 1} [8].

The mapping may be represented as a set of ordered pairs, with exactly one ordered pair present for each element of S. The first element of the ordered pair is an element of the set S, and the second element is an element of the set (0, 1). Value zero is used to represent non-membership, and the value one is used to represent complete membership. The truth or falsity of the statement. The 'X is in U' is determined by finding the ordered pair whose first element is X. The statement is true if the second element of the ordered pair is 1, and the statement is false if it is 0.

3 From Fuzzification to Slam

3.1 Online Analytical Process

In order to collect and provide quality data for business intelligence analysis, the use of decision support system (DSS) becomes crucial to assist managers within problem solving critical area (Dunham, 2002). Online analytical process (OLAP) is a decision support system (DSS) tool which allows accessing and parsing data in a flexible and timely basis. Moreover, OLAP enables analysts to explore, create and manage enterprise data in multidimensional ways (Peterson, 2000). The decision maker, therefore, is able to measure the business data in different deeper levels and aggregate them depending on his specific needs. According to Dayal and Chaudhuri (1997), the typical operations performed by OLAP software can be divided into four aspects: (i) roll up, (ii) drill down, (iii) slice and dice and (iv) pivot. With the use of OLAP, the data can be viewed and processed in a real time and efficient way. Artificial Intelligence (AI) is one of the techniques that support comprehensive knowledge representations and practical manipulation strategy (Robert, 1990). By the use of AI, the system is able to learn from the past experiences and handle uncertain and imprecise environment (Pham et al., 1996). According to Chen and Pham (2006), fuzzy logic controller system comprises three main processes: fuzzification, rule base reasoning and defuzzification. Petrovic et al. (2006) argued that fuzzy logic is capable to manage decision making problems with the aim of optimizing more than one objective. This latter proved that fuzzy logic could be adopted to meet the multi-put-away objective operation in the warehouse industry. Lau et al. (2008) proposed a stochastic search technique called fuzzy logic guided genetic algorithms (FLGA) to assign items to suitable locations such that the required sum of the total travelling time of the workers to complete all orders is minimized. With the advantages of OLAP and AI techniques in supporting decision making, an intelligent put-away system - namely Fuzzy Storage Assignment System (FSAS) - for the novel real world warehouse operation is proposed to enhance the performance of WMS system. Two key elements would be embraced: (1) Online Analytical Processing (OLAP) in the Data Capture and Analysis Module (DCAM); (2), a fuzzy logic system in the Storage Location Assignment Module (SLAM), with objective to achieve the optimal put-away decision minimizing the order cycle time, material handling cost and damage of items.

3.1.1 Fuzzy Storage Assignment System

The Fuzzy Storage Assignment System (FSAS) is designed to capture the distributed item's data (warehouse status included) from different organizations along the supply chain. The crucial passage concerns the conversion of data into information to hone the correct put-away decision for SLAP [10]. The tangible influence on the warehouse performance is immediately recognizable. In fact, it also allows warehouse worker to visualize a report regarding the status of SKUs in real time, both in arriving both already stocked in the warehouse. The architecture of the FSAS is illustrated in Figure 1. Generally, FSAS consists of two modules: (1) Data capture and analysis Module (DCAM) and (2) Storage location assignment module (SLAM). These are used to achieve the research objectives through a fully automated recommendation storage system.

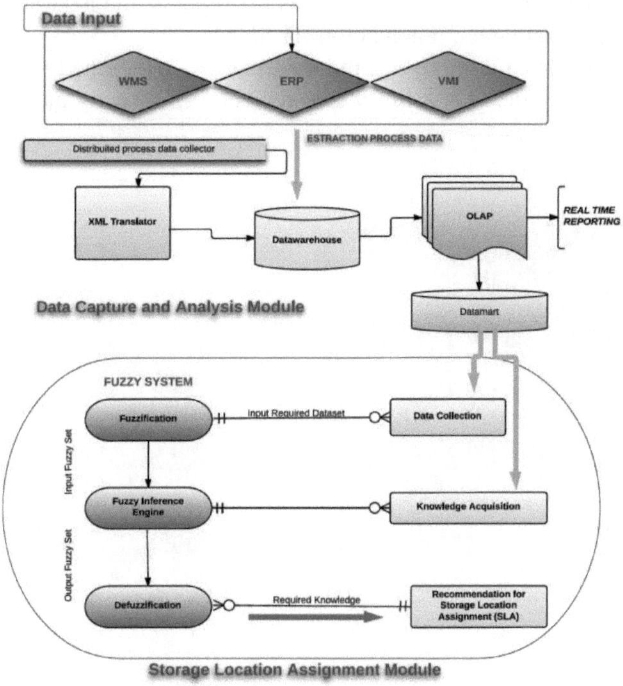

Fig. 1. Fuzzy Storage Assignment System Algorithm

4 Leverage the Model

4.1 Data Capture and Analysis Module (DCAM)

Three major components of Data capture and analysis Module (DCAM) are central data warehouse, OLAP and data mart. The first one is used for connecting various systems with different data sources, including the WMS, ERP and VMI, which are independent systems but may hold valuable data of the storage SKUs to a central data warehouse in order to fulfill SLAP. It ensures the availability of required data for further analysis. DCAM plays the role of decision support system that offers warehouse engineer or manager relevant specific information, including the item's activities information, demand correlation and warehouse information.

The second component, OLAP, provides calculation and multi-dimension structure of data. Warehouse management bases on these information in SKUs and the warehouse to make strategic decision formulating the Fuzzy rules for SLAP. Through holistic manipulation of quality information, the warehouse engineers are able to develop a set of specific rules or algorithms to fit their unique daily operations, warehouse configuration and their operational objective. DCAM offers the refined parameter of SKUs and warehouse that act as the input of the next module-SLAM for generating the automatic recommendation for SLAP.

The last but not least component, the data mart, is developed to store the refined parameter and fuzzy rules (as a fuzzy rule repository), directly and specifically support the SLAP.

(a) (b)

(a): Portion of a six-tiers warehouse
(b): Software to design customized warehouse in 3D

4.2 Storage Location Assignment Module (SLAM)

The storage location assignment module is used to decide the correct storage location for arrival SKUs, based on the analyzed information and the fuzzy rules set from DCAM. Its major component is the fuzzy logic system that consists of a fuzzy set, a fuzzy rule and fuzzy inference.

The fuzzy rules is a set of rules that integrate the elements of the selected storage strategies, experience and knowledge of expert, and regulations. It is characterized by an IF (condition) THEN (action) structure.

The set of rules determines each item storage location; the system will match the characteristics of the SKU and current warehouse (conditions) with the fuzzy rule and then find out the action (where it should be shored). Finally the automatic put-away solution is generated. The SLAM start from the data mart in the former DCAM, it provides the parameters that are format compatible to the fuzzy system, than the parameters will be the input into the fuzzy system that is specifically developed to support the SLAP. The output of fuzzy system will be explained as the recommendation of final storage location for the inbound cargo, than the warehouse workers will store the inbound cargo as the recommendation, finally the storage information will be updated to the WMS system.

4.2.1 The "Golden Zone" Partition

There are golden zone (the most accessible), silver zone (middle accessible) and bronze zone (the least accessible). Therefore there are 3 subzones inside each of storage zone, A, B and C in the sequence of the accessibility in which zone A with the highest accessibility.

5 The Real Case

5.1 Problem Identification

Generally, the Espresso Siena & Co. is characterized by a handle large amount of requests in warehouse operation. Efficient storage location assignment may result in minimizing the cost as well as the damage rate in order to increase the customer satisfaction. However, the current practice of SLAP in deciding storage department, location and the suitable tier relies on the warehouse manager is based on his knowledge. Problems may be raised as the wrong storage environment offers the storage item (resulting in deterioration of item quality) and the long storage location process (resulted in longer inbound shipment processing cycle). This is caused by the insufficient data availability and the lack of systematic decision support system in the decision process. According to the past experience, cargo storing in high tier of a pallet rack or the item with higher loading weigh have more probabilities of getting damage or higher loading height, because of the difficulty to control the pallet truck well. The more expensive cargo, the higher loss the warehouse suffers from the damage.

To ensure the accurate and real-time data can be used, the proposed FSAS for integrating data, extracting quality data from different data source and assigning appropriate storage location for the inbound items, in the way to minimize the risk of getting damage and the loss from it during the put-away and storage process.

5.1.1 Deployment of Online Analytical Process in DCAM

SKU data and warehouse data are captured and transferred into the centralized data warehouse from the data source systems. Through the OLAP application it's possible to build up a multidimensional data model called a star schema. This is composed of a central fact table and a set of surrounding dimension tables and each table has its own attributes in variety of data type. The users are able to view the data in different levels of detail, so the warehouse engineer can generate real-time report for decision-making. In fact the OLAP function allows finding out the statistics of SKUs activities for a specific period of time, representing the SKUs dimension, storage environment and information of warehouse etc [11]. This gives the possibility to master the critical decision support data by the warehouse operator. To ensure the OLAP approach functions properly, the OLAP data cube needs to be built in advanced in the OLAP sever. The cube is developed in a star schema (Figure 2) consisting of dimensions, measures and calculated members.

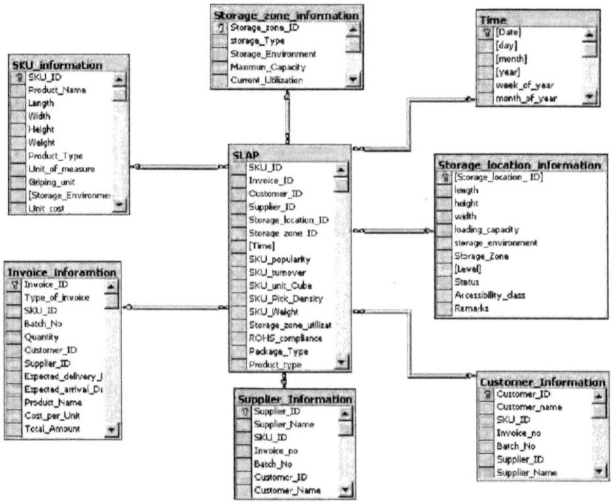

Fig. 2. The relational database structure of DCAM

5.1.2 Relevant Properties Considered

Dimensions

In SKU dimension, the "SKU_ID", and "Product Type" fields are used to find out the dimensions of the SKU and the other characteristics for the storage department selection. In the "Invoice" dimension, the "Invoice ID", and "SKU_ID" and "Invoice _Type" fields are used to find the activity patterns of SKU's for deciding the location inside the department for the SKU.

In "Time" dimension, the "Delivery Date" and "Arrival Date" field are used to find the expected storage time for the SKU and the number of transaction during the specific period.

Measures

"Loading Item Height", "Loading Item Width", "Unit Cost" and "Unit_Cube" etc. are all used to provide critical information for the warehouse manager, in order to realize fuzzy rule composition and perform as a fuzzy input for implication process.

Calculated Member

The calculated member calculates the mean of "Popularity", "Turnover", "Cube_Movement" and "Pick_density" etc., needed for fuzzy rule composition and implication process.

5.2 Deployment of Fuzzy System in SLAM

The fuzzy rules and membership function of the company have to be first formulated in the fuzzy system for each parameter. The parameters (Table 1) and the fuzzy rules of others rule sets are specifically set by the warehouse manager, in order to truly reflect the operational conditions of such product families. The formulation is worked out by the knowledge of experts with the revision on the past experience on the warehouse daily operation; the historical revision could be achieved by the help of the OLAP report, in the former module-DCAM. Different sets of fuzzy rule, with particular parameters, make the decision to determine the storage zone/department, storage location and tier level for the item storage. The fuzzy rules are stored in the knowledge database and defined as a conditional statement in IF-THEN form [10], [12].

Some examples of fuzzy rules are shown in Table 2. The warehouse manager ranged from 0-1 determines the membership function of each parameter. There is more than one type of membership functions existing, some with Gaussian distribution function, others with sigmoid curve, or quadratic and cubic polynomial curves. For this case study, since it's possible to demonstrate the manager's knowledge through the trapezoid and triangular membership functions, the graphic formats of the membership functions of the example parameters are demonstrated as the Figure 3.

The MATLAB-Fuzzy Logic Toolbox needs to create and execute fuzzy inference systems. With the above fuzzy rules and required data, the final storage location for the incoming item would be automatically generated from the Fuzzy Toolbox for SLAP.

In order to demonstrate the feasibility of the system, one supplier delivery input is selected into the FSAS system. When the market-operating department fulfill the relevant data into the ERP, these will be extracted by central data warehouse and then go to the OLAP module. At the same time, the warehouse department is informed and starts to go through their slotting decision tree.

Table 1. The parameters taken into account to optimize the put-away decision process

SYMBOL	PARAMETER	RANGE
C	Cube Movement (m3/Month)	0 – 1830
D	Expected Storage Days (Days)	1 / 365
IH	Loading Item Height (m)	0.3 / 3
IW	Loading Item Width (m)	0.2 / 2.5
IL	Loading Item Length (m)	0.5 / 6
IWt	Loading Item Weight (kg)	20 / 5000
IVol	Loading Item Volume (m3)	0.03 – 26
IV	Loading Item Value (€)	750 / 12000
PWt	Pallet Department Loading Item Weight (kg)	75 / 1000
PH	Pallet Department Loading Item Height (m)	0.5 / 2.5
Pp	Pick Density (Request/ m3)	0 / 60
P	Popularity of Storage (Times/Month)	0 / 60
T	Turnover Rate (Units/Month)	0 / 450
SC	Storage Department Capability (m3)	0 / 46
SA	Storage Location Accessibility (m)	20 – 140
ST	Storage Tier	1 – 6

Table 2. Fuzzy Association Decision Rule

RULE 1:		
IF	Loading Item Height → Low	AND
	Loading Item Width → Short	AND
	Loading Item Length → Short	AND
	Loading ItemWeight → Small	AND
	Loading Item Cube → Small	
THEN	**Capability of Storage Depart-ment is LOW**	

RULE 2:		
IF	Popularity → High	AND
	Turnover Rate → High	AND
	Cube Movement → Hight	AND
	Pick Density → not Hight	AND
	Expected Storage Days → Short	
THEN	**Accessibility of Storage Zone is Good**	

RULE 3:		
IF	Loading Item Value →High	AND
	Loading Item Height → High	AND
	Loading Item Weight → High	
THEN	**Tier Selection is Medium**	

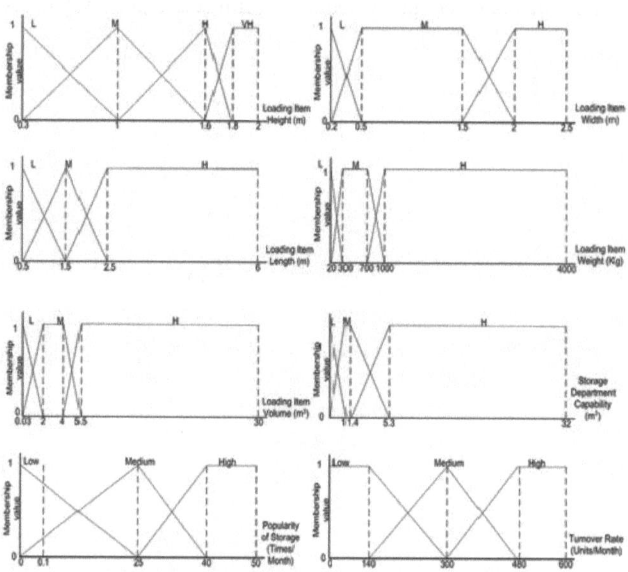

Fig. 3. The MATLAB graphic function model

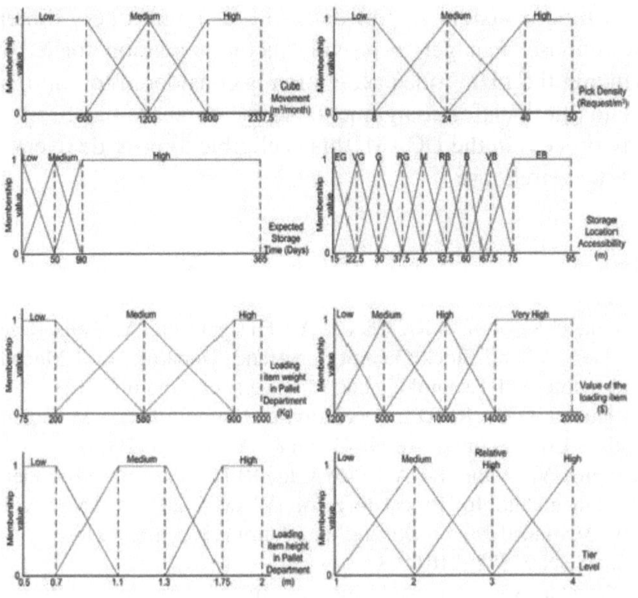

Fig. 3. *(continued)*

6 Conclusions

This research tries to introduce the design and the implementation of a FSAS, which embraces the fuzzy theory to achieve warehouse capacity improvement and optimize the put-away process. The implementation of the proposed methodology, in the aspect of warehouse management through simulation, has been succeeded. Incorporating the error measurement and the complexity of the process into the fitness evaluation, the generalized fuzzy rule sets can be less complex and more accurate. In the matter of generation of new fuzzy rules, the membership functions are assumed to be static and known. Other fuzzy learning methods should be considered to dynamically adjust the various parameters of the membership functions, to enhance the model accuracy. Future contribution of this endeavor goes to validate the decision model in a way to be launched in case companies. Despite increasingly manufacturers and retailers emphasize the just-in-time inventory strategy, the delivery orders will become more frequent with smaller lot size. This creates considerable demand for put-away processes in warehouses, since put-away process is able to match the characteristics of the storage item and the storage location. In order to achieve this standard, the warehouse operators first need to master the characteristics of the incoming items and storage location and then correctly match the storage location, minimizing the material handling cost, product damage and order cycle.

An OLAP based intelligent Fuzzy Storage Assignment System (FSAS) becomes suitable to integrate day-by-day operational knowledge from human's mind, supporting a key operation in warehouse-put-away process, minimizing product damage and material handling cost. FSAS enables the warehouse operators to perform put-way decision: (i) real-time decision support data with different query dimensions (ii) mimicking the warehouse manager to provide recommendation for SLAP. Further research on enhancing the fuzzy rules generation is considered to improve the accuracy in the suitable storage location assignment. As the database has been well developed for the put away process in the DCAM, this is eligible to provide overview on the past performance of the warehouse.

References

1. Chede, B., Jain, C.K., Jain, S.K., Chede, A.: Fuzzy Logic Analysis Based on Inventory Considering Demand and Stock Quantity on Hand. Department of Mechanical Engineering, Mahakal Institute of Technology and Management, Ujjain (2012)
2. Blecker, T., Huang, G.Q.: RFID in Operation and Supply Chain Management – Research and Application. Erich Schmidt Verlag Gmbh & Co., Berlin (2008)
3. Chirici, L., Wang, Y., Wang, K.: A Tailor Made RFID-fuzzy Based Model to Optimize the Warehouse Managment. In: Proceedings of IWAMA 2012 - The Second International Workshop of Advanced Manufacturing and Automation, pp. 491–498. Tapir Akademisk Forlag (2012) ISBN 9788232101108
4. Castro, L., Wamba, S.F.: An inside look at RFID Technology. Journal of Technology Management & Innovation 2(1), 128–141 (2007)
5. Convery, T.: RFID Technology for Supply Chain Optimization: Inventory Management Applications and Privacy Issues. Northwest Pump & Equipment Co. University of Oregon (USA) (2004)
6. Ho, G.T.S., Choy, K.L., Poon, T.C.: Providing decision support functionality in warehouse management using the RFID-based fuzzy association rule mining approach. Hong Kong Polytechnic University (2010)
7. Wang, K., Zhang, Z.: Application of Radio Frequency Identification (RFID) to Manufacturing, SINTEF Report, Norway (2012)
8. Zadeh, L.A.: Fuzzy sets. Information and Control 8(3), 338–353 (1965)
9. Zou, C., Jiang, C.: The Applications of RFID Technology in Logistics Management and Constraints. Northwestern Polytechnical University, Trans. Tech. Publications, China, Switzerland (2006)
10. Lam, C.H.Y., Chung, S.H., Lee, C.K.M., Ho, G.T.S., Yip, T.K.T.: Development of an OLAP Based Fuzzy Logic System for Supporting Put Away Decision. International Journal of Engineering Business Management 1(2), 7–12 (2009)
11. Laurent, A., Lesot, M.J.: Scalable Fuzzy Algorithms for Data Management and Analysis: Methods and Design. IGI Publishing (2009)
12. Li, D., Laurent, A., Poncelet, P.: Mining Belief-Driven Unexpected Sequential Patterns and Implication Rules in Rare Association Rule Mining and Knowledge Discovery: Technologies for Infrequent and Critical Event Detection. IGI Publishing (2009)

Improving the Usability of Collaboration Methods and Technologies in Engineering

Lars Wolter[1], Haygazun Hayka[2], and Rainer Stark[2]

[1] Chair of Industrial Information Technology (IIT)
Department of Machine Tools and Factory Management, Technische Universität Berlin
lars.wolter@tu-berlin.de
[2] Division Virtual Product Creation
Fraunhofer Institute for Production Systems and Design Technology (IPK)
rainer.stark@ipk.fraunhofer.de

Abstract. This paper illuminates certain needs and solutions in collaborative engineering like the ability to initiate collaboration sessions without preparation steps, the acceptance of collaboration solutions due to heterogeneous IT landscapes and the protection of intellectual property rights. These problems are observed together with current solutions in commercial products and research projects. An overview of those solutions is presented, including their workings, problems and current state of development adding some in depth information about current research work conducted by the authors to address these problems.

Keywords: Collaboration, collaborative engineering, PLM, heterogeneous IT landscape, intellectual property rights.

1 Introduction

The industry sets new requirements for collaborative engineering due to technological improvements on products and product development methods, increasing complexity of supply chains and the trend to establish virtual teams. The paper discusses these requirements and the resulting fields in need of action. Each field shows different opportunities for the industry. Technology can give better integration and usability, the processes can be more transparent and standardized and the human factors can receive more attention to increase the stimulation of the stakeholders. To utilize all the opportunities in these fields, the collaborative engineering processes, methods and tools can not only focus on the technical goals and how to achieve them, they must also consider the human factor, containing the other stakeholders and there reasons. Only then the task work can be united with teamwork for successful collaboration.

1.1 The Need for Collaborative Engineering

The increasing complexity of consumer products and industrial goods also increases the complexity of their development. This is given by a raising number of parts in

G.L. Kovács and D. Kochan (Eds.): NEW PROLAMAT 2013, IFIP AICT 411, pp. 73–84, 2013.

each individual product and especially the combination of multiple engineering domains into a single product. Additional complexity in today's product development originates from release cycles that need to get shorter to stay competitive. To address the complexity a company needs to involve more people in the development process, each being an expert in his domain. This also includes the engineers which are experts for specific domains, parts, functionalities or steps in the development and manufacturing process.

As with common meetings collaboration is a task that does not produce anything but needs to be done to be successful. Therefore the collaboration needs to be efficient and natural for all participants. The collaboration in product development is happening on many different levels starting from asynchronous groupware systems, telecooperation solutions, viewing collaboration and full featured interactive collaboration. Additionally the collaboration can be done locally or across long distances, it can be with or without the use of digital tools. Then there is also the difference of collaboration inside companies which happens intensively and the collaboration with partners and suppliers which is performed less intensively [1] and is also shown in figure 1. This increases the complexity of collaboration management which only needed to supply meeting rooms and a telephone number to allow collaboration. To manage all the collaboration scenarios in companies today, additional people need to be involved being experts in the area of collaboration.

The focus during this paper lies in engineering scenarios during virtual engineering. It does not address collaboration during concept development, production or selling.

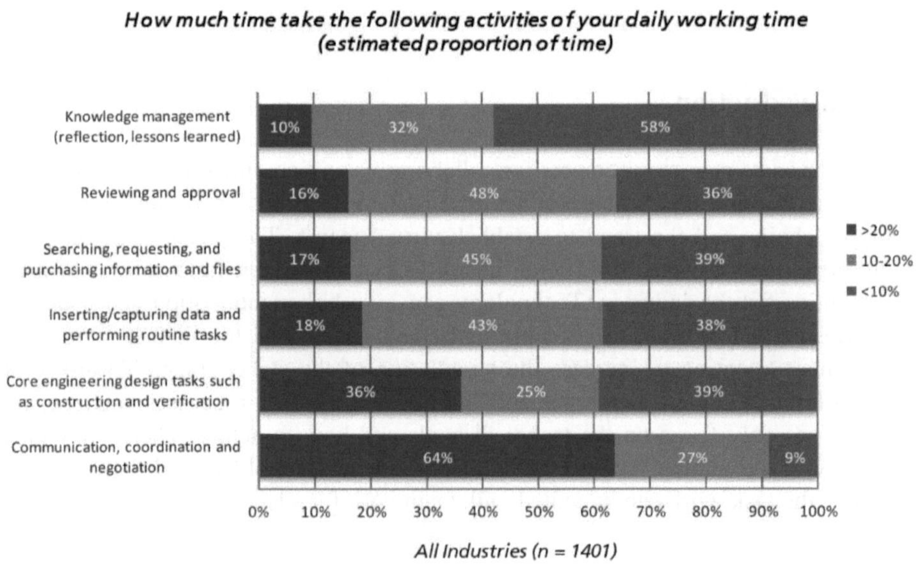

Fig. 1. Distribution of Engineering Working Time [1]. Communication, coordination and negotiation happen intensively in the industry.

1.2 Problems of Collaborative Engineering

In collaboration there is a large field of problems that can be analyzed and addressed. This article focuses on collaboration supported by digital tools that an engineer uses during his product development tasks. For an engineer the types of tasks have changed over time. There is much more information needed to fulfill his day-to-day tasks forcing him to spend a lot of his time to search and acquire this information, not only from IT-Systems but also from other colleagues or partners. This also means he has to supply information he generates during his work either by storing them in an IT-System or by directly communicating them to other people. This additional overhead is already part of the collaboration happening in a company. Looking at basic collaboration tasks without the use of digital tools already introduces a lot of problems. If you leave someone a message on a post-it it can get lost during cleaning, your block of post-its could be empty or you try to find someone for a talk but don't know the room where that person currently is. Those problems can be directly translated to collaboration using digital tools. For example the email getting lost in the spam-folder, no free space on the network drives or the Web-Ex session that cannot initiate because of a too restrictive firewall. These kinds of problems can be addressed very effectively by rules and a good organization, which are necessary to achieve robustness in using digital engineering technology [2]. But digital tools for collaboration introduce new kind of problems that need to be addressed separately.

1.2.1 Heterogeneity

Due to the increasing fragmentation of companies, either through outsourcing or by integrating other companies leads to the use of multiple IT-Systems dedicated to the same kind of task. This is also true for the collaboration tools, either stand-alone or integrated, because most of them **can only be connected to tools of the same vendor.** This is not a problem with the telephone for example; it can be used with every other telephone from different vendors.

1.2.2 Tool Acceptance

An engineer today has to use multiple digital tools to solve his engineering tasks, this includes CAx, PDM, ERP, MES, Excel, Outlook and various other tools. The number of tools increases in a collaboration situation because an extra collaboration tool needs to be used. Same goes for engineering tools, which normally only the collaboration partner uses. This is a frustrating situation for the engineer because it maybe a tool he only uses irregular when contacting a supplier for example and sometimes differs from his regular tools in usage and even methodic concepts. **The engineer is naturally blocking to use this extra tool** for collaboration resulting in reduced efficiency or lesser collaboration.

This also happens for local collaboration scenarios, where multiple people discuss a situation with a specific tool. This happens in design reviews for example, where a dedicated operator is necessary to operate the tool introducing an extra layer into the interaction.

1.2.3 Protection of Intellectual Property

The ideas of new products as well as the methods and processes to produce them are a constant interest for all competitors in the market. Therefore this knowledge needs to be kept secret. Collaboration using digital tools is normally associated with sharing your own digital data. This results in trust problems with any kind of new or specialized tool which connects multiple stakeholders for collaborative purposes. Opposite to that generally used tools like emails have a very high trust even if they are used insecurely. The problem is to make any kind of collaboration tool used be trusted to **protect the intellectual property**; otherwise it cannot be effectively used.

1.3 How Are These Problems Currently Addressed

The market has a multitude of solutions to achieve collaboration in engineering. The system vendors of larger software products started to integrate collaboration features in their products. This includes features common to current social media products like text, audio and video chat combined with the sharing of product data. One example of those solutions is 3D-Live in Catia V6 from Dassault-Systemes. Such a system works very well in a homogenous environment, in which all participants use the same CAD program which tightly integrates into the PDM of the same vendor to make the product data sharing available for collaborative engineering. Using a different CAD system breaks all the collaboration features.

This problem can be solved by deploying all applications to every user, forcing the engineer to use multiple IT-Systems in collaborative scenarios. Another way to address the heterogeneity problem is to use stand-alone collaboration solutions. Standalone solutions are a less complex extra piece of software, but also need a context switch and the conversion of the data from the engineering tool. Multiple vendors developed standalone tools to allow collaboration without the need of supplying the whole authoring tool set to all users.

There are two kinds of stand-alone solutions for collaboration. Screensharing solutions are the easiest to use in this category and don't need any data conversion. They work by capturing the contents of the screen or a specific application window and transmit that as a video stream to the other participants. These solutions are well known in web conferencing and are widely used because of their ease of usability. The user just starts the screensharing solution and decides which screen to share. After the setup, he only operates his preferred tool. Products in this category are WebEx from Cisco or WebConf from Adobe. But the concept of screensharing **lacks the possibility of an equal participation** in the collaboration scenario, because only one participant can present his screen to all others. Even when using multiple monitors, a monitor for each participant would be needed to present all the different views.

The second type of solution is applications that use their own visualization engine to display geometric models. Most of those solutions import data from the engineering tool and share them across multiple participants. Because these tools are separate from the authoring tools, they normally only import common exchange formats. This means the collaborating engineer needs to export the product data from his application in a suitable format for the collaboration tool and all the data need to be transmitted to all participants. This conversion and transfer of product data induces **long setup times**. This kind of tools can have a lot of functionality to alter the viewing, leaving annotations and therefore support the collaboration very well, but requires in additional tool-knowledge for the engineer. Changes that can only be done in the authoring system would need an export of the changed product data and redistribution of those changes to all participants.

The most basic way of collaboration thru sharing data is handled asynchronously. Data management solutions allow locking of complete files by different users but there is no technique available that can merge two differently changed CAD files like Microsoft Word is offering for documents. Most PDM and other data storage systems ease the asynchronous collaboration with large files by duplicating them to different sites making them rapidly available from different locations. This is still limited in speed, but is already well established among the industries.

The last kind of well-established collaboration solutions consists of web based groupware solutions. They are often based around whole communication solutions for email, task and workflow management. They allow the users, to exchange tasks, documents and other information using wikis forums or blogs. Newer systems also incorporate so called social office functions to allow commenting, rating and sharing of information across the company intranet. Examples for these kinds of systems are Microsoft Office with Sharepoint and Outlook or open source groupware solutions like Liferay, Tiki-Wiki or other portal solutions. These solutions are independent of product development but can be customized to fit specific products and companies. The type of collaboration is limited, but due to the inclusion of real time communication and web based editors for documents these solutions are not only used for cooperation but also full collaboration.

1.4 Research to Increase the Collaboration Efficiency

Some research is done to support the engineering collaboration. One approach is to let different CAD systems communicate with each other, to allow collaborative design even with heterogeneous CAD systems. One of those approaches in [3] uses a common set of commands. Every CAD system translates its own authoring commands to this common set which is distributed to all participants and at each target platform converted to a command of the specific CAD system. Therefore it does not need to exchange any product data beforehand. This approach allows concurrent design from the beginning; it does not allow modification of existing product data.

Other Research is done to allow better network communication even in firewalled scenarios across companies for communication applications [4]. This is a very fundamental type of research affecting nearly all collaboration attempts which communicate thru the internet. Also very fundamental is research to resolve conflicts in collaboratively authored CAD. This is addressed for example by research activities to integrate Boolean operations in CAD systems [5].

There is also research activity to enhance distributed design reviews. There are solutions that adapt to different kind of devices allowing the use of large VR-Cave-Systems together with participants only using desktop computers during the collaboration. One such approach is documented in [6]. The setup they are using uses VRML models to visualize 3D data but can also show the screens of the participants using screensharing.

Many research projects focus on the protection of the intellectual property. Everything related to encryption is only protecting the data while it is traveling between the collaboration participants. If the aim is to not let one of the participants misuse data from other participants, any encryption is of no use. Therefor some research projects describe methods and algorithms to watermark geometric models [7]. This way it can be traced back where copies originated allowing the owner to sue them. Other methods reduce the details in certain areas by using multiresolution meshes and specifying the detail priorities of certain areas [8].

On the other hand there is also research going on for 3D model reconstruction from image sequences. With flaws this was possible 1996 from camera images [9]. Better algorithms, the good quality of the rendered images and faster computers allow much better reconstruction [10]. But for all reconstruction algorithms it is important that there is at least one image from every feature necessary to reconstruct the original model. If the back side or inside of an object is never shown, it cannot be reconstructed.

1.5 Areas in Need of Action

1.5.1 Technology

The technology for collaboration is in a very good state, but there are technological problems when different solutions need to be connected.

The rapid increase in information and the expectation of its global availability introduces a new field of information management that does not require a central distribution point but intelligent information containers that can manage the containing information and is able to route those information to systems and participants in a collaborative scenario that need them.

Besides the technical areas that are in need of action, the collaboration between engineers needs to address human factors to make the collaboration to something an engineer wants to do instead of something he needs to do. This can be seen in interaction with tools, where a user is pleased to use a tablet-pc to read its newspaper using simple gestures. This kind of technology common to consumer products needs to be adapted to solutions for the industry.

1.5.2 Processes

The business processes need to address the specialty in collaborative processes. When working in a collaborative manner, each step has a meaning for the task the group does together, like the patterns in [11]. This can also lead to additional tasks for collaborative processes, because after a collaborative session the gathered information needs to be converged from multiple participants before it can be evaluated.

1.5.3 Methods

The different methods of collaboration can be as coarse has to choose from personal meeting, phone call or use of email to very fine grained methods that define the format of the emails to send for specific tasks. Having a method for handling specific collaboration tasks is a must, to ensure correct flow of the information.

2 New Solutions for Collaborative Engineering

To fill some of the gaps presented above the following solutions are described. These solutions handle different collaboration scenarios like local collaboration on multitouch tables as well as remote collaboration over the network.

2.1 Using Current Touch Technology for Local Cooperation

The technology for touch is well established in the consumer area. The technology is also very mature in its use. To be of a real benefit in the industry not only the technology needs to be used, but also the methods must adapt to the scenarios, where multitouch environments can be used to raise the efficiency of engineering processes. During the research at the Fraunhofer IPK methods where developed to visualize product structures on multitouch tables [12]. The requirement was a good usability with touch devices but also be understandable und usable in a multi-person-scenario where the participants have different views onto the multitouch device. This resulted in a Voronoi [13] based structure as seen in figure 1. This special structure was analyzed to see if it can fulfill some typical tasks in engineering like search and compare operations with this structure [14].

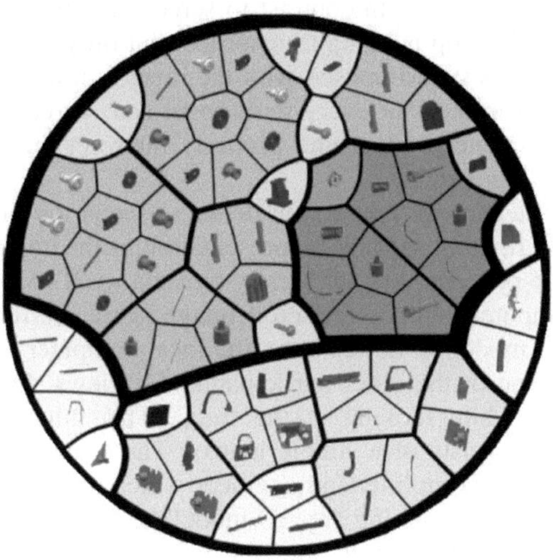

Fig. 2. Touch optimized visualization of a product structure

Fig. 3. Multi-User Multitouch environment for design reviews

In figure 3 this special application can be seen in a multi user environment. It allows multiple participants in a meeting to either work in their own workplace or to cooperate with some or all of the other participants. This allows part of the cooperating group to prepare their content while others are discussing a previous item. The example is the door of a car. In this example the expert for the door opener can discuss some details with the chassis expert while the experts for the window-lifter-system illuminate the use of an extra-ordinary expensive part to the management.

2.2 Technology for Secure and Instant Collaboration

The here presented solution constitutes a combination of screen sharing and the local visualization at each participant. In contrast to screen sharing not the whole program window or the whole desktop is being transmitted, but only the 2D image of the rendered 3D model, which is superimposed with the images of all participants. A correct superposition is necessary so that every participant can correctly perceive the visual impression of the complete product and properly interpret the correlations and distances between the components.

This collaboration technique focuses on different scenarios shown in figure 4. All participants shown in figure 4 see a 3D representation of the object being reviewed, in this case a truck. The parts in blue are locally existent as 3D-Modells and are locally rendered on the computer and the rendered image is transferred to all participants. The gray parts of the model do not exist on the local computer. They are just 2D images streamed from one of the other participants. All views share the same point of view and orientation while looking at the truck. This information is also shared among the users and consists of a simple matrix. The scenario can also incorporate special participants like the mobile lead engineer which only needs a web browser to join the session. He is not supplying any 3D model, he just consumes the images. The opposing case is the PDM System at site B which just renders it locally stored data and sends it to the others. This participant does not consume any information.

All the other participants, the OEM at site east and the two suppliers deliver their own data and consume from the others. The OEM holds the 3D models of the chassis, while supplier A holds the cabin and supplier B the wheels. They only deliver their

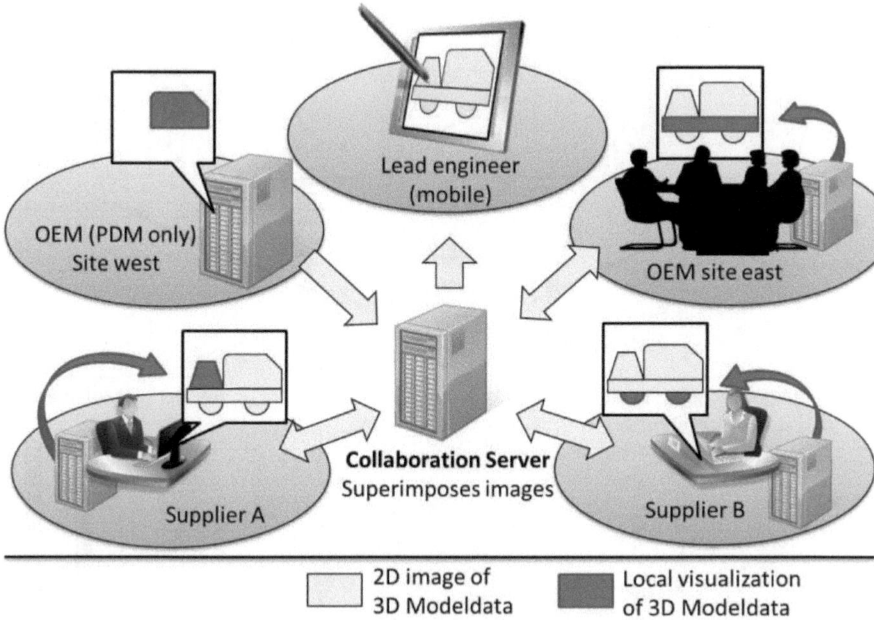

Fig. 4. The different collaboration scenarios condensed in a single collaboration

own property as images without being afraid that for example supplier B can steal the 3D model data from supplier A.

To achieve the correct superposition a so-called depth image is transmitted additionally. The depth images can be used, to decide for every pixel from which 2D image the respective pixel should be used (figure 5, bottom right).

To create, transmit or show it on a screen, a 2D image of a 3D model must be rendered. This process is called image synthesis, where for each pixel on the screen must be determined, which part of the 3D model it represents. The color of that point is being shown at this pixel (figure 5, bottom left). The depth image is constructed on the same principle. This is achieved by storing a distance value instead of a color value. If you interpret this distance value as a color, so is the value that represents the pixel brighter for larger distances and darker for shorter distances.

At multiple participants a color- and depth rendering is created that are all collected. These can now be used to construct a joint image. Condition is that all participants have the same point of view. This joint point of view acts as a zero point of the viewed scene and is reached by using the aforementioned matrix that is shared among the participants. This aspect corresponds to the technique of local visualization where likewise just the view matrix needs to be exchanged.

Assuming there are three participants with color renderings C1-C3 and depth renderings D1-D3. Then for each pixel it is analyzed which of the three Dx is the darkest. If it is D3 then the color value of C3 is being used. That way a new image is assembled where all models of all three participants are integrated. An interpolation allows combining images with different resolutions as long as the aspect ratio is maintained.

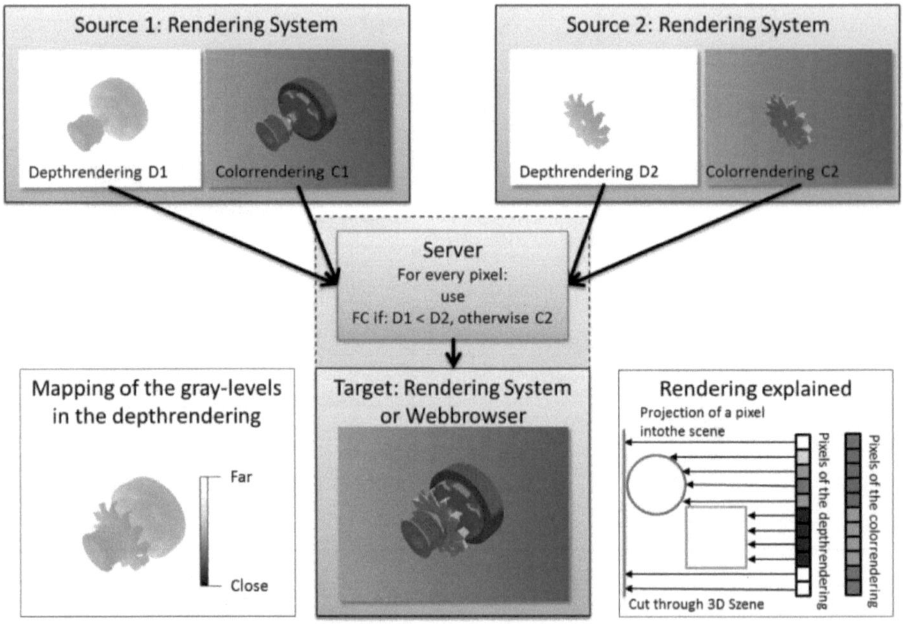

Fig. 5. Description of the image merging algorithm that merges two rendered images displayed at the top into a single image in the bottom middle. Therefor it uses a depth image explained on bottom left. Bottom right describes the basic principle of rendering a 3D scene into a 2D image in conjunction to the rendering of a depth image.

This is of advantage when generating images on hardware with different performances. The images from different sources can still be combined. The realization of this technology was executed in a prototypic collaboration tool.

2.3 Increased Acceptance through Collaboration Using CAD Systems

To approach the problem of acceptance, methods were evaluated to include collaboration functionality in existing CAD systems. Through the strong link with the rendering an easy connection via a plugin is difficult. In the course of work the CAD systems Spaceclaim and NX were extended via plugins in a way that they communicate their viewing location with other participants and could react to commands. Doing that the CAD systems showed limitations of different magnitude. The realization of the plugin proved more difficult for NX than for Spaceclaim. An automatic reacting e.g. to the network commands could not be realized with the NX API until the end of the project. Nonetheless the realization of the plugins proved that a collaborative coupling of two very diverse CAD systems is possible and a consequent commitment across manufacturers towards an open adaption of the API according to the CPO could create new possibilities.

Similar as in [3] it was also tested to fulfill some modeling tasks using heterogeneous CAD systems. But the target was different as in the paper because the focus is still located in the design review scenario where existing CAD Models need to be examined and possibly changed. Therefor the problem of identifying the same parts in the different systems needs to be managed. The research in this area is still ongoing, to deliver a solid solution.

3 Conclusions

The increasing need for effective collaboration solutions is a challenge for today's companies. But system vendors and research facilities are continuing to generate better solutions. Useful cross vendor programs like the Codex of PLM Openness can ensure that future collaboration solutions will not suffer because of heterogeneous IT landscapes. The continuous evolvement of user interaction technologies from the consumer market into the industry gives the opportunities for user friendly and easy to use solutions. The presented solutions and ongoing research work introduce new accents for collaboration. These accents can be used by the industry to bring new concepts of collaboration into the companies to increase their collaboration abilities.

References

1. Müller, P., Pasch, F., Drewinski, R., Bedenbender, H., Hayka, H., Stark, R.: Study on Collaborative Product Development and Digital Engineering Tools. In: Rivest, L., Bouras, A., Louhichi, B. (eds.) PLM 2012. IFIP AICT, vol. 388, pp. 389–399. Springer, Heidelberg (2012)
2. Stark, R., Kim, M., Woll, R., Wolter, L.: The way forward of Virtual Product Creation – how to achieve robustness in using digital engineering technology? In: 15 Seminario International de Alta Tecnologia, pp. 41–52 (2010)
3. Li, M., Gao, S., Wang, C.C.L.: Real-Time Collaborative Design With Heterogeneous CAD Systems Based on Neutral Modeling Commands. Journal of Computing and Information Science in Engineering 2007(7), 113–125 (2007)
4. Stark, R., Gärtner, H., Wolter, L.: Verteilte Design Reviews in heterogenen Systemwelten. ProduktDaten Journal 15(1), 45–49 (2008)
5. Zheng, Y., Shen, H., Xia, S., Sun, C.: Conflict resolution of Boolean operations by integration in real-time collaborative CAD systems. In: Meersman, R., Tari, Z. (eds.) OTM 2007, Part I. LNCS, vol. 4803, pp. 238–252. Springer, Heidelberg (2007)
6. Daily, M., Howard, M., Jerald, J., Lee, C., Martin, K., McInnes, D., et al.: Distributed Design Review in Virtual Environments. In: Churchill, E., Reddy, M. (eds.) Proceedings of the Third International Conference on Collaborative Virtual Environments (CVE 2000), pp. 57–63. ACM, New York (2000)
7. Kuo, C.-T., Cheng, S.-C., Wu, D.-C., Chang, C.-C.: A Blind Robust Watermarking Scheme for 3D Triangular Mesh Models Using 3D Edge Vertex Detection. Asian Journal of Health and Information Sciences (1), 36–63 (2009)
8. Zyda, M., Capps, M., Pausch, R., Bishop, G., Pojar, E., Schmalstieg, D.: User-controlled creation of multiresolution meshes. In: Proceedings of the 2003 Symposium on Interactive 3D Graphics, SI3D 2003, p. 127. ACM Press (2003)
9. Beardsley, P., Torr, P., Zisserman, A.: 3D model acquisition from extended image sequences. In: Buxton, B., Cipolla, R. (eds.) ECCV 1996. LNCS, vol. 1065, pp. 683–695. Springer, Heidelberg (1996)
10. Snavely, N.: Scene Reconstruction and Visualization from Internet Photo Collections: A Survey. IPSJ Transactions on Computer Vision and Applications 3, 44–66 (2011)
11. de Vreede, G.-J., Briggs, R.O.: Collaboration Engineering: Designing Repeatable Processes for High-Value Collaborative Tasks. In: Sprague, R.H. (Hg.) Proceedings of the 38th Annual Hawaii International Conference on System Sciences. IEEE Computer Society Press, Los Alamitos (2005)

12. Woll, R., Wolter, L.: Kollaborativer Design Review am Multitouchtisch. In: Reiterer, H., Deussen, O. (Hrsg.) Mensch & Computer 2012 – Workshopband: Interaktiv Informiert – Allgegenwärtig und Allumfassend!?, pp. 439–443. Oldenbourg Verlag, München (2012)
13. Balzer, M., Deussen, O., Lewerentz, C.: Voronoi treemaps for the visualization of software metrics. In: Proceedings of ACM Symposium on Software Visualization, pp. 165–172 (2005)
14. Schulze, E.E., Wolter, L., Hayka, H., Röhlig, M.: Intuitive Interaktion mit Strukturdaten aus einem PLM-System. In: Konferenzband des 10. Gemeinsames Kolloquium Konstruktionstechnik, pp. 71–87 (2012)

Towards Product Avatars Representing Middle-of-Life Information for Improving Design, Development and Manufacturing Processes

Karl Hribernik, Thorsten Wuest, and Klaus-Dieter Thoben

BIBA - Bremer Institut für Produktion und Logistik GmbH,
Hochschulring 20, 28359 Bremen, Germany
{hri,wue,tho}@biba.uni-bremen.de

Abstract. In today's globalized world, customers increasingly expect physical products and related information of the highest quality. New developments bring the entire product lifecycle into focus. Accordingly, an emphasis must be placed upon the need to actively manage and share product lifecycle information. The so-called Product Avatar represents an interesting approach to administrate the communication between intelligent products and their stakeholders along the product lifecycle. After its initial introduction as a technical concept, the product avatar now revolves around the idea individualized digital counterparts as targeted digital representations of products enabling stakeholders to benefit from value-added services built on product lifecycle information generated and shared by Intelligent Products. In this paper, first the concept of using a Product Avatar representation of product lifecycle information to improve the first phases, namely design, development and manufacturing will be elaborated on. This will be followed by a real life example of a leisure boat manufacturer incorporating these principles to make the theoretical concept more feasible.

Keywords: PLM, Product Avatar, BOL, Intelligent Products, Digital Representation, Information, Data.

1 Introduction

In today's globalized world, customers increasingly expect physical products and related information of the highest quality. New developments bring the entire product lifecycle into focus, such as an increased sensibility regarding sustainability. Accordingly, an emphasis must be placed upon the need to actively manage and share product lifecycle information.

The so-called Product Avatar [Hribernik, Rabe, Thoben & Schumacher, 2006] represents an interesting approach to administrate the communication between intelligent products and their stakeholders. After its initial introduction as a technical concept, the Product Avatar now revolves around the idea individualized digital counterparts as targeted digital representations of products enabling stakeholders to benefit from value-added services built on product lifecycle information generated and shared by Intelligent Products [Wuest, Hribernik, Thoben, 2012].

G.L. Kovács and D. Kochan (Eds.): NEW PROLAMAT 2013, IFIP AICT 411, pp. 85–96, 2013.

During the middle of life phase (MOL) of a product a broad variety of data and consequently information can be generated, communicated and stored. The ready availability of this item-level information creates potential benefits for processes throughout the product lifecycle. Specifically in the beginning-of-life phase (BOL) of the product lifecycle, opportunities are created to continuously improve future product generations by using item-level MOL information in design, development and manufacturing processes. However, in order to make use of the information, its selection and presentation has to be individualized, customized and presented according to the stakeholders' requirements. For example, in the case of design processes, this means taking the needs of design engineers into account, and during manufacturing, the production planner.

In this paper, first the concept of using a product avatar representation of product lifecycle information to improve the first phases, namely design, development and manufacturing will be elaborated on. This will be followed by a real life example of a leisure boat manufacturer incorporating these principles to make the theoretical concept more feasible.

2 Product Lifecycle Management and Intelligent Products

The theoretical foundation for the Product Avatar concept is on the one hand Product Lifecycle Management. This can be seen as the overarching data and information source from which the product Avatar retrieves the bits and pieces according to the individual needs of a stakeholder. On the other hand, this depends on Intelligent Products being able to gather and communicate data and information during the different lifecycle phases. In the following both areas are introduced as a basis for the following elaboration on the Product Avatar concept.

2.1 Product Lifecycle Management

Every product has a lifecycle. Manufacturers are increasingly becoming aware of the benefits inherent in managing those lifecycles [Sendler, 2009]. Today's products are becoming increasingly complicated. For example, the amount of component parts is increasing. Simultaneously, development, manufacturing and usage cycles are accelerating [Sendler, 2009] and production is being distributed geographically. These trends highlight the need for innovative concepts for structuring and handling product related information efficiently throughout the entire lifecycle. On top that, customer demand for more customisation and variation stresses the need for a PLM at item and not merely type-level [Hribernik, Pille, Jeken, Thoben, Windt & Busse, 2010].

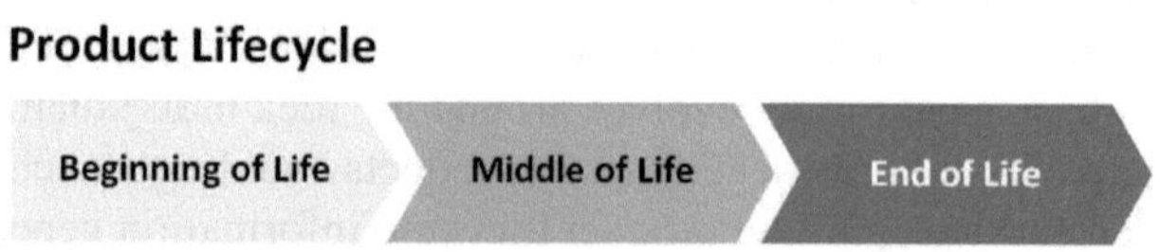

Fig. 1. Phases of the Product Lifecycle

PLM expands on the concept of Product Data Management (PDM) to include information generated and used beyond design and manufacturing [Paul & Paul, 2008; Fasoli, Terzi, Jantunen, Kortelainen, Sääski & Salonen, 2011]. Besides merely handling product and process related data, PLM also has to take into account the interdependencies of information and communication between all of the stakeholders involved in the product lifecycle.

Common graphical representations of the product lifecycle encompass three phases – beginning of life (BOL), Middle of Life (MOL) and End of Life (EOL) –arranged either in a circle or in a linear form (see Figure 1). The linear form represents the product lifecycle "from the cradle to the grave".

The social web offers a number of opportunities for item-level PLM. For example, Web 2.0-based product information acquisition could contribute to the improvement of the quality of future products [Merali & Bennett, 2011; Gunendran & Young, 2011].

2.2 Intelligent Products

Intelligent Products are physical items, which may be transported, processed or used and which comprise the ability to act in an intelligent manner. McFarlane et al. [McFarlane, Sarma, Chirn, Wong, Ashton, 2003] define the Intelligent Product as "...a physical and information based representation of an item [...] which possesses a unique identification, is capable of communicating effectively with its environment, can retain or store data about itself, deploys a language to display its features, production requirements, etc., and is capable of participating in or making decisions relevant to its own destiny."

The degree of intelligence an intelligent product may exhibit varies from simple data processing to complex pro-active behaviour. This is the focus of the definitions in [McFarlane, Sarma, Chirn, Wong, Ashton, 2003] and [Kärkkäinen, Holmström, Främling & Artto, 2003]. Three dimensions of characterization of Intelligent Products are suggested by [Meyer, Främling, Holmström, 2009]: Level of Intelligence, Location of Intelligence and Aggregation Level of Intelligence. The first dimension describes whether the Intelligent Product exhibits information handling, problem notification or decision making capabilities. The sec-ond shows whether the intelligence is built into the object, or whether it is located in the network. Finally, the aggregation level describes whether the item itself is intelligent or whether intelligence is aggregated at container level. Intelligent Products have been shown to be applicable to various scenarios and business models. For instance, Kärkkäinen et al. describe the application of the concept to supply network information management problems [Kärkkäinen, Holmström, Främling & Artto, 2003]. Other examples are the application of the Intelligent Products to supply chain [Ventä, 2007], manufacturing control [McFarlane, Sarma, Chirn, Wong, Ashton, 2003], and production, distribution, and warehouse management logistics [Wong, McFarlane, Zaharudin, Agrawal, 2009].

A comprehensive overview of fields of application for Intelligent Products can be found in survey paper by Meyer et al [Meyer, Främling, Holmström, 2009].

Thus, an Intelligent Product is more than just the physical product – it also includes the enabling information infrastructure. Up to now, Intelligent Products are not "socially intelligent" [Erickson, 2009] in that they could create their own infrastructure to communicate with human users over or store information in. However, Intelligent Products could make use of available advanced information infrastructures designed by socially intelligent users, consequently enhancing the quality of information and accessibility for humans who interact with them.

3 Product Avatar

One approach to representing the complex information flows connected to item-level PLM of an intelligent product is the Product Avatar. This concept describes a digital counterpart of the physical Intelligent Product which exposes functionality and information to stakeholders of the product's lifecycle via a user interface [Hribernik, Rabe, Thoben, & Schumacher, 2006; Wuest, Hribernik & Thoben, 2012; Wuest, Hribernik & Thoben, 2013].

3.1 Concept Behind the Product Avatar

The concept of the Product Avatar describes a distributed and de-centralized approach to the management of relevant, item-level information throughout a product's lifecycle [Hribernik, Rabe, Thoben, & Schumacher, 2006]. At its core lies the idea that each product should have a digital counterpart by which it is represented towards the different stakeholders involved in its lifecycle. In the case of Intelligent Products, this may also mean the implementation of digital representations towards other Intelligent Products. Consequently, the Avatar concept deals with establishing suitable interfaces towards different types of stakeholder. For Intelligent Products, the interfaces required might be, for example services, agents or a common messaging interfaces such as QMI. For human stakeholders, such as the owner, producer or designer, these interfaces may take the shape, e.g., of dedicated desktop applications, web pages or mobile "apps" tailored to the specific information and interaction needs. This contribution deals with the latter.

Examples of embryonic Product Avatars are already emerging on the market. For example, each new Smart Fortwo electric drive (Smart ED) car has its own web page, which shows e.g. maximum range capable with current battery charge [N.N., 2012]. The cur-rent charge status or the SmartCharging charge configuration can be controlled and managed via a web portal from a home computer or with any modern smartphone. In the future, several other features will be controlled remotely through a smart drive application for the iPhone [N.N., 2011].

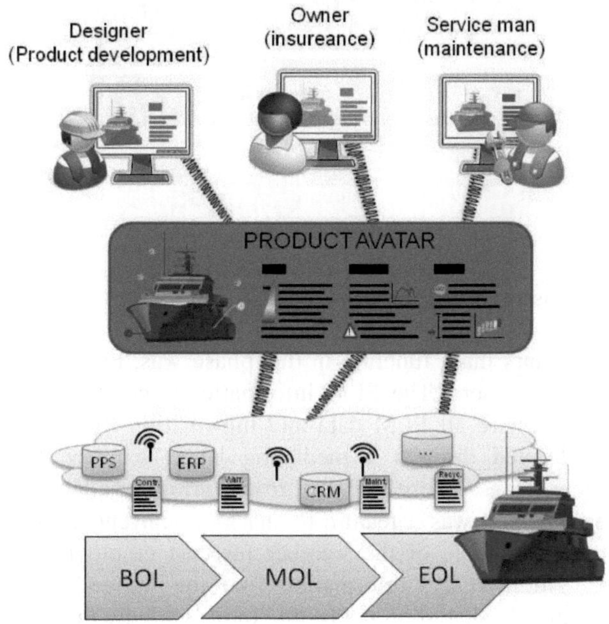

Fig. 2. Digital, Stakeholder specific Representation of a Product through a Product Avatar

3.2 Example for Product Avatar Application during the MOL Phase

In order to make the theoretical concept of a Product Avatar more feasible, a short example based on a real case will be given in this section.

The authors successfully implemented a Product Avatar application for leisure boats in the usage (MOL) phase of the lifecycle for the stakeholder group "owner" by using the channel of the popular Social Network Service (SNS) Facebook. The goal was to create additional benefits for users by providing services (e.g. automatic log-book with location based services). The rationale behind using the popular SNS Facebook was that users are already familiar with the concept and that the inherent functions of the SNS expanded by the PLM based services increase the possibilities for new services around the core product of a leisure boat (see Figure 3).

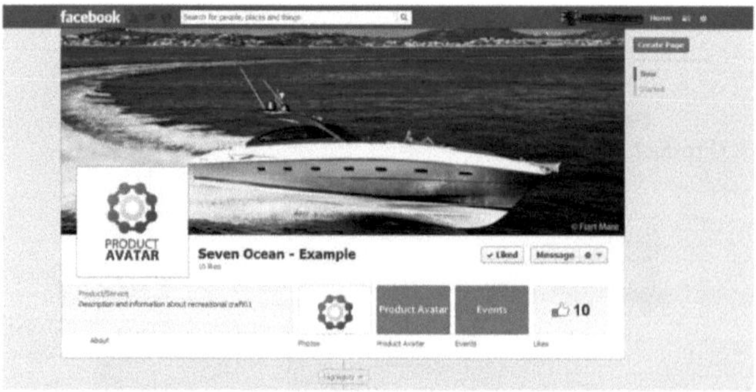

Fig. 3. Screenshot of the Facebook Product Avatar for the Usage phase (MOL)

The Product Avatars main function in this phase was, to provide pre-defined (information) services to users. The PLM information needed were either based on a common data base where all PLM data and information for the individual product were stored or derived through a mediating layer, e.g. the Semantic Mediator (Hribernik, Kramer, Hans, Thoben, 2010), from various available databases. Among the services implemented was a feature to share the current location of the boat including an automatic update of the weather forecast employing Google maps and Yahoo weather. Additionally, information like the current battery load or fuel level were automatically shared on the profile (adjustable by the user for data security reasons). (see Figure 4)

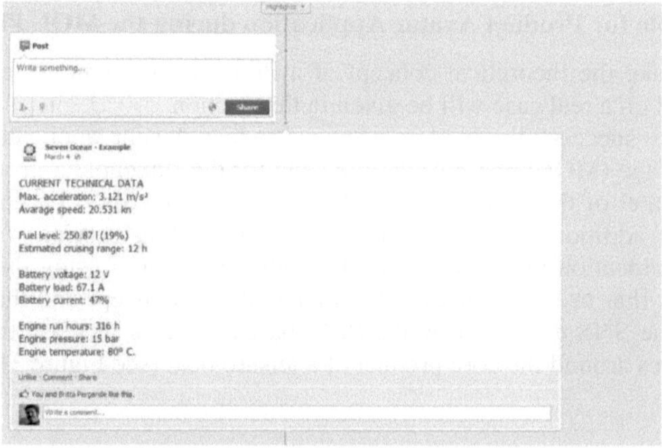

Fig. 4. Screenshot of an excerpt of information provided by the Product Avatar

Throught the digital, stakeholder specific representation of the individual product the available data and information allows the users and their friends, through social features) to benefit from their boating experience even remotely.

4 Product Avatar Application in the Bol Phase

In this section the practical example of a Product Avatar for a leisure boat, shortly introduced with a focus on MOL in the section before, will be described in more detail focusing on the BOL phase. First the stakeholders with an impact on the BOL phase will be presented and discussed briefly. Afterwards, some insights on MOL data capturing through sensor application and the different existing prototypes are introduced. The last sub-section will then give three examples of how MOL data can be applied during the BOL phase in a beneficial way for different stakeholders.

4.1 Stakeholders

The stakeholders having an impact on BOL processes can be clustered in two main groups: data producing (MOL) and data exploiting (BOL).

The group of data producing stakeholders during the MOL phase is fairly large and diverse. The main stakeholders with the biggest impact are:

- Users (owners): This stakeholder controls what data will be communicated (data security). Furthermore, they are responsible for the characteristic of the data captured through the way they use the boat.
- Producers: This group has an impact on updates (software), what sensors are implemented and what services available all influencing the data availability and quality.
- Maintenance: This group on the one hand produces relevant data themselves when repairing the boat but also ensures the operation readiness of the sensors etc.

In the BOL phase, the stakeholders are more homogenious as all have a common interest of building the boat. However, they have different needs towards possible MOL data application. There are two main groups to be identified:

- OEMs: This stakeholder is responsible for the overall planning and production of the boat and later the contact towards the customer. He has the strongest interest in learning about the "real" usage of the boat based on MOL data.
- Suppliers: This group is mostly integrated in the planning process through the OEM. However, even so indirectly included in planning activities, MOL data can be of high value for their operations.

Depending on the Customer Order Decoupling Point, the user might also fall into this category of important stakeholders during the BOL phase. However, the user at this stage is mostly considered not to be directly involved in the product developing activities and has to rely on the OEMs communication.

4.2 Capturing of MOL Data

Today's technological development, especially in the field of sensor technology, presents almost unlimited possibilities of data capturing. Of course this is limited by common sense and economic reasons.

To capture MOL data of a leisure boat, the development included three stages of prototypes.

The first stage of the so called Universal Marine Gateway (UMG) were three sensors (humidity, pressure and temperature) connected to a processing unit (here: BeagleBone) and mounted in an aquarium. In this lab prototype (see Figure 5) first hands on experience was gathered and the software interfaces with PLM data infrastructure was tested.

The next stage, the UMG Prototype MK.II (see Figure 6) incorporated the findings of the first lab prototype on a miniature model of a boat in order to learn about the effects of a mobile application and wireless communication on the data quality and impact on capturing still in a secure environment.

Fig. 5. Lab Prototype "Universal Marine Gateway" (UMG) with Example Sensors

Fig. 6. UMG Prototype Mk. II „in Action"

The final stage,UMG Prototype MK. III (see Figure 7), consists of a fully functional and live size boat where a set of sensors, based on the findings of the earlier stage tests, is implemented. This prototype will be tested under realistic settings and different scenarios.

Fig. 7. Prototype boat for sensor integration and testing

The practical implementation of the sensor equipment implies a series of challenges. E.g. the sensors need to be protected against damage coused by impact when debarking on shore. On the other hand they have to be "open" to the sourrounding environment to measure correctly. Other challenges include how the captured data is communicated "in the wild" to the data base.

4.3 Application, Limitation and Discussion

In this sub-section three exemplary cases of utilization and application of MOL data of a leisure boat by the BOL stakeholders through the Product Avatar are presented below. It is however evident that the use cases are just a short description without going into detail as this would exceed the purpose of this paper.

The first use case of MOL data is based on the Product Avatar supplying data of location, temperature, humidity in combination with a timestamp to boat designers. Ideally they can derive information on not only suitable material (e.g. what kind of wood can withstand high humidity and sun) and dimensioning of certain details (e.g. sunroof more likely to be used in tropical environments) but also on equipment needed under the circumstances (e.g. heating system or air conditioning).

Whereas the benefit of the first use case could also be realized utilizing other methods, the second one is more technical. The Product Avatar provides information directly to the suppliers of the Boat OEM, namely the engine manufacturer. Through aggregated data of, on the one side, the engine itself, e.g. rpm or heat curve and on the other side supplying information about the conditions it is used, e.g frequency, run-time, but also outside temperature etc the engine designers can reduce the risk of over-engineering. When a boat is just used a few times a year, the durability of the engine module might not be as important.

The third use case is in between the former two. Whilst it is unlikely that MOL data can influence manufacturing processes directly, it definiately can influence them indirectly through the process planning. An example is that through location based data and accompanying legal information for that location, both provided by the Product Avatar, the production planner can change the processes. So can it be necessary to e.g. add Shark-Inspired Boat Surface on the hull instead of using toxic paint as for the region the boat is mostly used the toxic paint is illegal. This could also be an application for the MOL phase again, notifying boat users not to enter a certain area (e.g. a coral reef) as they might inflict damage to the environment, which might be valued by environmelntal conscious users.

5 Conclusion and Outlook

This paper presented an introduction on the basic principles of PLM and Intelligent Products as a basis for the concept of a Product Avatar as a digital representation of a physical product. After Introducing the theoretical concept and giving an example of application of PLM data during the MOL phase, the usage of MOL data during the BOL phase was elaborated. To do so, the main stakeholders of both phases were derived and the process towards data capturing on leisure boats was briefly introduced. This was followed by three hypothetical use cases on how MOL data provided by a Product Avatar can be beneficial for the stakeholders.

In conclusion, the Product Avatar can only be as good as the existing data and information and, very importantly, the knowledge on what information and data is needed in what way (e.g. format) through which channel by which individual stakeholder.

In the next steps the Product Avatar concept will be expanded and evaluated further through scenarios as described in the use cases.

Acknowledgement. This work has partly been funded by the European Commission through the BOMA "Boat Management" project in FP7 SME-2011-1 "Research for SMEs". The authors gratefully acknowledge the support of the Commission and all BOMA project partners. The results presented are partly based on a student project at the University of Bremen. The authors would like to thank the participating students for their significant contributions: Anika Conrads, Erdem Galipoglu, Rijad Merzic, Anna Mursinsky, Britta Pergande, Hanna Selke and Mustafa Severengiz.

References

1. Hribernik, K.A., Rabe, L., Thoben, K.-D., Schumacher, J.: The product avatar as a product-instance-centric information management concept. International Journal of Product Lifecycle Management 1(4), 367–379 (2006)
2. Wuest, T., Hribernik, K., Thoben, K.-D.: Can a Product Have a Facebook? A New Perspective on Product Avatars in Product Lifecycle Management. In: Rivest, L., Bouras, A., Louhichi, B. (eds.) PLM 2012. IFIP AICT, vol. 388, pp. 400–410. Springer, Heidelberg (2012)
3. Sendler, U.: Das PLM-Kompendium. Referenzbuch des Produkt-Lebenszyklus-Managements. Springer, Heidelberg (2009)
4. Hribernik, K.A., Pille, C., Jeken, O., Thoben, K.-D., Windt, K., Busse, M.: Autonomous Control of Intelligent Products in Beginning of Life Processes. In: International Conference on Product Lifecycle Management (2010)
5. Paul, R., Paul, G.: Engineering Data Management and Product Data Management: Roles and Prospects. In: Proceedings of the Fourth International Bulgarian-Greek Conference Computer Science 2008, Kavala, Greece, pp. 614–619 (September 2008)
6. Fasoli, T., Terzi, S., Jantunen, E., Kortelainen, J., Sääski, J., Salonen, T.: Challenges in Data Management in Product Life Cycle Engineering. In: Hesselbach, J., Herrmann, C. (eds.) Glocalized Solutions for Sustainability in Manufacturing, pp. 525–530. Springer, Heidelberg (2011)
7. Merali, Y., Bennett, Z.: Web 2.0 and Network Intelligence. In: Warren, P., et al. (eds.) Context and Semantics for Knowledge Management, pp. 11–26. Springer, Heidelberg (2011)
8. Gunendran, A.G., Young, R.I.M.: Methods for the capture of manufacture best practice in product lifecycle management. Int. J. of Production Research 48(20), 5885–5904 (2011)
9. Wuest, T., Hribernik, K., Thoben, K.-D.: Digital Representations of Intelligent Products: Product Avatar 2.0. In: Abramovici, M., Stark, R. (eds.) Smart Product Engineering. LNPE, vol. 5, pp. 675–684. Springer, Heidelberg (2013)
10. McFarlane, D., Sarma, S., Chirn, J.L., Wong, C.Y., Ashton, K.: Auto ID systems and intelligent manufacturing control. Eng. Appl. of Artif. Intell. 16, 365–376 (2003)
11. Kärkkäinen, M., Holmström, J., Främling, K., Artto, K.: Intelligent products - a step towards a more effective project delivery chain. Comput. in Ind. 50, 141–151 (2003)
12. Meyer, G.G., Främling, K., Holmström, J.: Intelligent Products: A Survey. Comput. in Ind. 60, 137–148 (2009)
13. Ventä, O.: Intelligent and Systems. Technology Theme - Final Report. VTT, p. 304. VTT Publications, Espoo (2007)
14. Wong, C., McFarlane, D., Zaharudin, A., Agarwal, V.: The Intelligent Product Driven Supply Chain. In: IEEE Int. Conference on Systems, Man and Cybernetics, vol. 4 (2002)
15. Erickson, T.: Social systems: designing digital systems that support social intelligence. AI & Soc. 23, 147–166 (2009)
16. N.N.: Stromer suchen Anschluss. Die ZEIT (2012), `http://www.zeit.de/2012/31/Elektroauto-Reichweite-Kosten` (received September 15, 2012)

17. N.N.: Third-generation smart fortwo electric drive to launch worldwide in spring 2012. Green Car Congress (2011), http://www.greencarcongress.com/2011/08/smart-20110816.html?cid=6a00d8341c4fbe53ef014e8ab1feeb970d (received August 13, 2012)
18. Hribernik, K.A., Kramer, C., Hans, C., Thoben, K.-D.: A Semantic Mediator for Data Integration in Autonomous Logistics Processes. In: Poppelwell, K., Harding, J., Poler, R., Chalmeta, R. (eds.) Enterprise Interoperability IV - Making the Internet of the Future for the Future of Enterprise, pp. 157–167. Springer Publications, London (2010) ISBN 978-1-84996-256-8

Challenges and Solutions in Customer Process Based on Collaborative Engineering

Sergej Bondar[1], Leo Potjewijd[2], and Josip Stjepandic[1]

[1] PROSTEP AG, Dolivostraße 11, 64293 Darmstadt, Germany
{sergej.bondar,josip.stjepandic}@opendesc.com
[2] SKF AB, Kelvinbaan 16, 3430 DT Nieuwegein, Netherlands
leo.potjewijd@skf.com

Abstract. Due to the higher-level business trends, such as increasing globalization and outsourcing, the collaborative engineering gets more and more important. This increases the challenge for the suppliers to accomplish their customer processes optimally with minimal effort. Efficient data communication in all phases of the product development process is a prerequisite for lean and flexible collaboration processes.

In this paper, specific concepts and solutions are described, as this problem can be solved from the perspective of a tier-1 supplier by using advanced IT technology based on contemporary standards. As a case study, the example of the company SKF is shown, one of the major bearing manufacturers for the automotive sector that develops on 14 locations worldwide and communicates with over 50 OEMs (original equipment manufacturer) synchronously. The recently implemented scalable solution fosters the customer process by ensuring the data communication in the right format, quality and time and achieving significant cost savings and error reduction.

Keywords: Collaborative engineering, distributed development, CAD data exchange, CAD data translation.

1 Introduction

Nowadays modern manufacturing industries like automotive are acting in the dynamic global marketplace that demands instant responses to customer requirements, short time-to-market and high flexibility in production. In the past decades the manufacturing industries were even more shifting to a distributed environment like an extended enterprise with increasing agility. We set the customer's satisfaction on the first place in order to get the assignment against many competitors. This has caused the mass customization at high level and even more complex development, manufacturing as well as logistics processes along the manufacturing supply chain. Thus the upcoming outsourcing has derived a multi-tier supply network structure involving numerous enterprises around the globe. The product development as well takes place in global development partnerships. OEMs accomplish the development of new products at many locations in several countries across the world [1]. Furthermore a variable

G.L. Kovács and D. Kochan (Eds.): NEW PROLAMAT 2013, IFIP AICT 411, pp. 97–107, 2013.

number of external service providers and suppliers take part in individual projects. The most relationships in this supply network are temporary, exceed by the end of project and today's project partner can quite become tomorrow's harshest competitor (Fig. 1).

The suppliers were involved in the product development processes as early as possible, because they mostly possess a greater depth of domain expertise, which is necessary for best product development. The OEM-supplier relationship is characterized by a sequential interaction whereby the OEM gives clear product and production requirements to the supplier and the supplier delivers the product or service to the OEM. Supplier integration is a crucial method for incorporating a supplier's innovativeness in the product development process and reduce the costs and risk [2]. It also creates synergy through mutually interacting deliverables and decisions between OEM and suppliers. Both sides take advantage of each other's capability to develop the product as well as to obtain feedback from the other party to improve the product development. Due to the complex development cycle, the OEM took the lead and has begun to adopt supplier integration into its product development process. To respond to this trend, the collaboration and partnership management between the OEM and suppliers need to be continuously improved to reduce costs and time. Regarding the depth of collaboration, the integration of suppliers into the OEM process chain can be defined in many ways, depending on corresponding work package and type of collaboration [3]. To enable the success of supplier integration, this work describes how to shape the collaboration between the OEM and its suppliers, through determining an appropriate supplier integration method and using the proper tools.

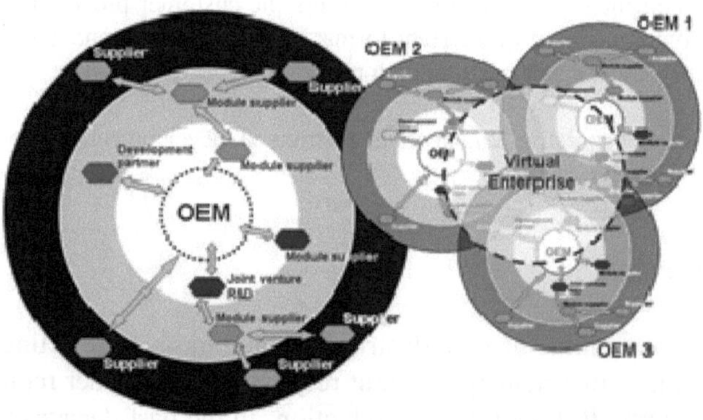

Fig. 1. Virtual enterprise in automotive industry

In the context of concurrent and collaborative engineering, the validity and consistency of product information become important. However, it is difficult for the current computer-aided systems to check the information validity and consistency because the engineers' intent is not fully represented in a consistent product model. Due to the different approaches and IT systems in the automotive OEM industry, a unified solution is not possible at this time. In particular, automotive suppliers that develop system components for a number of different OEMs or tier-1 suppliers, face the challenge of ensuring that they make the CAD data available in the format required by their customers and with a high level of reliability and, if data translation is involved, that they take the system configuration of the respective customer into consideration. Special requirements like the flexibility for further changes in business relationships and the intellectual property protection are also taken into account.

2 Cooperation Models

As the outsourcing became standard business approach in the manufacturing industries, many industry associations like the German automotive manufacturers association (VDA) accomplished the basic development work to define and classify the typical collaboration models and corresponding processes [4] (Figure 2). There are 6 supplier types defined according to the criteria: production technical integration, process integration, functional integration, and geometrical (spatial) integration of the whole product (car). Beside of the prime contractor who can be defined as a clone of the OEM without the product management, sales and marketing function, all other types of suppliers (system supplier, module supplier, component supplier, part supplier, and engineering service provider) maintain a high level of independency in their corresponding processes. Taking the fact into account that a supplier supplies many customers, who have their own, various processes and infrastructures, there yields a strong need for a comprehensive integration approach based primarily on standards and serving the relationships to all the customers.

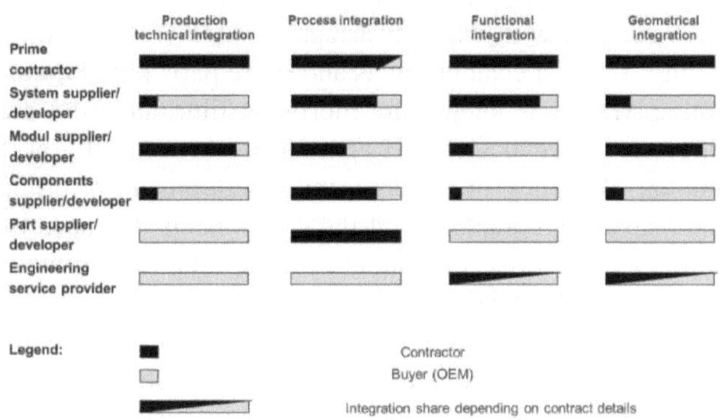

Fig. 2. Virtual enterprise in automotive industry

The responsibility of suppliers to provide the suitable product documentation in each phase of product emerging process is self-evident for an OEM. Thus the supplier is forced to take the corresponding measures to adapt his data to OEM's requirements.

3 Case SKF

The SKF Group is one of the leading manufacturers of precision bearings, seals, mechatronic components and lubrication systems. It also offers customers a wide range of services including maintenance services and condition monitoring. Founded in 1907 in Sweden, the company is a global player with 110 manufacturing sites distributed around the world and represented in over 130 countries. In the 2012 business year, the Group, which has over 46,000 employees, generated sales of 64.6 billion Swedish krona or the equivalent of 7.1 billion euros.

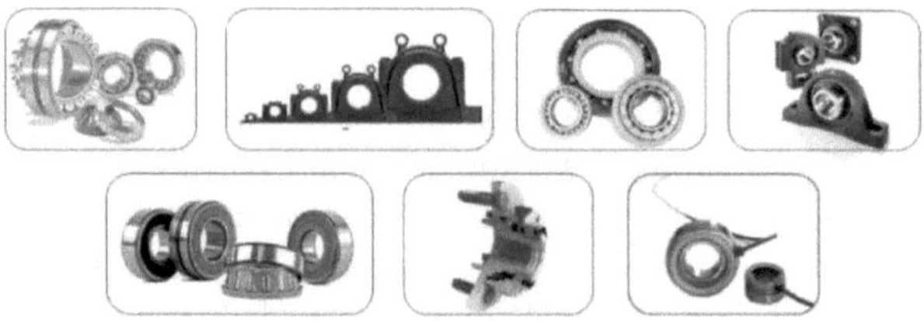

Fig. 3. Product platform "Bearing and units" at SKF

SKF is organized into 2 Markets servicing more than 40 customer segments. While the Industrial Market takes care of industrial original equipment manufacturers (OEMs) in the energy, railway vehicle, machine tool, medical, food and beverage and other industries, as well as the respective aftermarkets, the Automotive Market supplies manufacturers of passenger cars, trucks, buses, two-wheeled vehicles, and their authorized service centers, with wheel bearings, tapered roller bearings, small deep-groove ball bearings, seals, special products and spare parts kits (Figure 3).

The Automotive Market is subject to enormous time and cost pressures due to global competition. In order to set itself apart from the competition, SKF invests a considerable amount of money in research and development, resulting in numerous innovations. The increasing complexity of the components that need to be embedded in an overall solution means that the amount of coordination required between SKF and its customers naturally also increases.

Many SKF's products for the Automotive Market are designed based on customer specifications, which is why developers have to work closely with OEMs and tier-1 suppliers. To ensure the greatest possible proximity to customers, application development is decentralized to a great extent (Figure 4). It involves teams comprising of rarely more than 12 members of staff that are distributed over 14 development sites in ten countries. Cross-site collaboration is facilitated by the fact that SKF uses the CAD software Pro/ENGINEER throughout the corporation and manages the CAD data centrally using the PLM solution Windchill PDMLink. As far as external data communication is concerned, however, this means that data constantly needs to be converted into other formats.

Fig. 4. SKF design centers worldwide

4 Requirements and Weakness in the Customer Processes

Where data communication is concerned, automotive suppliers have to take numerous requirements into consideration that are not always easy to fulfil. Not only do they have to make their CAD data available in the format of the authoring systems used by their customers, but they also need to maintain their respective system environments (system configuration, version, starting model, etc.), which sometimes vary from one vehicle program to another. They have to log exchange processes so that they can provide proof of who received which version of the data when. The changes in the development process require a high degree of automation to make the increasing number of exchange processes manageable. At the same time, aspects such as intellectual property protection (IPP) must be given due consideration when converting data since it is possible that the project partners in the distributed development network are competitors. Last but not least almost each OEM prefers the use of a rich PDM client of his own PDM system on the supplier site for the data export and import. Such client must be bought and licensed, and presupposes an expensive internet access with high bandwidth. That is the reason why many suppliers employ additional resident engineers on the corresponding OEM's sites who do nothing else but data management.

Even if the engineers at SKF are developing bearings and seals for tier-1 suppliers, they often have to send the CAD data to the automotive OEM in question because the components involved are safety relevant and have to be simulated in the context of the whole vehicle. As a rule, the OEM's own data exchange specifications are used, which means that suppliers have to make the data available in the format used by the OEM's CAD system. As far as SKF was concerned, this meant that the company had to support approximately 20 different CAD configurations with several versions of a single CAD system. Data exchange was not even straightforward with customers who, like SKF, used Pro/ENGINEER.

The requirements relating to the quality of the data to be exchanged are not uniform throughout the entire development process, something which must be taken into consideration when designing the exchange process. As long as the design is being harmonized and optimized in a dialog with engineers at the OEM or tier-1 supplier, requirements relating to data quality are not excessively high. Subsequent to formal release, however, the procurement managers insist on the data being made available according to the contractually agreed guidelines. These guidelines fill volumes since they include not only the data format to be supplied but also the standards, protocols and methods that must be adhered to for the actual data exchange [5]. In many cases a significant manual rework is necessary to supply the data according to those requirements.

A great deal of know-how is embedded in the parametric design models of the bearings as features, formulas and dependencies, and this know-how must not end up in the wrong hands. In the past, the different translation processes meant that a great deal of manual intervention was required to simplify the data during translation and avoid revealing too much intellectual property [6]. The assemblies could not simply be converted into a single part because certain attributes of the individual parts such as, for example, mass or center of gravity, are needed for the calculations and simulations [7]. Certain details do, however, have to be removed as they could reveal how the engineers at SKF solved certain problems from a technical point of view. Thus an additional step became necessary to derive the customer models from the original models.

As is the case with many automotive suppliers, data communication at SKF in the past was characterized by heterogeneous translation and exchange processes, which made traceability more difficult and time-consuming. Translation and data exchange were not integrated, which meant that the files passed through numerous hands. Although the quality of the converted data normally satisfied customer requirements, the costs involved in preparing and transferring the data were relatively high and the data was not always delivered on time. Many weak points were identified.

In addition to Pro/ENGINEER, several business units of SKF Automotive also maintained other CAD systems whose sole purpose was to convert the Pro/ENGINEER data into the respective target format and adapt/rework the converted models and drawings in particular. Maintaining this menagerie of CAD systems alone was costly, without including the time required for user training. And if the specialist for a certain CAD system went on holiday or changed jobs, there was always the danger of not being able to deliver data on time.

Most business units in the Automotive Market had the local engineering service provider, who maintained a close relationship with the respective OEM and was familiar with its CAD requirements, convert the data. However, the collaboration with different partners was not ideal – neither with regard to cost nor in lead time. Bearings do not involve a large volume of data, which meant that the amount of data to be converted for each partner was not very big. As a result, the translation orders never qualified for the more favorable terms and were never processed the fastest.

Another cost driver was data communication itself due to the fact that different, sometimes country-specific, protocols and OEM-specific exchange processes had to be supported [8]. SKF normally uses the OFTP protocol in conjunction with ENX, but several customers ask for ENGDAT, which required additional software and support capacity from the outsourced IT department [9]. In addition, customer requirements sometimes change from one project to the next [10]. In view of the large number of parallel projects, it was becoming increasingly difficult to organize the data exchange processes in line with current requirements (Figure 5).

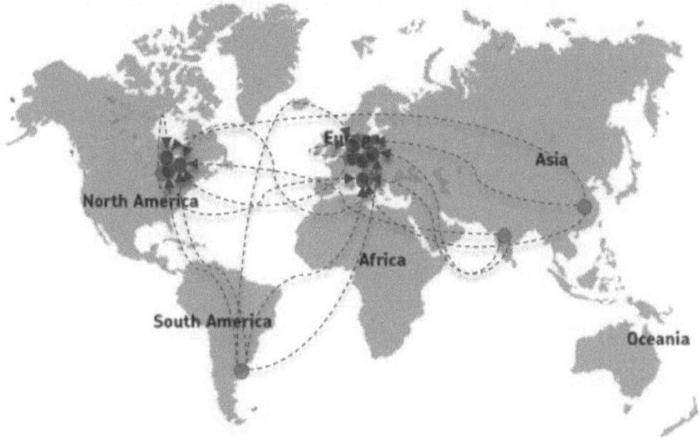

Fig. 5. Data flow required for Global data exchange

5 Solution Approach

Since the data communication is not one of the core competences of an automotive supplier, there the question arises as to whether the companies want to take this task into their own hands or whether they should outsource it to an external service provider [11]. The answer depends on the number of exchange partners involved, the variety of the exchange scenarios to be supported, the data exchange frequency and volume of data to be exchanged [12]. In the case of SKF Automotive, outsourcing data translation and exchange promised a return on investment of less than three years. The bearing manufacturer reached an agreement on a cost model that comprises a flat rate for maintaining the services and a rate for the converted and exchanged data volume. This typical pay-per-use scenario makes the costs predicable and accountable for each development project.

To make data communication more efficient, SKF Automotive chooses a uniform solution for all the application areas and sites. A key requirement of the solution and one of the primary reasons for selecting the dedicated service provider was the provision of both data translation services and the organization of data exchange processes with SKF's customers (OEM and tier-1 suppliers) (Figure 6). SKF selected a central service provider for the customer process so that it could avoid having to develop competences of its own if possible. The web-based OpenDESC.com portal offered the added advantage that it required almost no infrastructure nor initial implementation effort and could therefore be rolled out at the various SKF sites very quickly. The basic prerequisites for use of OpenDESC.com are the web browser and fast internet access.

As technological base the OpenDXM software is used which allows workflow automation of typical data management activities in engineering like import, export, convert, check, send, receive by embedding the corresponding software tools (CAD, PDM, check tools, EDI.). The communication logic of SKF including all communication partners and corresponding persons is stored in the OpenDXM database and controls the data exchange processes preserving a high level of traceability. The software is installed at PROSTEP site including all communication components and devices (ENX, EDI). The access by the end-users at SKF site is given with a web-interface.

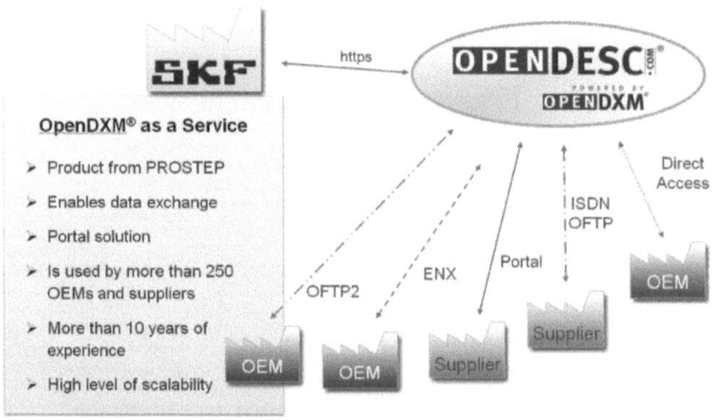

Fig. 6. Data exchange CAD via www.OpenDESC.com

Following a short period of preparation, SKF Automotive started using the central translation and data exchange service at the beginning of 2008. Mapping the exchange partners with their customer-specific requirements regarding translation, quality assurance and data availability took just a few weeks since the data exchange specialists at PROSTEP were able to take advantage of a large pool of existing profiles. Data communication functioned straightaway, with just one exception. There was initially a slight coordination problem with one OEM, which the PROSTEP specialists were able to solve quickly and to the users' complete satisfaction. Since then, the solution has been running smoothly.

To quickly bring users up to a uniform level of learning, key users were named throughout the company. These users were given half a day's training in how to use the solution, and today they are the people which the broad mass of users contact, if they have problems. In addition, PROSTEP offers webinars that allow users to improve their knowledge of the solution and reduce the number of operating errors. In principle, any user at the company today can initiate an exchange process. All they have to do is log on to www.OpenDESC.com, upload the data and select the exchange partner – all the other processes are executed automatically in the background, or if problems occur, they are solved by the translation specialists at PROSTEP within the framework of the agreed terms of delivery [13] (Figure 7). Despite of high progress of the information technology there are still task which must be done manually (repair of grave data quality problems, accurate comparison of drawings). Users at SKF have readily accepted the solution because the amount of time and effort required for data communication has been noticeably reduced.

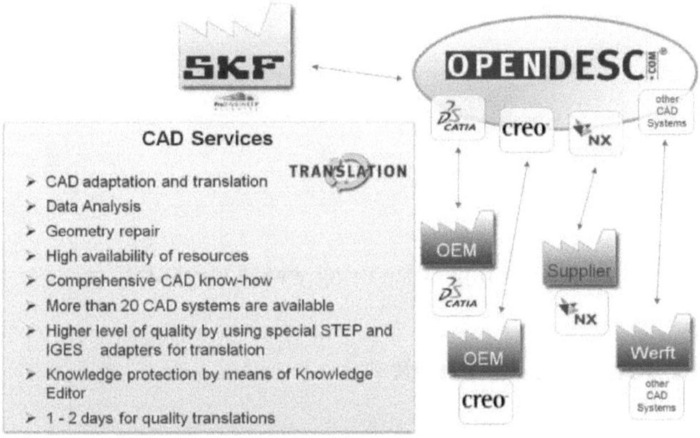

Fig. 7. CAD translation via www.OpenDESC.com

Today, the portal solution is being used at all relevant development sites. A total of more than 100 exchange relationships with the corresponding translation requirements are currently being stored, not all of which, however, are always active. The solution is currently being used for communication with almost 50 OEMs and five tier-1 suppliers at almost 90 locations (Figure 8).

The relatively low number of suppliers can be explained by the fact that SKF uses the normal supplier portal to exchange data with smaller manufacturing partners since the data exchange requirements involved are not as high. The suppliers often also receive drawing data. Last year, SKF converted approximately 4 GB of data via the PROSTEP portal, which was a greater increase than originally planned.

With the reduction of the number of tools for Data Exchange in SKF the company has achieved an ROI of 2 years. This is excluding the increased efficiency by using dedicated trained resources. Therefore the outsourcing is a success for SKF (Figure 8).

The most important benefit of the solution is knowing, that customers will receive the desired data at the agreed point in time and with the stipulated level of quality. Compliance with delivery dates has improved perceptively, which means that customers can be promised delivery on a certain date with a clear conscience. The company expects to reap additional benefits from linking the portal solution to the PLM system Windchill, which will allow data exchange to be better integrated in the customer's engineering change processes.

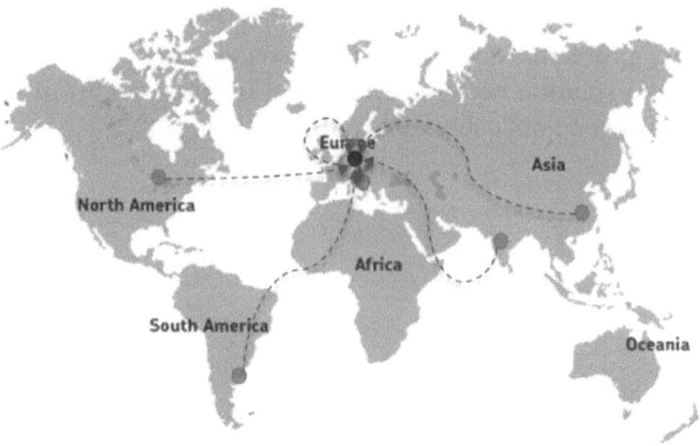

Fig. 8. Data flow when using www.OpenDESC.com

6 Conclusions and Outlook

In a dynamic collaborative environment like global automotive industry the working conditions are undertaken continuous change. Suppliers who work together with different OEMs and tier-1 suppliers have to constantly cope with new requirements relating to exchange partners, data formats, system environments to be supported, quality and security requirements, etc. If they take data communication with their customers into their own hands, this means that they have to constantly adapt their data translation and exchange processes to the ever-changing requirements. This involves considerable administrative overhead in terms of time and money, which can on occasion have a negative impact on quality and adherence to deadlines. Collaboration with a competent service provider is therefore an interesting alternative as it not only cuts costs but also facilitates making the exchange processes uniform and ensures a higher level of reliability and traceability.

Good example for the long-term stability of SKF's approach is the recent move of Daimler by replacing CATIA with NX. While the other suppliers which provide the data communication by their own are forced to adapt their infrastructure, processes and methods to new environment at Daimler by spending a significant amount of money and time, SKF has no need for action because this scenario is included in their service contract with OpenDESC.com.

Whether outsourcing is worthwhile depends on various factors such as the number of exchange partners involved, the volume of data, requirements regarding data quality, etc. The example provided by SKF, however, makes it clear that the ROI for such an investment can be calculated relatively well.

The future development belongs the further automation of whole communication process and provision of "communication plugin" [14] for each OEM based on recent standards (STEP AP242 and JT) to avoid the expensive point-to-point connection.

References

1. Stark, J.: Global Product. Strategy, Product Lifecycle Management and the Billion Customer Question. Springer, London (2007)
2. Dyer, J.H.: Collaborative advantage: winning through extended enterprise supplier networks. Oxford University Press (2000)
3. Tang, D., Chin, K.-S.: Collaborative Supplier Integration for Product Design and Development. In: Wang, L., Nee, A.Y.C. (eds.) Collaborative Design and Planning for Digital Manufacturing, pp. 99–116. Springer, London (2008)
4. VDA, VDA Empfehlung 4961/3, Abstimmung der Datenlogistik in SE-Projekten, VDA, Frankfurt (2012)
5. Fischer, T., Martin, H.P., Endres, M., Stjepandic, J., Trinkhaus, O.: Anwendungsorientierte Optimierung des neutralen CAD-Datenaustausches mit Schwerpunkt Genauigkeit und Toleranz. VDA, Frankfurt (2000)
6. Liese, H., Rulhoff, S., Stjepandic, J.: Securing Product Know-how by Embedding Intellectual Property Protection into the Organization. In: Proceedings of ICE 2010 Conference (2010)
7. Wendenburg, M.: Kein Datenaustausch ohne Datensicherheit. Digital Engineering, 1 (2008)
8. Hering, U.: Beim Datenaustausch macht Masse die Klasse. In: CADCAM 2008, 3 (2008)
9. Barkholz, D.: Trans-Atlantic networks link up to ease automotive data sharing. Automotive News, 26 (Juli 2010)
10. Wendenburg, M.: Ein System für alle Datenaustauschfälle. In: CADCAM 2006, 7–8 (2006)
11. Sendler, U.: Komplexität ist die größte Herausforderung. In: CADCAM 2007, 9–10 (2007)
12. Wendenburg, M.: Konvertierung als Dienstleistung. In: CADCAM 2007, 9–10 (2007)
13. Reim, R., Cordon, P., Hund, A., Stjepandic, J.: CAD-Konvertierung: Motivation, Probleme, Lösungen. In: CADCAM 2007, 9 (2007)
14. Mervyn, F., Kumar, A.S., Nee, A.Y.C.: A 'Plug-and-Play' Computing Environment for an Extended Enterprise. In: Li, W.D., et al. (eds.) Collaborative Product Design and Manufacturing Methodologies and Applications, pp. 71–91. Springer, London (2006)

Additive Manufacturing as Integral Part of the Digital Solution Process - An Industrial Short Note

Klaus Müller-Lohmeier

Festo AG&Co. KG, Esslingen, Germany
kum@de.festo.com

Abstract. Within the last years additive manufacturing established itself in the industrial environment as cutting-edge technological possibility to reduce time-to-market and manufacturing costs and to take in account even the chance of reducing resources consumption.

The paper figures that the digital development and production process chain also continues by stepping from virtual to physical world by using additive manufacturing means. This means, after all kind of sophisticated simulation based part and product optimisation, there is a time where you need parts to have a hand on, to test and verify them and even produce them in a new way to face actual economical and ecological challenges.

The additive manufacturing technology bears also a paradigm shift: from manufacturing-driven design to design-driven manufacturing. Based on a long-time practical in-house experience state-of-the-art applications with already proven practical relevance (among them is the bionic handling assistant awarded with the German Prize for Achievements in Technology and Innovation) highlight this new freedom of design.

Subsequent the current state of additive manufacturing as possibility for serial production will be discussed. There shows up both: real usable potentials and existing restrictions, too. The results of different research studies on PA-based specimen and parts manufactured by laser sintering (SLS) regarding mechanical properties, reproducibility, long-term behaviour and cost deployment will be presented. They pinpoint the huge progress of the last years but also further research needs to overcome mentioned obstacles for a continued and broader industrial application.

Keywords: Digital process chain, rapid prototyping, additive manufacturing, design-driven manufacturing, serial production, plastic laser sintering, long-term behaviour, cyclic loading.

1 Introduction

Festo is a leading world-wide supplier of automation technology, figure 1. As a globally-oriented and independent family company with its head office in Esslingen, Germany, it has become the performance leader in its industry over the last 50 years, thanks to its innovations and problem-solving competence in all aspects of pneumatics as well as its unique range of industrial training and education programs. Today,

G.L. Kovács and D. Kochan (Eds.): NEW PROLAMAT 2013, IFIP AICT 411, pp. 108–111, 2013.

the company provides pneumatic and electric drive technology for over 300,000 customers in 200 industries world-wide both for factory and process automation as well as cutting-edge training solutions.The Festo group achieved sales revenues € 2,24 billion in its fiscal year 2012 and with 16,200 employees is represented world-wide in 250 locations.

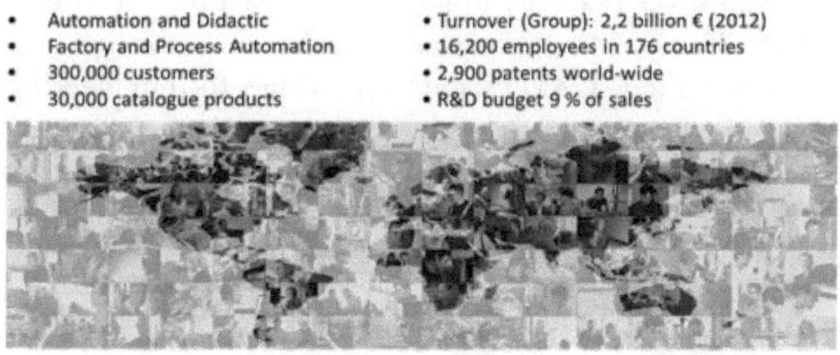

Fig. 1. Think global, act local: Festo at a glance

2 Status of Additive Manufacturing at Festo

Festo has a well founded practical and in-house experience in Additive Manufacturing, figure 2. Installing the first FDM machine in 1995, Festo invested continuously on this promising technology resulting in a company-wide network of currently 8 machines handling plastic materials like ABS, PC or PA12 and 2 laser melting systems for metal powders (tool and stainless steel, aluminium). The result is an annual inbound of more than 1.200 orders for additive manufactured parts (means nearly 5 orders per working day) with an admirable average door-to-door time of 4 days.

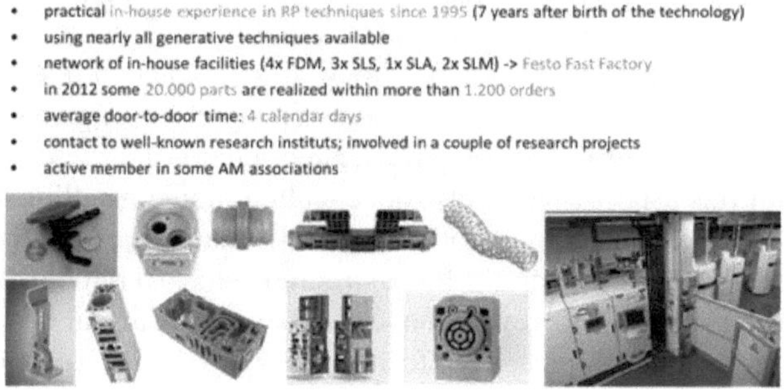

Fig. 2. Additive Manufacturing at Festo

3 Additive Manufacturing and the Digital Solution Process

Additive Manufacturing is a valid member of the digital solution process tools in new product development, figure 3. Starting in the virtual world ideas get first structure in

3D CAD drawings and are optimized using various FEM software programs (mostly this might be some kind of an iterative circle). But at a certain point, there rises the wish to have some hands-on experience with the new product, to shift to the real world and feel the haptics of the parts and to carry-out functional tests.

In this case Additive Manufacturing has the advantage to use direct the CAD data and convert this in thin slices which are sended to the AM machines where a repeated deposition of material takes place building up the physical part. If design changes are created on this parts or tests show the need of further optimization subsequent reverse engineering may create a new CAD model to (re-) start the engineering process again.

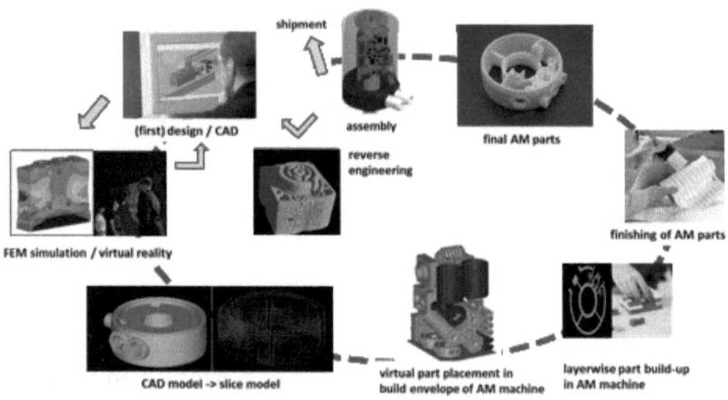

Fig. 3. Additive Manufacturing is a part of the digital solution process !

4 Paradigm Shift in Design

Another big benefit of applying additive techniques is the possibility to liberate design from manufacturing. The layer-wise and tool-less build-up of parts allows for example now internal structures and bending channels (to put it in other terms conventionally you will not be able to drill around 2 corners !) as well as a function integration (eg. direct co-build of hinges) and a reduction of parts and assembly efforts, figure 4. Now manufacturing follows design leading to complete new design solutions!

A highlight of this aspect is certainly the so-called bionic handling assistant. Flexible movement and precise gripping: at first glance, the bionic handling assistant appears to be a resilient gripper arm whose structure and overall functional principle imitate an elephant's trunk. However, above and beyond its actual benefits, the bionic handling assistant is a development platform combining a wide range of technologies and components - from manufacturing concepts to series products like sensors and valves, actuators and grippers, to control technology and software for developing applications and products. This platform, as other solutions of the bionic learning network of festo, shows future concepts for tomorrow's industry and new perspectives in hazard-free human-machine cooperation. The bionic handling assistant is thus an example of how structural flexibility can be realized and adaptive soft-gripping features can be included by using additive manufacturing technology. All parts are made by plastic laser sintering leading to this unique design.

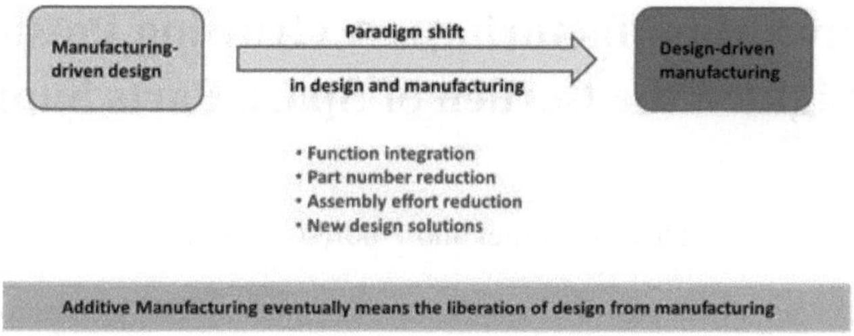

Fig. 4. Additive Manufacturing is more than just another manufacturing technology !

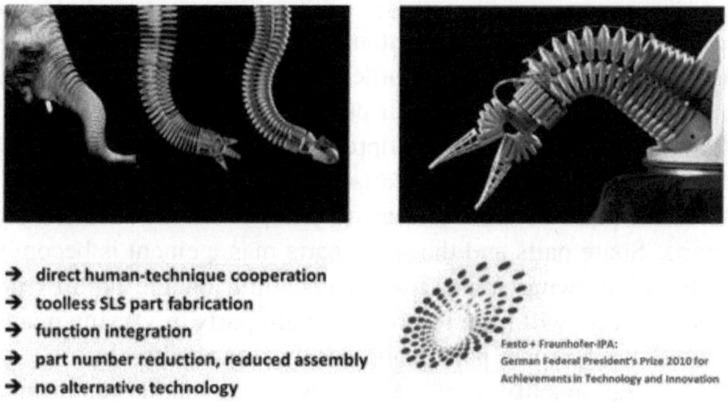

Fig. 5. Bionic Handling Assistant: Physical realisation of new handling concepts

Additive Manufacturing – A Growing Possibility to Lighten the Burden of Spare Parts Supply

Steve Rommel and Andreas Fischer

Fraunhofer Institute for Manufacturing Engineering and Automation,
Nobelstrasse 12, 70569 Stuttgart, Germany
{andreas.fischer,steve.rommel}@ipa.fraunhofer.de

Abstract. Corporations compete constantly in the market and for market share with their products and services. In order to do so one element is research for new areas of improvement within their products and processes. The focus is often on efficiency and cost effective improvements. The goal is often to provide a superior service or product in comparison to the competition. This is not only true for first time parts but also increasingly important for secondary products-the spare parts. Spare parts and the spare parts management is becoming a production factor of growing importance. Depending on the type of product or service manufacturers will have to provide spare parts in varying quantities and versions increasing the spare parts management complexity.

For the growing quantity of product types on the market requiring currently enormous warehouses to keep stock for spare parts, with corresponding high costs and complex logistics in combination with continuously decreasing product life time, decreasing time-to-market and increasing regulatory affairs Additive Manufacturing (short: AM) technology could be one solution to lighten the spare parts burden by enabling economically viable, one-off manufacturing on demand and on-location.

Keywords: Additive Manufacturing, Spare Parts, Spare Parts Management.

1 Introduction

Considering the global but also the local market and the competition of corporations within these markets constant improve of products and processes are required to search and find more cost effective solutions to manufacture products. At the same time services and products need to offer growing possibilities and ways for customers to individualize, specialize and improvise these products. This fact holds also true for spare parts and its market importance.

Today's spare parts industry is characterized by high volume production with sometimes specialized products, long distance transportation and extensive warehousing, resulting in huge inventory of spare parts. These spare parts even hold the risk of being outdated or not usable at the time of need so that they are often scrapped afterwards. On an industrial level, reacting to this burden, companies are competing or collaborating with OEMs, to provide a variety of maintenance services and products which in turn are limited with regards to the broadness and flexibility of their service

G.L. Kovács and D. Kochan (Eds.): NEW PROLAMAT 2013, IFIP AICT 411, pp. 112–123, 2013.

solutions, especially when design or feature changes to spare parts (copies or OEM-manufactured) are required.

Additive Manufacturing offers new sometimes unimaginable possibilities for manufacturing a product which have the potential to change the logistical and business requirements and therefore create the possibilities to lighten the burden. Being a new possibility the following aspects of business need to be further developed with the focus on additive manufacturing:

- standardization of manufacturing processes
- logistics
- product and process management
- certification process
- Product and business management.

The goal of funded and private research projects is to develop a model which incorporates the old and new requirements of manufacturing and the market demands to assist companies to better compete in the market settings.

2 Addititve Manufacturing and Spare Parts Management

2.1 Additive Manufacturing Technologies

Additive Manufacturing and its technologies involve all technologies used to manufacture a product by adding (placing and bonding) layers of the specific material to each other in a predetermined way. These so called layers are generally speaking 2D-cross-sections of the products 3D-model. AM therefore is creating the geometry as well as the material characteristics during the build predetermined by the material selected. The contour is created in the x-y-direction. The z-direction creates the volume and therefore the 3rd dimension.

Additive Manufacturing offers the possibility to optimize products after each run of parts being build and the lessons learned. There is generally speaking little to no limitations to the freedom in design given by this process. Complex shapes and functional parts can be realized by these innovative processes directly from CAD data. Two examples of technologies are Selective Laser Sintering (SLS) as shown in Figure 1 and Fused Deposition Modeling (FDM) as shown in Figure 2.

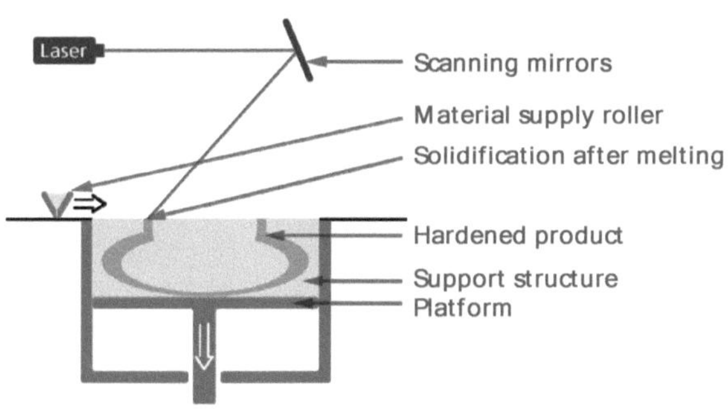

Fig. 1. Schematic diagram of SLS [following VDI 3404]

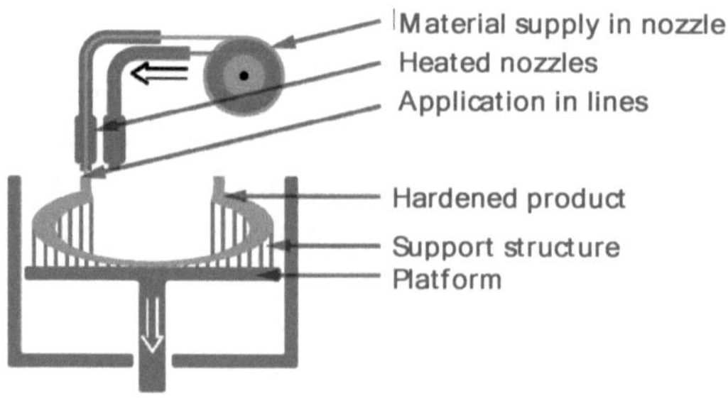

Fig. 2. Schematic diagram of FDM [following VDI 3404]

In order to hold a finished real product in hand two main process steps need to be performed (Gebhardt, 2007):

1. Developing of the cross-sections (layers) of the 3D-model
2. Fabrication of the physical product.

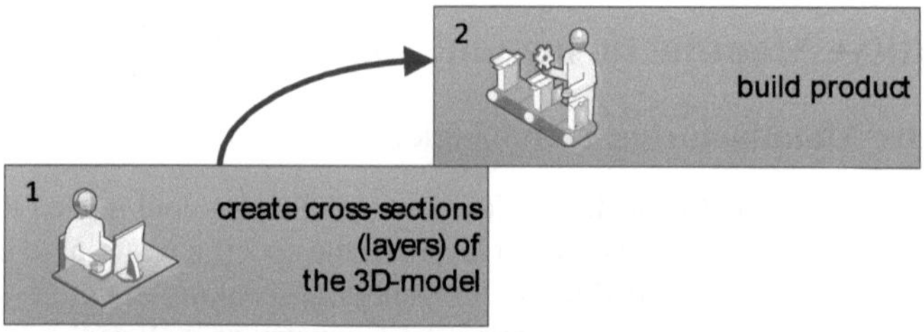

Fig. 3. Figure 3: Two main process steps of AM

These two main process steps either run simultaneously or in sequence depending on the technology. They are broken down into a series of steps following a sequence which is shown in Figure 4 (Gibson et al, 2010):

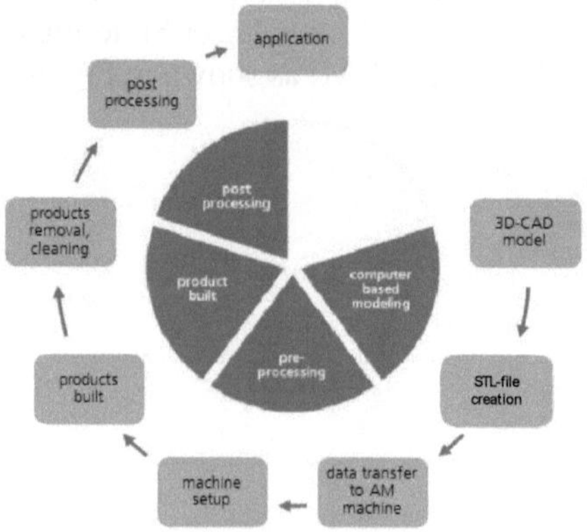

Fig. 4. Process Steps of AM

In order to withstand the test of being a true addition or even alternative to conventional manufacturing technologies additive manufactured products are looked at possessing the same mechanical and technological characteristics as comparable conventionally manufactured products. This does not equal the fact that its material characteristics have to be exactly the same to the once of conventional technologies. This can be a false point of view and a limiting factor to the use of additive manufacturing, because the new freedom of design offered also offers the possibility to create new products which may look completely different but perform the required function equally well or better.

The thinking has to shift from "get exactly the same product and its material characteristics with another technology, so I can compare it" to "get the same performance and functionality of the product regardless of the manufacturing technology" used.

Besides the benefit of using the 3D-model data directly for the manufacturing process of a product exist additional benefits listed below (Gebhardt, 2007; Hopkinson et. al., 2005):

Table 1. Benefits of Additive Manufacturing

Benefits of AM
Integration of functions, increase in complexity of products and components and including of internal structures for the stability of the product
Manufacturing of products very difficult or not possible to manufacture with traditional/conventional manufacturing technologies (e.g. undercuts …)
Variation of spare products – ability to adopt "local" requirements to the same products and therefore supply local markets with the product and its expected features, low effort and no true impact on the manufacturing
Customization – two forms of customization are possible - Manufacturer customization (MaCu), and Client customization (CliCu)
No tooling, and reduction in process steps
One-piece or small volume series manufacturing is possible,
Product-on-demand
Alternative logistic strategies based on the current requirements give the AM an enormous flexibility with regards to the strategy of the business model

2.2 Spare Parts Management

Spare parts nowadays have become a sales and production factor, especially for any manufacturing company with a highly automated, complex and linked machinery park and production setup. Any production down-time caused by the failure of a component of any equipment within production lines will lead not only to capacity issues but also to monetary losses. These losses may come from (Biedermann, 2008):

- Distribution: sales lost due to products not being manufactured
- Production: unused material, increase in material usage due to reduced capacity, additional over-time for employees to balance the inventory and make up for the lost production, additional maintenance costs of 2-30% of overall production costs
- Purchasing and supply chain: increase in storage for spare parts and costs incurred due to the purchase of spare parts

These points alone illustrate the need for each corporation to choose the right spare parts strategy in order to reduce to risk to the business to a minimum by determining the right balance between the minimum inventory and the availability to delivery spare parts intime to prevent production down-time or dissatisfied customers. When choosing this strategy on important aspect is the type of spare parts. According to DIN 13306 a spare part is an "item intended to replace a corresponding item in order to retain or maintain the original required function of the item". Biedermann is defining the items in the following way (Biedermann, 2008):

- Spare part: item, group of items or complete products intended to replace damaged, worn-out or missing items, item groups or products
- Reserve/Back-up item: item, which is allocated to one or more machines (anlagen), and therefore not used individually and in disposition and stored for the purpose of maintenance, Back-up items are usually expensive and are characterized by a low inventory level with a high monetary value
- Consumable item: item, which due to its setup will be consumed during use and has no economically sound way of maintenance

Another aspect in determining the strategy is the type of maintenance the company is choosing or offering. There are three basic strategies:

- Total Preventive Maintenance (TPM): characterized by the performance of inspections, maintenance work and replacement of components prior to the failure of the equipment.
- Scheduled Maintenance or Reliability Centered Maintenance: strategy where the replacement of an item is as the term says planned ahead of time.
- Corrective Maintenance or Repair which is also called Risk Based Maintenance: an item fails and will be replaced in order to convert the installations or equipment back into production mode.

Besides the mentioned type of spare part and the maintenance strategy the following aspects play an equally important role when selecting the strategy

- failure behavior and
- failure rate of the item
- reliability requirements
- level of information available and receivable
- back-up solutions
- Alternatives amongst others.

The decision on the strategy and the type of spare parts determines the logistics and supply chain model to be chosen and therefore the cost for the logistics portion.

2.3 Spare Parts Logistics

The current spare parts logistics strategies are typically focusing on the procurement of spare parts from an already established supplier. In many cases this supplier is responsible for manufacturing the initial primary products. This bares the benefits of an already established business relationship, defined and common understanding of the requirements for the products and services offered, clear definition of the responsibilities, established logistic, established payment modalities and a common understanding of the expectations of either party.

On the other hand there are some downfalls like lack of innovative ideas, unwanted dependence of each other, or an increase in logistics costs to name a few.

The logistics strategy itself is determined by two groups of factors:

1. Exogenous Factors: social and political environment and settings, market situation and competition, type of spare part and customer requirements and expectations [Michalak 2009].
2. Endogenous Factors: company internal factors, in this case inbound, production and outbound logistics.

With the focus on additive manufacturing the parameters for spare parts supplied using Additive Manufacturing are shown in the tables 5-7:

Table 2. Parameter Selection inbound logistics I Rommel, Fraunhofer IPA (following Michalak 2009)

<table>
<tr><th colspan="2">Parameter</th><th colspan="4">Parameter Value</th></tr>
<tr><td rowspan="4">Sourcing</td><td>Place of spare parts production</td><td colspan="2">Internal</td><td colspan="2">External</td></tr>
<tr><td>Location of spare parts manufacturer</td><td>Local</td><td colspan="2">Domestic</td><td>Global</td></tr>
<tr><td>Number of possible spare parts manufacturers</td><td colspan="2">Single</td><td colspan="2">Multiple</td></tr>
<tr><td>Vertical production integration</td><td colspan="2">Components</td><td colspan="2">Modular</td></tr>
<tr><td colspan="2">Allocation concepts</td><td>Stock</td><td colspan="2">JIT</td><td>Postponement</td></tr>
</table>

Table 3. Parameter Selection outbound logistics I Rommel, Fraunhofer IPA (following Michalak 2009)

<table>
<tr><th colspan="2">Parameter</th><th colspan="3">Parameter Value</th></tr>
<tr><td rowspan="2">Outbound logistics structure</td><td>Vertical (steps of distribution)</td><td colspan="2">single-step</td><td>multiple-step</td></tr>
<tr><td>Horizontal (number of distribution units)</td><td colspan="2">single</td><td>multiple</td></tr>
<tr><td colspan="2">Sales strategy</td><td>intensive</td><td>selective</td><td>exclusive</td></tr>
<tr><td colspan="2">storage location structure (if needed)</td><td colspan="2">central</td><td>local</td></tr>
</table>

Table 4. Parameter table for selecting the storage location strategy I Rommel, Fraunhofer IPA (following Schulte 2005)

Parameter	*Degree of centralization*		
	Trending towards centralized storage		*Trending towards decentralized storage*
Assortment	broad	⊂•••••⊃	limited
Delivery time	sufficient	⊂•••••⊃	fastest delivery (specific time …)
Product value	high	⊂•••••⊃	low
Level of concentration of manufacturing sites	one source	⊂•••••⊃	multiple sources
Customer structure	few big size companies	⊂•••••⊃	many small sized companies
Specific storage requirements	yes	⊂•••••⊃	no
Specific national/regional requirements	few	⊂•••••⊃	many

3 New Process Design and Business Model

3.1 Proposed Process Flow

Derived from the limitations and effects of the current supply strategies of spare parts, customer feedback from questionnaires and based on the processes of AM in combination with the product figure 5 illustrates a generic conventional process flow model and figure 6 the proposed preliminary process flow model.

As with standard process models the proposed process flow model covers all the process steps starting with the input from the market in the form of customer orders and customer feedback up until the delivery of the finished product to the customer.

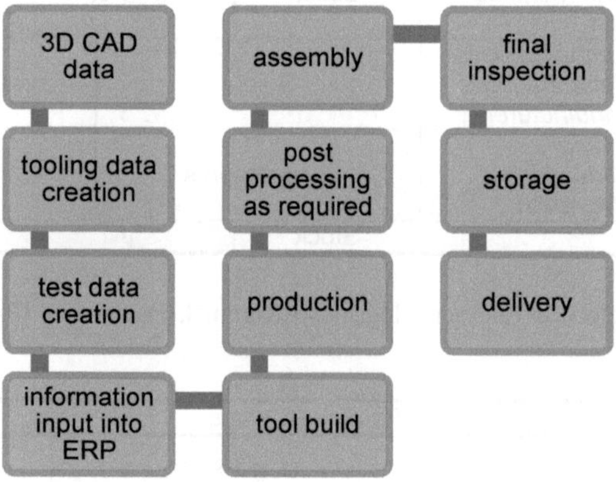

Fig. 5. Conventional process flow

The process model will be in a permanent updating stage for some time due to the development stage of the technologies.

Producing parts using Additive Manufacturing technologies has an impact on multiple levels and multiple areas of a business' operation.

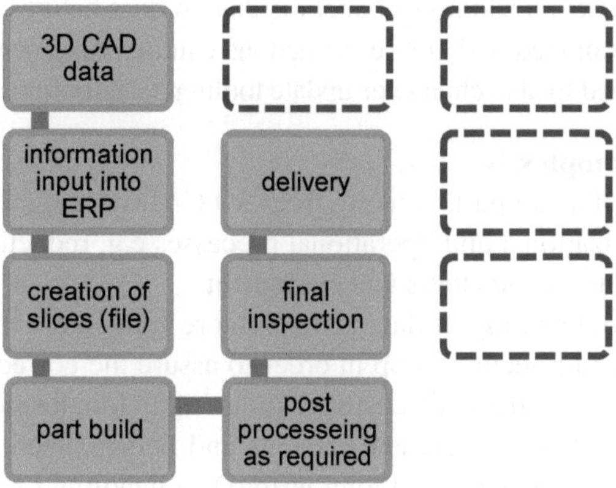

Fig. 6. Preliminary Process Model

3.2 Impact of AM Technologies on Spare Parts Manufacturing

The impact of using AM technologies to manufacture spare parts will be described in the following subchapter. This chapter presents only an overview of the main benefits.

3.2.1 Reduction or Elimination of Tooling

Conventional manufacturing like injection molding requires various tools in order to fabricate a product from start to finish. This results not only in costs for the tool build and tooling material, but also in time for the tool build, setup procedures during production periods and maintenance activities in order to keep the tools and therefore production running. Additionally tooling often has to be stored for a defined time after the end of production (EOP) to be able to produce spare parts when needed.

There are two possible alternatives to the conventional way of manufacturing spare parts being proposed: one being the fabrication of products including its spare parts strictly using Additive Manufacturing technologies from the start, thus eliminating tooling completely. The other alternative is to manufacture primary products using conventional technologies including tooling but manufacturing spare parts using Additive Manufacturing technologies.

In order to make a decision which alternative is to be preferred the suggestion is to perform an analysis of the spare part in question to determine the potentials and risks of using Additive Manufacturing. Depending on the spare parts characteristics and the spare parts strategy the following benefits can be achieved:

- Reduction or elimination of tooling
- Freeing up storage space for not needed tooling
- Freeing up storage space for already produced products
- Reduction in costs for logistics
- Freeing up production time for not required products to be made ahead of time after EOP
- Reductions of obsolete or excessive spare parts being produced at EOP and being disposed if not required.

In the case of spare parts an additional benefit is that product failures causing the need for spare parts can be examined and corrective actions can be implemented into the product design without the need to also change or update tooling data, tooling and processes.

3.2.2 Reducing Complexity

The manufacturing of spare parts directly from 3D CAD data significantly reduces the complexity in organizational and operational processes e.g. reduction of data transfers and conversion for the various tools and equipment.

On the other hand, handling of data is much more convenient than handling of real parts, but also required a secured loop in order to assure the correct data handling and storage. Within the mega trend of customization / individualization of products it is very easy to produce lots of different versions and personalized products with very little additional effort short-term and long-term. Data handling of all the versions will be the limitation.

3.2.3 Manufacturing "On Demand" and "On Location"

The main advantage of Additive Manufacturing spare parts is the possibility to produce these parts on demand. Two alternative models of this process are possible. First the spare parts will be kept on stock in very small numbers and the customer demand will trigger the delivery of the parts from the stock and the immediate production of the desired number of parts to refill the stock. Second is to eliminate the stock and produce directly the number and version of parts that the customer demands. The timing demand will be longer but no capital will be tied up in the spare parts sitting in storage. Another advantage is the future production on location: Production on location envisions sending the 3D-CAD part data with additional information regarding building process, materials and tolerances to a production site close to the customer. The parts could be manufactured in independent or dependent production facilities that have clearly defined and certified Additive Manufacturing capacities. This model could have a large impact on the logistics that will be evaluated. The impact of a production on demand, on location and with local material is recapped in the following table 5:

Table 5. Impact of Additive Manufacturing on Spare Parts

On demand	On location	Local material
No more warehousing for spare parts including space, building maintenance, energy for climate control, workers…	Worldwide service without limitations	Reaction on local requirements
No more logistics of scrapping unused old spare parts	No more logistics for end products	Environmental friendly
No more time limitations for spare parts support	Faster response time over long distances	Much less raw material logistics
		Social benefits of job creation in the local area
		Cultural adaption

3.2.4 Business Model Opportunities

Staying competitive using traditional business model concepts is becoming more and more difficult. Customization and the response time to customer needs are two critical factors of being successful. 21st century companies have to focus on moving physical products as well as their information quickly through retail, distribution, assembly, manufacture and supply. This is part of the value proposition manufacturing and service providers offer to their customers. Using Additive Manufacturing can provide a significant competitive advantage to a company.

Business models which consist of a deeper cooperation between suppliers and receivers of on-demand parts with possible virtual networks will have to be developed. The stake holders involved vary depending on the type of spare part and the setup between the manufacturer and the user. The involved stake holders are shown in figure 7.

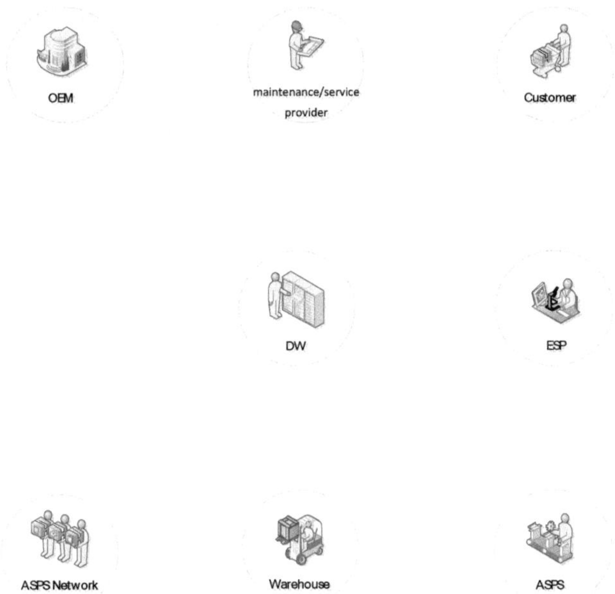

Fig. 7. Key stake holders of an Additive Manufactured Spare Parts logistics

The involvement of these stake holders can vary as shown in figure 8:

Considering the different possible scenarios of the involvement of the key stake holders one examples of the setup of such business models is illustrated in figure 9:

Scenario No.	OEM	Customer	ESP	DW	ASPS	Warehouse	ASPS Network	Maintenance/Service
	Key Stakeholders							
1	◑	●	○	○	●	○	○	●
2	◑	●	◑	○	●	○	○	●
3	◑	●	◑	●	●	○	◑	●
4	◑	●	○	○	●	●	○	●
5	○	●	◑	○	●	○	○	●
6	●	●	○	○	○	○	○	●
7	●	●	○	○	●	◑	○	●

● full involvement ◑ partial involvement

○ no involvement

Fig. 8. Involvement of key stake holders

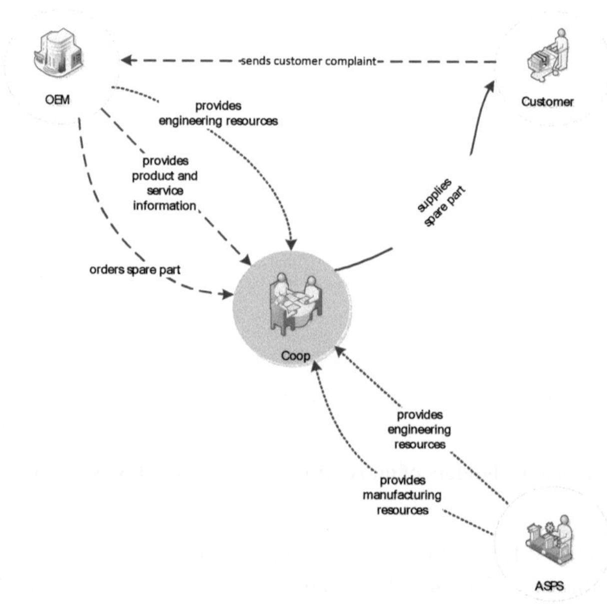

Fig. 9. Additive Manufacturing Coop

Depending on the business model each player has a different level of involvement and therefore a different level of value creation he adds to the overall product.

4 Conclusions and Outlook

Implementing and using Additive Manufacturing in order to manufacture spare parts offers a viable option not only to familiarize one with a new emerging manufacturing technology but also presents opportunities to offer products and services to the customer which fit their desire and requirements regarding time and cost effective deliver.

It is however important to take in account that Additive Manufacturing has also its current limitations like size, surface finish quality and the production volume (number of parts). Additive Manufacturing and its benefits has the potential of an enormous economic impact by reducing inventory levels to an absolute minimum as well as reducing logistics cost significantly.

References

1. Becker, R.: Midterm Report – Direct Spare. Fraunhofer IPA, Stuttgart (2010)
2. Biedermann, H.: Ersatzteilmanagement: Effiziente Ersatzteillogistik für Industrieunternehmen (VDI-Buch). Springer, Heidelberg (2008)
3. Gebhardt, A.: Generative Fertigungsverfahren: Rapid Prototyping - Rapid Tooling - Rapid Manufacturing. Carl Hanser Verlag GmbH & CO. KG, München (2007)
4. Gibson, I., Rosen, D.W., Stucker, B.: Additive Manufacturing Technologies: Rapid Prototyping to Direct Digital Manufacturing. Springer, New York (2009)
5. Hopkinson, N., Hague, R., Dickens, P.: Rapid Manufacturing: An Industrial Revolution for the Digital Age: The Next Industrial Revolution. John Wiley & Sons, West Sussex (2005)
6. Pfohl, H.: Logistiksysteme – Betriebswirtschaftliche Grundlagen. Springer, Heidelberg (2010)
7. Wirtz, B.: Business Model Management: Design - Instruments - Success Factors. Gabler Verlag, Heidelberg (2011)
8. Rommel, S., Becker, R.: Supplying Spare Parts using AM Technologies. In: Proceedings of the 1st Annual International Conference on Industrial Engineering Theory, Applications and Practice, Stuttgart, Germany

Laser Beam Melting for Tooling Applications – New Perspectives for Resource-Efficient Metal Forming and Die Casting Processes

Bernhard Müller[1], Ralf Hund[2], Roland Malek[3], and Nancy Gerth[4]

[1] Fraunhofer Institute for Machine Tools and Forming Technology IWU, Chemnitz, Germany
[2] BRAUN CarTec GmbH, Schwalbach, Germany
[3] Volkswagen AG, Wolfsburg, Germany
[4] Druckguss Heidenau GmbH, Dohna, Germany

Abstract. Applying additive manufacturing technologies in the tooling sector is reaching a new level with Laser Beam Melting, since this technology allows layer-by-layer manufacturing of completely dense tool and die inserts in standard high-alloyed tool steel. The technology is now ready to go beyond applications in low impact processes like plastic injection moulding and enters metal working process applications like metal forming and die casting. The potential of additive manufacturing for added value in tooling applications has now been investigated for various metal working processes.

The paper presents results of research and pilot application projects to apply laser beam melting to manufacture tooling for metal forming and aluminium die casting. The paper describes the shortcomings of conventional cooling channels in metal working tools and the resulting inadequate cooling effect in critical areas. The paper shows how innovative cooling systems can be implemented in metal working dies through laser beam melted die inserts. Cooling of specific die areas has been realized by placing specially designed cooling channels very close to the die cavity, targeting shorter cycle times, structural and dimensional quality improvements of manufactured metal parts and a reduction of energy consumption for cooling and idle times of forming presses and die casting machines. The paper will present the achieved results for both metal working applications and point out the general potential of additive manufacturing in tooling.

Keywords: Additive Manufacturing, Laser Beam Melting, Selective Laser Melting, tooling, metal forming, die forging, press hardening, die casting, conformal cooling, resource-efficiency.

1 Introduction

Metal forming and casting are important production technologies in Europe and worldwide. Virtually any product on the market contains formed metal parts or cast components, especially in the automotive sector. A serious challenge that manufacturers are facing today is that clients expect ever-rising complexity and functional

G.L. Kovács and D. Kochan (Eds.): NEW PROLAMAT 2013, IFIP AICT 411, pp. 124–137, 2013.

integration in all products, pushing business to providing global solutions and the conventional manufacturing technologies to their limits. From a technological point of view, main problems the industry is facing are reproducibility of the production process conditions, part quality and cycle time. This asks for new knowledge of material behaviour and results in a lot of extra effort to continuously adjust processing parameters.

The emerging additive manufacturing technology laser beam melting has been identified to have the potential to realize tooling features out of standard tool steels due to its virtually unlimited freedom of design that can fulfil the desired functions in metal forming tools and casting moulds. The additive approach in building up tooling components layer by layer is the key factor to provide this freedom of design without compromises for manufacturability, cost for complexity and tool-modularity.

2 State of the Art

2.1 Press Hardening

To implement lightweight, high-strength steel parts in car bodies, the press hardening (hot sheet metal forming) technology is an emerging technology. The sheet metal is heated above the austenitization temperature (more than 800 °C) and rapidly cooled down during the forming process below 200 °C, whereby a martensitic microstructure is created [1]. This process is especially advantageous when different demands are placed on one component, for example a component that needs to have one area of higher strength or hardness and another area with higher elongation [2]. The cycle time in press hardening is determined to as much as 30% (see Figure 1) by the cooling time (holding time of closed die after forming operation before re-opening for part extraction).

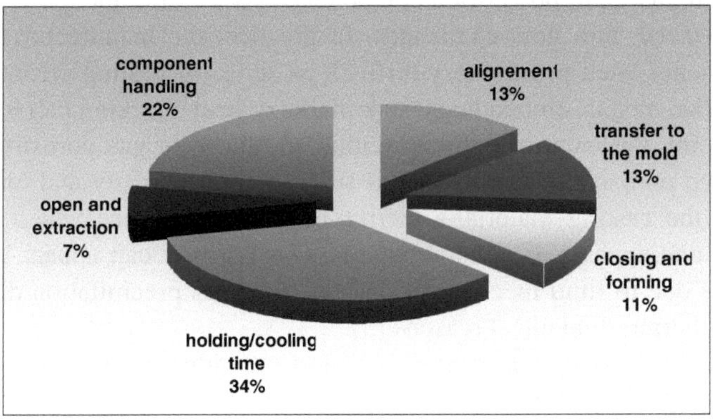

Fig. 1. Exemplary illustration of the cycle time in press hardening

So it was assumed that, through an optimized cooling system manufactured by laser beam melting, cycle time of the hot forming process can be reduced significantly. Furthermore, it was assumed that, by increasing the cooling rate, an improvement of the component's strength can be achieved. Subsequently, through the improved mechanical properties, a further reduction of wall thickness is getting possible. That results in a reduced need of raw material and lesser component weight and therefore in resource savings in manufacturing and use of the component. Due to the reduced cycle time, reduced energy demands and potential material savings, savings in resource consumption of up to 10 per cent per produced part were expected to be possible.

The setup of a hot forming tool (see Figure 2) is more complex than that of a conventional one. Mainly this is due to the fact that cooling channels must be implemented into punch and die. The implementation of the channels is usually done by deep drilling or a segmentation of the tool. Due to complex geometry of the tools, the cooling system design is especially demanding for tool manufacturers. The added complexity of cooling bores increases the expenses for hot forming tools. Current production effort is estimated with about one hour per meter borehole and a high consumption of resources (energy, drilling oil, compressed air, etc.). The mostly angled cooling bores also require additional preparation like mirroring and/or generation of a pilot hole. Therefore, the effort of work preparation like creating CAM tool paths and drilling programs as well as defining the workflow (e.g. re-clamping) is significant.

2.2 Aluminium Die Casting

High performance aluminium casting components need high performance quality. This applies especially to innovative engine concepts which offer substantial improvement in performance and efficiency and leads to high thermo-mechanical loads, especially on lightweight powertrain components. As a result, the die casting industry is constantly faced with new challenges. In practice, the manufacturing of complicated components often proves very difficult as series of casting errors such as cold shots, die marks, mends, embeddings, sink marks or heat checking can lower the quality. At this point, the issue of porosity (either shrinkage or gas porosity) plays a key role. Shrinkage porosity is also known as solidification porosity and blowholes. It is the result of the casting material's thermo-physical properties which occur during solidification and appears as volume deficits. Gas porosity can appear in the form of gas inclusions due to fluid mechanical reasons or as gas precipitation during solidification due to thermo-dynamical reasons [3].

Up to now, a selective temperature control of individual areas close to the die cavity could only be realized with enormous effort and limitations. Hence, the result: an insufficient attainability of the target temperature and often inadequate heat dissipation in the critical areas.

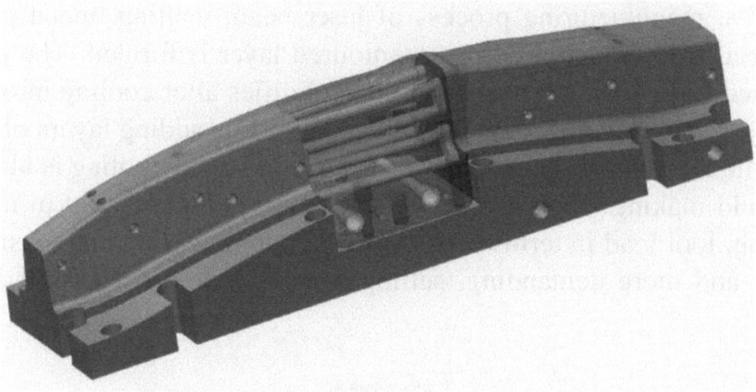

Fig. 2. Tool for hot sheet metal forming

2.3 Laser Beam Melting

A way to reduce the manufacturing effort while increasing the freedom in cooling system design is the application of additive manufacturing in tool and mould making. So called rapid tooling applications emerged very soon after introduction of first layer-based rapid prototyping technologies like stereo lithography or laminated object manufacturing (LOM). Since no metallic or other durable material could be directly processed in an additive manufacturing process at that time, the need for metallic or series-material prototypes lead to first applications of layer manufactured parts as tooling like sand and investment casting patterns [4], as well as prototype and pre-series moulds for various moulding processes [5]. Direct rapid tooling was limited to very low volume production, for higher volumes only indirect processes like Keltool were applicable [6]. First research was also done toward metal forming operations with layer-manufactured tooling [7].

With further development of the selective laser sintering process (SLS) towards direct metal laser sintering (DMLS), it became possible to directly layer manufacture metallic tooling. Limitations of that technology were initially to be found in the necessary second process step of finish sintering with significant shrinkage ratios or otherwise the infiltration with a low melting bronze alloy. This infiltrated material was able to survive complete pre-series or low-series production up to a couple of thousand shots in plastics processing like injection moulding [8]. Material properties were still far away from those of standard tooling materials like hot-work steel.

This was changed with the emergence of laser beam melting technologies. Standard tooling materials like 1.2709 can now be processed and completely melted rather than only superficially fused to an almost 100 per cent dense microstructure. Now it has become possible to use laser beam melting technologies to manufacture full series tooling for mass production without tool life limitations compared to conventional tool making by machining or EDM.

Additive manufacturing allows overcoming the limitations of today's common manufacturing technology and opens up new ways for cooling of metal forming tools, e.g. in die forging [9]. Components and tools can be manufactured directly based on 3D CAD data from powdered materials such as hot work steel and built up layer-wise.

For the additive manufacturing process of laser beam melting, metal powder is the starting material from which a defined contoured layer is formed. The powder is selectively melted layer by layer by a laser and solidifies after cooling into a solid body (see Figure 3). Therefore, the component is produced by adding layers of material and not by removing. Thanks to laser beam melting, conformal cooling is already state of the art in mould making when it comes to injection moulding. But in metal forming and die casting, tool load in terms of temperature, compression and tension is considerably higher and more demanding, setting a new challenge for laser beam melted tooling.

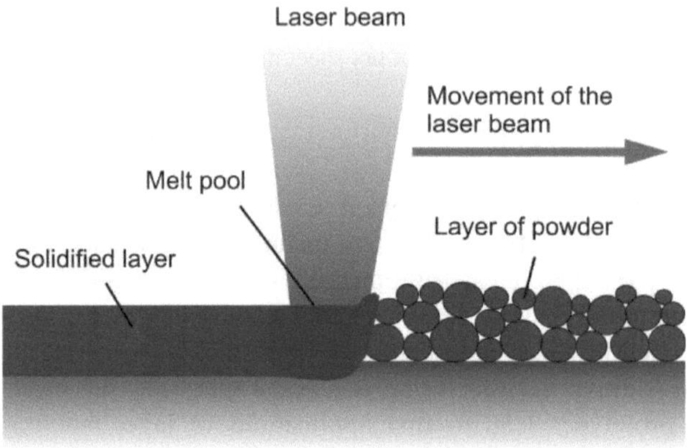

Fig. 3. Laser beam melting principle

3 The Projects

3.1 Press Hardening

Within the so-called Innovation Alliance "Green Carbody Technologies", funded by the German Federal Ministry of Education and Research and combining 60 companies and research institutes in joint research to increase resource efficiency along the entire car body production chain, it has been investigated how the hot sheet metal forming process of press hardening can become more resource efficient by using innovative laser beam melted tooling inserts. Thereby, specific die areas were getting locally tempered by arranging cooling channels very close and conformal to the cavity shape. Research aims were: the reduction of process cycle time, further enhancement of mechanical properties of hot formed metal sheets, further reduction of wall thickness and a general reduction of amount of energy used per component in its manufacturing.

The aim of the presented Innovation Alliance project was to develop and manufacture tool inserts with a new type of cooling system to improve resource efficiency in hot sheet metal forming. Therefore, thermo-fluidic simulation and laser beam melting were used.

After investigating current mass production, the project partners have jointly developed a representative demonstrator (see Figure 4). To enable easy transfer of the project's results into mass production, the demonstrator's geometry is very similar to a serial component. The design reflects a typical hot forming component and its difficulties and potential problems. It incorporates geometric features such as curved surfaces and cavities to demonstrate limitations of conventional, deep hole drilled cooling channels in terms of rapid and homogeneous cooling of the sheet metal component.

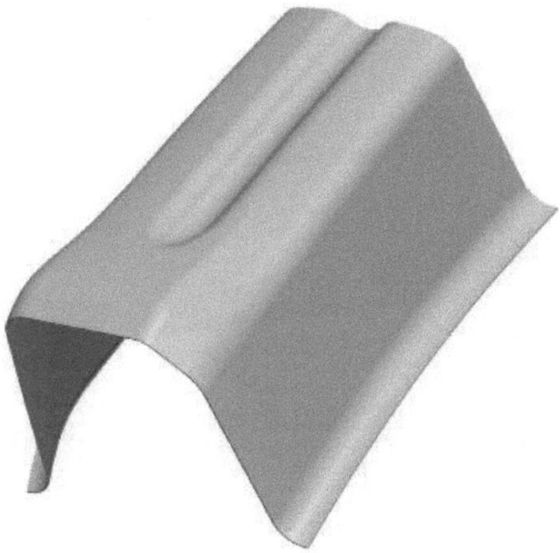

Fig. 4. Demonstrator

The tool design and the cooling system design were done based on conventional manufacturing methods such as milling and deep drilling. Parallel to this, the development of the innovative, conformal cooling system began. Various iterations of the cooling system were designed. First proof of positive effects of the optimized tool temperature control was provided by numerical simulation (see Figure 5).

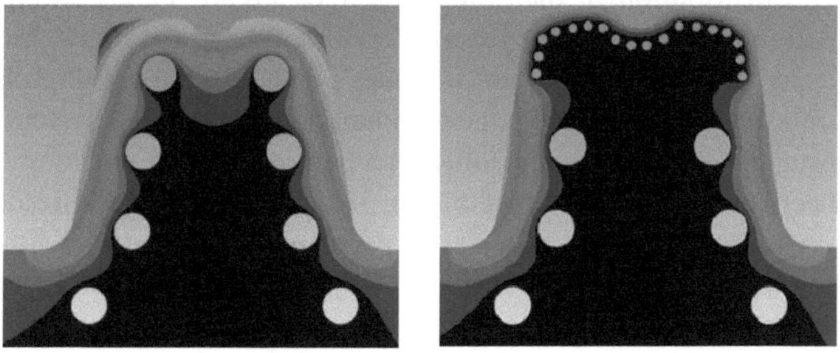

Fig. 5. Thermal simulation: comparison of conventional drilled cooling channels (maximum temperature in the tool 191 °C, left) and optimized cooling channels with a significantly lower thermal load (81 °C, right)

In the project, thermal behaviour of the tool as well as coolant flow was analysed and different cooling geometries were compared. The input variables such as compression force, work piece temperature, coolant temperature, flow rate, pump power and surface roughness of cooling channels were adopted from the mass production system. In order to assure comparability of simulation results with reality, thermal conductivity of specific materials in use were determined experimentally on the basis of material samples. The optimum cooling channel geometry was designed based on simulation results, considering technical characteristics of the laser beam melting technology (see Figure 6).

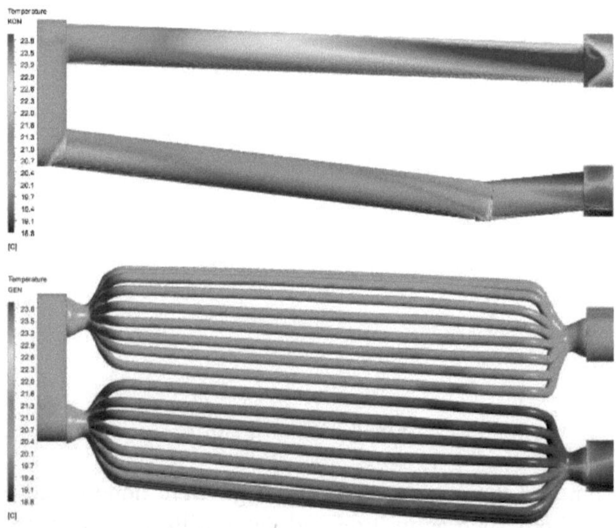

Fig. 6. Comparison of the temperature gradient in conventional cooling system (top) and in new cooling system (bottom)

Temperature distribution in the component (see Figure 7), according to thermo-fluidic simulation, showed inhomogeneous cooling due to the conventional die and limitations in its manufacturing when it comes to getting the cooling channels very close and conformal to the surface. Due to simulation, the cooling system's efficiency could be constantly improved and resulted in a homogeneous temperature distribution within the sheet metal component (see Figure 7). Thanks to the optimized cooling it is possible to cool down the parts more evenly and more rapidly. The simulation suggested the holding time to be shortened by 45 per cent from initially 11 down to 6 seconds, using the same temperature profile like in conventional cooling.

The most critical area was localized in the die's deepest cavity, where – due to limitations of conventional drilling – standard cooling channels have the longest distance to the cavity surface. Therefore, a complete re-design of the die's entire cooling

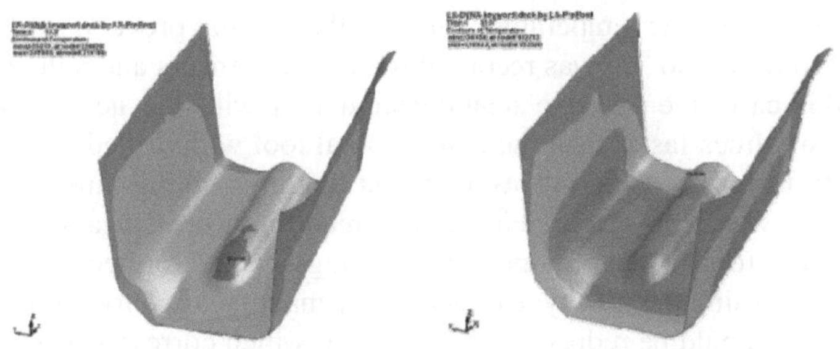

Fig. 7. Comparison of the temperature distribution in the sheet metal component with conventional cooling system (component temperature in the critical area 335 °C, left) and with the optimized cooling system (177 °C, right)

system was not even necessary. The base body was left untouched and so the re-design focused only on the critical areas around the deep cavity. Because of that and in order to achieve the best synthesis of greatest value, short production time and low costs it was decided to manufacture the tooling insert by so-called hybrid tooling, a combination of conventional manufacturing technologies like milling, drilling, turning with additive manufacturing like laser beam melting.

In this case, the laser beam melted functional structure with optimized cooling channels was applied on a conventionally milled base body (see Figure 8). Only rough machining and heat treatment needed to be done on the base body to prepare it for laser beam melting of the top section. To get the best possible bonding between base body and functional structure with conformal cooling system, the upper base body surface was grinded and afterwards sand blasted. The base body was then placed and fixated in the laser beam melting machine. After the functional structure was applied by laser beam melting, the tool insert (see Figure 8) was removed from the machine and heat treated for hardening and stress relief within the laser beam melted section.

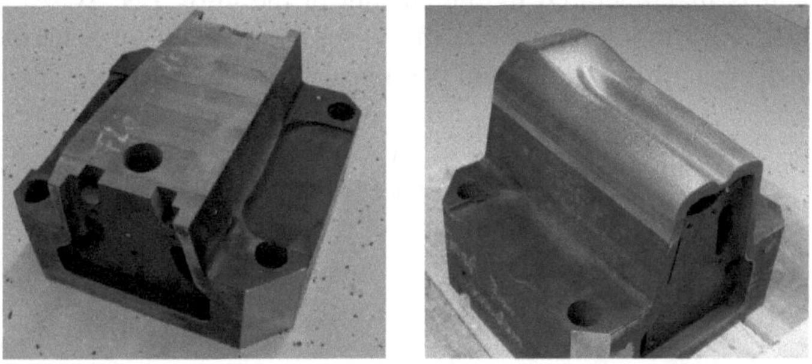

Fig. 8. Tool insert – milled base body (left) and ready for finish-machining (right)

To confirm simulation results, extensive forming trials were done. The trials took place on a standard hot forming press, under production-like conditions and applying a variety of different parameter settings. With the help of latest equipment like thermo-camera, temperature sensors and computer-assisted analysis, all relevant data from the trials were recorded and afterwards analysed.

In a first test series, the temperature curve of the cooling process, starting from an initial temperature of 200 °C, was recorded by thermal sensors and a thermo-imaging camera. In this case, the additive manufactured tool with the new cooling system cooled down six times faster than the conventional tool with drilled cooling channels (see Figure 9). In further experiments, different holding/cooling times were run with varying cooling water flow. For temperature recording, once again thermocouples were used in the tool and thermographic imaging for the formed component. The results show that, using the new type of additively manufactured tool inserts, the holding (cooling) time could be reduced by 50 per cent, which corresponds in this particular case to a total cycle time reduction of 20 per cent.

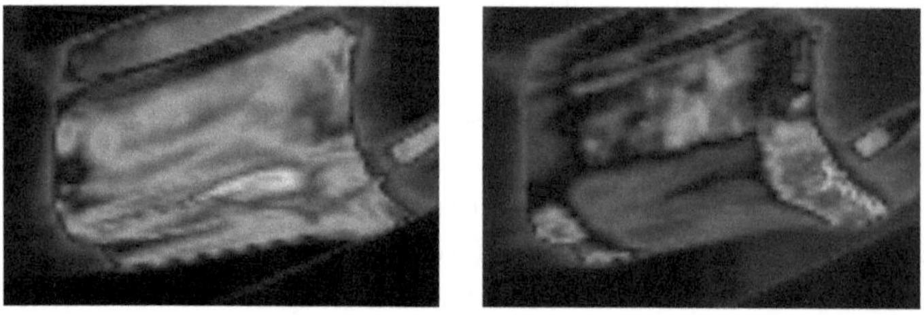

Fig. 9. Thermal image of the die after 5 seconds of cooling with the conventional cooling system (temperature in the top section of the tool 142 °C, left) and with the additively manufactured cooling system (68 ° C, right)

3.2 Aluminium Die Casting

The newly developed V8-TFSI engine of Audi combines efficiency and high performance. The Biturbo presents its Premium character with a torque of up to 650Nm in a range of 1.700 up to 5.500 1/min and power outputs of up to 382kW (520PS). The engine consists of a 90°V-cylinder block and the beneath attached bedplate [10]. The component is manufactured by die casting in the alloy EN AC-Al Si9Cu3 (Fe) and is equipped with moulded crankshaft bearing caps made of spheroidal graphite iron to optimize strength, acoustics and main bearing (Figure 10).

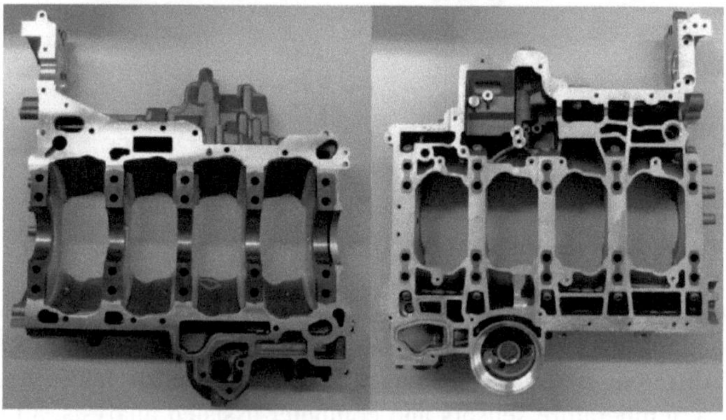

Fig. 10. Audi bedplate V8-TFSI (view of both sides)

In the development stages, the bedplate has already been classified as a highly complex component. High customer requirements on the die cast component lead to a particularly subtle approach in terms of design and manufacturing of the component with regard to the main criteria die casting technology, die casting alloy and casting die. Therefore, the influencing factors such as material properties, process and technological parameters of the die casting machine, die design under usage of an optimal gating and ventilation system were mutually modified.

The complicated bedplate was on a basic state after first series of manufacturing which has been continually developed and modified since then. The quality control, documentation and evaluation provided the basis for the on-going process analysis as well as for creation and realization of the action plan in order to optimize the component. Defects like local weak spots, shrink marks, cold runs and hot cracking are mostly thermal in nature [10] and have been improved via adaptations of the thermal management and also by precise adjustments of process parameters. Particular difficulties were, due to critical porosities, specifically in the area of the oil filter cartridge. Figure 11 shows the porosity's position in the component and its magnitude.

Fig. 11. Porosity in bed plate at oil filter cartridge (right: enlarged view)

The micro porosity could not be specified unambiguously – neither by x-ray methods nor computer tomography (CT) scan imaging of the whole bedplate. Material analysis – electron microscopy (REM) and micro CT [12] – provided insight (Figure 12). The pores were a hybrid of gas and solidification porosity. A minimization of the weak spot was partially attained on the conventional way by melt treatment, optimization of process parameters, optimization of process technology and periphery, optimization of process monitoring and management and optimization of the casting die, in particular the gating and venting system.

The threshold value of pore sizes, especially in the oil filter cartridge zone, should prevent a possible damage of the cartridge incl. sealing ring when it comes to mounting or removing of the filter, a material fatigue or damage of the component and exclude the leakage oil. A geometrical adaptation of the component by the Audi designer was limited by the given installation space and was no option. Therefore, new production technologies had to be applied in order to ensure production output, quality and efficiency of the bedplate.

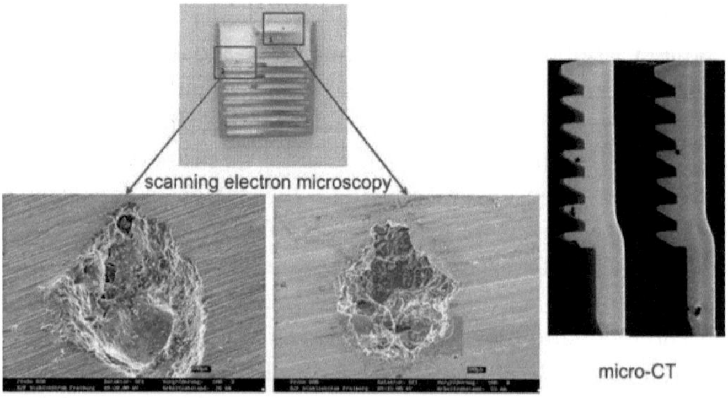

Fig. 12. Material analysis in the field of oil filter pot REM (left), micro-CT (right) (source: TU Freiberg – Foundry Institute)

The idea was to implement an additional conformal cooling system besides the common tempering of the mould which was allowed to consist of only minute cooling channels due to the wall thickness. The conflict was to maintain the energy input (temperature) by the melt to guarantee an even die filling and simultaneously dampening the momentum of the melt at the sprue to ensure a homogenous material solidification. Conformal cooling was implemented by individual laser beam melted tool inserts (Figure 13).

In this case, the tool insert was manufactured in hybrid design to realize a preferably economical production. It represents the combination of a conventional manufactured base body and an additively built-on functional geometry. This allows the best possible synthesis of greatest added value, short production time and lowest cost.

Fig. 13. Laser beam melted tool insert (left) and installed as a component in the die (right)

By using the additive laser beam melting technology, it became possible to focus on functionality and efficiency of the cooling system with no regard to manufacturability. With the help of numerical simulation and with consideration of existing cooling channels and connectors, an application-specific, innovative cooling system was developed. On the one hand, numerical simulation served as a tool to ensure the shortest possible distance of the cooling to the cavity, while still allowing a sufficient strength of the tool insert (FE simulation). On the other hand, simulation allowed a step-by-step optimization of the cooling system (thermal simulation) ranging from a simple spiral design to a complex system of small channels (see figure 14). In general, special care has been taken in cooling system design with regard to the total cross section of all individual cooling channels combined, which should correspond to the cross section of the coolant intake and outlet for a homogeneous flow distribution across all channels and therefor the most homogenous temperature distribution possible. Creating such a complex cooling system, including specific channel layout with wide contours and generously rounded transitions, in order to keep pressure losses to a minimum, became possible by using additive manufacturing technology only.

The first additively manufactured, optimised tool insert was run in the casting die of the bedplate and subsequently the follow-up version in series production. Test runs already proved function of tool insert and innovative cooling system. Thereupon, the insert was permanently attached to the die and the cooling channels could be connected to a specially water temperature control unit via conventional hoses. The cooling was gradually adjusted in sensitive tempering phases and a recognizable influence could be recorded. In a range of only 10 Kelvin, the component area showed no critical overheating or hypothermia. This temperature range needed to remain constant throughout the whole casting process. Immediate analyses of different tempering steps at the produced components via x-ray analysis, quick sample processing and additional materials analysis showed that gas and shrinkage porosity was significantly reduced and minimized to a level of permissible porosity.

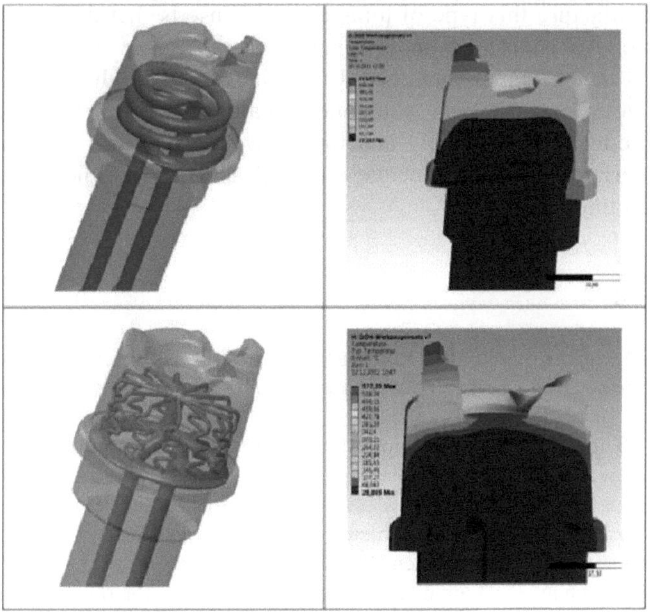

Fig. 14. Laser beam melted tool insert (left) and installed as a component in the die (right)

Meanwhile, all measures have been applied in series production. The achieved results could be proved and resulted in a scrap rate reduction of 10% referring to the error pattern of porosity. The innovative, additively manufactured tool insert enabled 5% less susceptibility of the process and a 3% cycle time reduction. Sustainability tests throughout the casting die's complete life cycle have not been finalized yet. Nevertheless, the results enabled a more stable production process of the bedplate without any part design changes and therefore assured a high focus on the component's functionality: a compact power unit, overall length limits and a high degree of integration. A rapid launch in series production could be realized [13][14].

4 Conclusions

The paper has described how metal forming and casting tool inserts can be manufactures by laser beam melting. A detailed in-sight has been given through case studies of innovative cooling systems, for hot sheet metal forming as well as for aluminium die casting. In the projects presented in this paper, it could be proved that laser beam melting is a well-suited technology for manufacturing highly complex moulds and tools which go beyond the limits of conventional production technologies. The unique laser beam melting technology opens up ways for new design approaches of cooling systems in metal forming tools and casting dies. This paper's main conclusions are to be found in increasing the cooling rate in hot sheet metal forming and in improving the part quality in light metal die casting and pointed out the superiority of laser beam melting to manufacture this type of tools and tool inserts that made these promising results possible. An enormous improvement in temperature distribution within the tool as well as in the components could be achieved for both manufacturing processes. Due to laser beam melting, component quality can be improved, process cycle times can be reduced significantly and therefore it is possible to increase resource efficiency of series production processes as well as to reduce the amount of energy used to manufacture each part.

References

1. Engel, B.: Hochfeste Stähle. IBU Vortragsreihe. Universität Siegen (2008), http://www.uni-siegen.de/fb11/fw/lehrstuhl/publikationen/pdf/ibu_hochfest.pdf
2. Beck, M.: Mehr als erwärmen und umformen. Industrieforum.net (2008), http://www.industrieforum.net/de/blechonlinede/oktober062008/rubrik/umformen/mehr-als-erwaermen-und-umformen
3. VDG Verein Deutscher Gießereifachleute: VDG-Merkblatt P201 – Volumendefizite von Gussstücken aus Nichteisenmetallen (Mai 2002)
4. Mueller, B., Kochan, D.: Laminated object manufacturing for rapid tooling and patternmaking in foundry industry. Computers in Industry 39/1(4), 47–53 (1999)
5. Chua, C.K., Hong, K.H., Ho, S.L.: Rapid Tooling Technology Part 2 a Case Study Using Arc Spray Metal Tooling. Advanced Manufacturing Technology 15(8), 609–614 (1999)

6. Jetley, S., Low, D.K.: A Rapid Tooling Technique Using a Low Melting Point Metal Alloy for Plastic Injection Molding. Journal of Industrial Technology 22(3), 2–8 (2006)
7. Voelkner, W.: Untersuchung der Möglichkeit des Einsatzes der Stereolithographie zum Bau von Blechumformwerkzeugen. DFG-Abschlussbericht. Institut für Produktionstechnik der TU Dresden, Dresden (1997)
8. Ferreira, J.C.: Rapid tooling of die DMLS inserts for shoot-squeeze moulding (DISA) system. Journal of Materials Processing Technology 155-156, 1111–1117 (2004)
9. Neugebauer, R., Mueller, B., Wagner, A.: Direct Rapid Tooling for Die Forging – a new challenge for Layer-Based Technologies. In: Bártolo, P.J., et al. (eds.) Innovative Developments in Design and Manufacturing – Advanced Research in Virtual and Rapid Prototyping, pp. 283–288. Taylor & Francis, Oxford (2010)
10. Königstedt, J., Aßmann, M., Brinkmann, C., Eiser, A., Grob, A., Jablonski, J., Müller, R.: Die neuen 4.0l-V8-TFSI-Motoren von Audi – die konsequente Verbindung von Effizienz und Performance. 33. Internationales Wiener Motorensymposium (2012)
11. Nogowizin, B.: Theorie und Praxis des Druckgusses. Fachverlag Schiele & Schön GmbH, Berlin (2011)
12. Eigenfeld, K., Jäckel, E.: Gutachten zur Bauteilanalyse. TU Bergakademie Freiberg, Gießerei-Institut (2012)
13. Gerth, N., Fischer, A., Hamann, I., Sauer, H., Müller, B., Rädel, T., Gebauer, M., Toeppel, T.: Prozessoptimierung im Druckgießverfahren – Laserstrahlgeschmolzener Werkzeugeinsatz im Praxistest. Giesserei 100(4), 34–41 (2013)
14. Gerth, N., Sauer, H., Gebauer, M., Rädel, T.: Prozessoptimierung in der Fertigung einer Lagertraverse. MTZ Motortechnische Zeitschrift 74(06), 488–493 (2013)

Design-Opportunities and Limitations on Additive Manufacturing Determined by a Suitable Test-Specimen

Karlheinz P.J. Hoeren and Gerd Witt

University Duisburg-Essen, Department of Mechanical Engineering,
Chair of Manufacturing Technologies
{karlheinz.hoeren,gerd.witt}@uni-due.de

Abstract. The key-feature all additive manufacturing processes have in common is parts being built up in layers. Due to this tool-less build-up principle, the degree of freedom in design is huge compared to conventional manufacturing processes. For instance rear sections, inner cavities, conformal cooling-channels and lightweight-structures can be realized without a significant rise of manufacturing-costs. However there are some specific limitations coming along with the layer-wise build-up. Depending on layer-thickness and the orientation of a surface, the so-called stair-step effect occurs with different levels of intensity. Furthermore some additive manufacturing processes have to use support-structures. These structures are used to hold the part in place or to lead the pro-cess heat away from the melting-zone. In order to convey these limitations and to show how de-tailed and filigree parts can be generated, a test-specimen has been designed that suits processes with- and without support-structures.

The geometry of the specimen`s bottom side contains chamfers and notches, so the amount of necessary supports is kept as low as possible and the specimen can be separated from its build-platform without complications. Furthermore, the specimen is designed as a web of test-areas, connected by bars. This way the curling effect that often leads to aborts in additive manufacturing processes is prevented and thus cannot have an influence on the geometries tested on the speci-men. The presentation is about showing first test results of Laser Beam Melted, Laser Sintered and Fused Layer Modeled specimens that have been evaluated by sight test and measured by a coordinate measuring machine. In addition the content of the individual test-areas will be ex-plained.

Keywords: Additive Manufacturing, Laser Beam Melting, Fused Layer Modeling, Laser Sintering, test-specimen.

1 Introduction

Additive manufacturing can be described as a direct, tool-less and layer-wise production of parts based on 3D product model data. This data can be based on Image-Generating Measuring Procedures like CT (computed tomography), MRI (magnetic resonance imaging) and 3D-Scanning, or, like in the majority of cases, a 3D-CAD construction. Due to the layer-wise and tool-less buildup-principle, additive manufacturing offers a huge amount of freedom for designers, compared to conventional

G.L. Kovács and D. Kochan (Eds.): NEW PROLAMAT 2013, IFIP AICT 411, pp. 138–152, 2013.

manufacturing processes. For instance rear sections, light weight constructions or inner cavities can be built up without a significant rise of manufacturing costs. However, there are some specific limitations on the freedom of construction in additive manufacturing. These limitations can partly be attributed to the layer-wise principle of buildup, which all additive manufacturing technologies have in common, but also to the individual restrictions that come along with every single manufacturing technology. [1] [2] [3]

Following, after a short description of the additive manufacturing technologies Laser Beam Melting (LBM), Laser Sintering (LS) and Fused Layer Mod-eling (FLM), the geometry of a test-specimen that has been designed by the chair of manufacturing technologies of the University Duisburg-Essen will be introduced. Based on this geometry, the design-opportunities and limitations of the described tech-nologies will be evaluated.

1.1 Laser Beam Melting (LBM)

Besides Electron Beam Melting, LBM is the only way of directly producing metal parts in a powder-based additive manufacturing process. [4] [5] In course of the LBM process, parts are built up by repeatedly lowering the build-platform, recoating the build-area with fresh powder from the powder-supply and selective melting of the metal powder by means of a Laser-Beam. For the schematic structure of LBM-Machine see Figure 1. [6]

Melting metal powder, a lot of energy is needed. Therefore a huge amount of thermal energy is led into the build-area. In order to lead this process-warmth away from the build-plane, and in order to keep the parts in place, supports (or support-structures) are needed in Laser Beam Melting. Sup-port-structures are to be placed underneath every surface that is inclined by less than about 45° to-wards the build-platform. They are build up simulta-neously with the part and consist of the same mate-rial. Thus parts have to be mechanically separated from their supports after the LBM-process, for example by sawing, milling or snapping off. As a re-sult, the surface-quality in supported areas is significantly reduced and LBM-parts are often reworked with processes like abrasive blasting, barrel finishing, or electrolytic polishing. [7] [8]

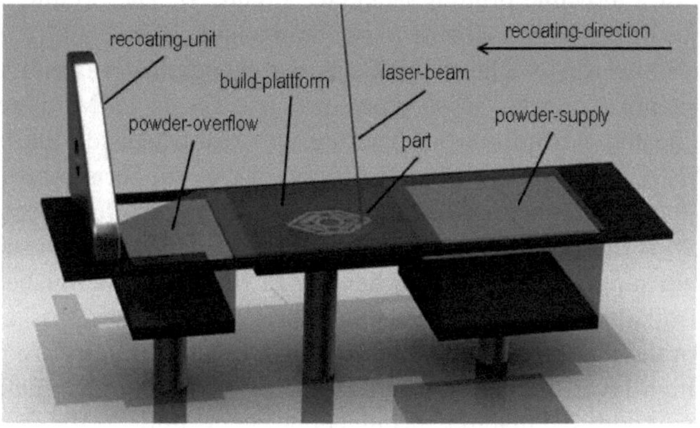

Fig. 1. Schematic illustration of the LBM process

1.2 Laser Sintering (LS)

In Laser Sintering, unlike LBM, plastic powder is used as basic material. Regarding the procedure, LBM and LS are very similar (see Figure 2). One of the main differences is that in LS no support-structures are necessary. This is because in LS the powder bed is heated to a temperature just below the melting point of the powder, so the energy that has to be introduced by the laser for melting the powder is very low. Therefore only little additional heat has to be led away from the build-area. On the one hand, this amount of energy can be compen-sated by the powder, on the other hand, due to the smaller temperature gradient, the curl-effect is less pronounced to occur. The curl-effect is the effect that causes a part to bend inside the powder bed and become deformed or even collide with the re-coating-unit. Latter would lead to a process break down. [9] [10] [4]

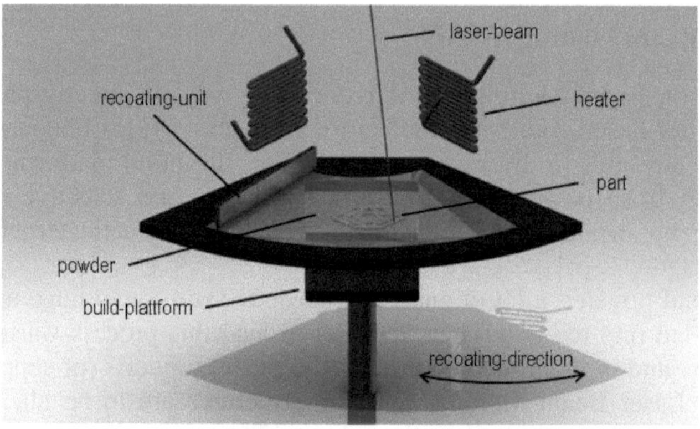

Fig. 2. Schematic illustration of the LS process

1.3 Fused Layer Modeling (FLM)

In FLM, slices of the part are built up by extruding an ABSplus wire through the heated nozzles of a movable printing head (see Figure 3). The printing head is moved in the x-, y-plane of the FLM machine to build up a layer. When a layer is finished, the build-platform is lowered by a layer-thickness and the next layer is built up. [4] [1]

Since FLM is not a powder-based procedure, like LS, or LBM, there is no powder to prevent the heated ABSplus from bending up or down inside the build-chamber. Therefore in FLM, supports are needed. In contrast to supports used in LBM, these supports only have the function, to hold the part in place. One special thing about supports in FLM is that a second, acid material is extruded through a second nozzle in order to build up supports. This way, when the FLM process is finished, the part can be put into an alkaline solution and the supports are dissolved. As a consequence, supports are not the main reason for the low finish-quality of parts produced by FLM. However, the high layer-thickness, which is one of the factors that make FLM cheap, compared to other additive manufacturing technologies, impacts the finish-quality in a negative way.

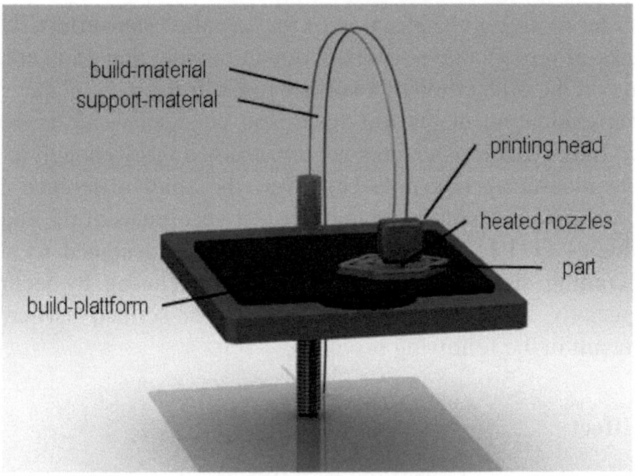

Fig. 3. Schematic illustration of the FLM process

2 Design of the Test-Specimen

The chair of manufacturing technologies of the University Duisburg-Essen has developed a test-specimen to convey the limits of additive manufacturing technologies. This specimen is designed to illustrate the smallest buildable wall-thicknesses, gap-widths and cylinder and bore diameters de-pending on their orientation inside the build-chamber. Thus diameters/thicknesses between 0.1 and 1 mm are built up at intervals of 0.1 mm and diameters/thicknesses between 1 and 2 mm are built up at intervals of 0.25 mm (see Figure 4).

In addition, the test specimen contains walls with different angles towards the build-platform (x-, y-plane), in order to show the change in surface-quality of down-skin surfaces with increas-ing/decreasing angles. Furthermore a bell-shaped geometry

Fig. 4. Test-specimen made of glassfilled polyamide 12 by LS

is built up in order to give a visualisation of the so called stair-effect. This effect characterises the lack of reproduction-accuracy due to the fact that parts are built up layerwise depending on the orientation of a surface towards the x-, y-plane.

For a further evaluation of the test-specimen, be-sides visual inspection, the distances between individual test-features are constructed large enough to enable the use of a coordinate measuring machine. However, the chief difference concerning the design of the test specimen compared to other test-specimens of the chair of manufacturing technologies [11] [12] is that this specimen is designed to suit the special requirements coming along with an additive manufacturing by technologies using sup-ports (especially LBM). Besides the features de-scribed earlier, these special requirements result in the following problems:

2.1 Curl-Effect

The curl-effect, which already was mentioned in the description of LBM, needs special focus. Since the production of large surfaces inside the x-y-plane is directly connected with a stronger occurrence of the curl-effect, this has to be avoided. Therefore the test-specimen is divided into eleven platforms, containing the individual test-features. The platforms are connected by small bridges, positioned in a z-level below the unpskin-surfaces of the platforms. This way, a large test-specimen can be produced without melting up large surfaces, especially in a z-level that may have an influence on the features to be evaluated.

2.2 Supports

As described before, in some additive manufacturing processes, supports have to be built up with the parts for several reasons. Since supports often have to be removed manually, the geometry of test-specimen should require few and little massive sup-ports. This way production costs for support material and post-processing requirements can be kept at a low level.

Fig. 5. Downskin-surface of the test-specimen produced by LBM after support-removal

In most cases the critical angle between build-platform and a surface to be built up is at about 45 degrees. Thus the downskin surface of each plat-form of the test specimen is equipped with a 60 degree groove. This way the amount of support that has to be placed under the platforms is significantly reduced without lowering process-stability (see Figure 5).

Additionally there are two kinds of test-features on the test-specimen which require supports. Since walls and cylinders that are oriented parallel to the build platform cannot be built without supports, the platforms containing these features are placed at the outside of the test specimen. By this means, the features are accessible for manual post-processing and visual inspection.

2.3 Recoating

In powder- or liquid-based additive manufacturing processes, different recoating-systems are used to supply the build-platform with fresh material. One thing, most recoating-systems have in common is some kind of blade that pushes the material from one side of the build-chamber to the other. Especially in LBM, the surfaces that just have been built up tend to have heightenings. In the majority of cases, these heightenings are the edges of the part, slightly bending up as a result of the melting-process. The combination of these circumstances can cause a scratching between the recoating-unit and the part to varying extends, depending on the build-material and the build-parameters used. In order to keep this phenomenon from affecting the features of the test-specimen, all platforms of the test specimen are oriented in an angle of 45 degrees towards the recoating-unit. This way, harsh contacts between recoating-unit the long edges of the platforms can be avoided.

However, there is another problem connected with the recoating-direction that may have an influence on the results, when small features are built up. As the test-specimen is designed to show the most filigree features that can be produced with each additive manufacturing process, the diameters and wall-thicknesses have to be decreased to the point, where they cannot be built up anymore. At this point, the features either are snapped by the recoat-ing-unit, or they cannot be built up as a connected object anymore. In both cases, fragments of the test-features are pushed across the powder-bed the recoating-unit. In order to prevent these fragments from having an influence on the build-process, by getting stuck between the recoating-unit and another area of the part or by snapping other test-features, the platforms and test-features are arranged in a suitable way. For instance, all diameters and wall-thicknesses decrease with the recoating-direction. Additionally, platforms with gaps are placed behind platforms with filigree features, so the space above them can be used as outlet zone.

3 Production of Test-Specimens with LBM, LS and FLM

Following, the results of visual inspections and measurements on test-specimen, build of Hastel-loy X (LBM with an EOSINT M270 Laser Beam Melting System), glass-filled polyamide (LS with a FORMIGA P 100 Laser Sintering System) and ABSplus (FLM with a Stratasys Dimension 1200es Fused Layer Modeling System) are discussed. For inspections and measurements, the test-specimen made of Hastelloy X and glass-filled polyamide have been exempted from powder adhesions by blasting with glass-pearls (LS), respectively corundum and glass-pearls (LBM). On the test-specimen produced by FLM, only supports have been removed by putting it into an alkaline bath.

3.1 Wall-Thicknesses

A look at the minimum producible wall-thicknesses conveys that in LBM the most filigree walls can be produced. Additionally walls in LBM show the slight-est deviation from the specified dimensions (see Figure 7). However, in LBM there is a considerable difference between walls oriented parallel the re-coating-unit and wall orientated orthogonal to the recoating-unit. The walls oriented parallel to the recoating-unit can only be built up down to a thick-ness of 0.7 mm. Thinner walls have been snapped by the recoating-unit (see Figure 6).

Fig. 6. Snapped walls, orientated parallel to the recoating-unit in LBM

Especially in FLM, but also in LS one can observe that from a certain threshold, in spite of decreasing nominal dimensions, the measured wall-thicknesses do not become any thinner. In FLM this can be explained by the fact that an object that is built up at least consists of its contours. Taking into account the diameter of an ABSplus wire and the fact that it is squeezed onto the former layer, it is clear that the minimum wall-thickness is situated at about 1 mm. In LS, the explanation is very similar. However the restricting factor is not the thickness of a wire, but the focus diameter of the LS-System in combination with the typical powder-adhesions.

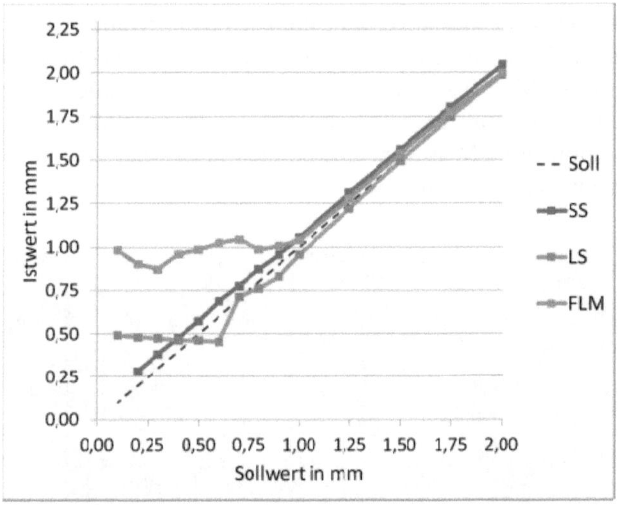

Fig. 7. Results of measuring minimal wall-thicknesses along the y-axis (parallel to recoating-unit in LBM)

The wall-thicknesses along the z-axis in powder-based manufacturing technologies (LBM and LS) are always slightly thicker than the nominal size (see Figure 8). This is to be explained by the fact, that melting the first layers of the walls, especially in LS, excess energy is led into the powder underneath the walls and melts additional powder particles. In LBM this effect can be observed less intense, since the manual removal of supports affects the results.

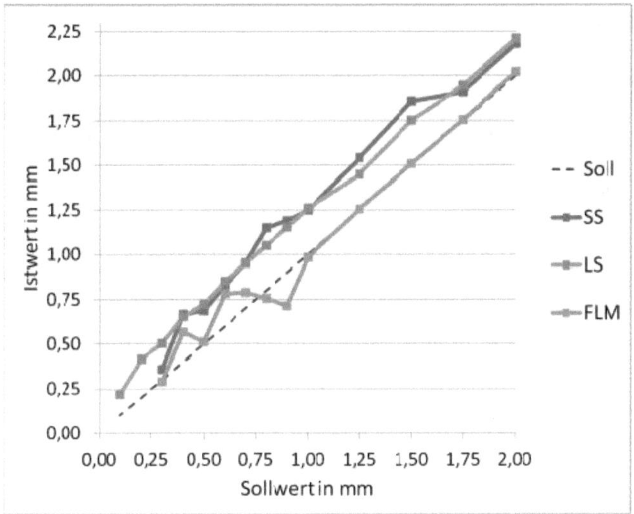

Fig. 8. Results of measuring minimal wall-thicknesses along the z-axis

In FLM, the course of measured wall-thicknesses is erratic within the range from 0.25 to 1.0 mm. This can be explained considering the layer-thickness in FLM. The layer-thickness in FLM is 0.254 mm. That way, a nominal thickness of 0.35 mm for example can either be represented by one, or two layers (0.254 mm or 0.508 mm). Since the resolution in FLM is very coarse, this effect can also be seen by a visual inspection of the walls (see Figure 9).

Fig. 9. Minimal wall-thicknesses along the z-axis in FLM

3.2 Cylinders

The test-specimen contains clinders with an angle of 0 degrees and cylinders with a polar angle of 45 and 90 degrees, each in negative x- and y- direction. The orientation along the x-axis has been chosen, since the process-stability in LBM is a lot higher, if the unsupported cylinders with a polar angle of 45 degrees dont grow against the re-coating-direction.

Comparing the cylinders with a polar angle of 0 degrees, shows again, that in LBM the most filligree featuers can be built up with the best accuarcy (see Figure 10). However, there are breaks visible in the cylinders in a hight of about 5 mm at cylinders with a diameter of less than 0.9 mm (see Figure 11). These breaks are a result of the scratching between the cylinders and the recoater-blade. This time the blade did not snap the cylinders since the geometry is more flexible. Thus the cylinders were able to flip back into their former positions be built on.

The results in LS are comparable tho those of the minimal wall-thicknesses along the x- and y- axis (see Figure 10). The smallest possible diameter in LS is 0.5 mm. In FLM, however, only cylinders with a diameter of 2 mm can be built up.

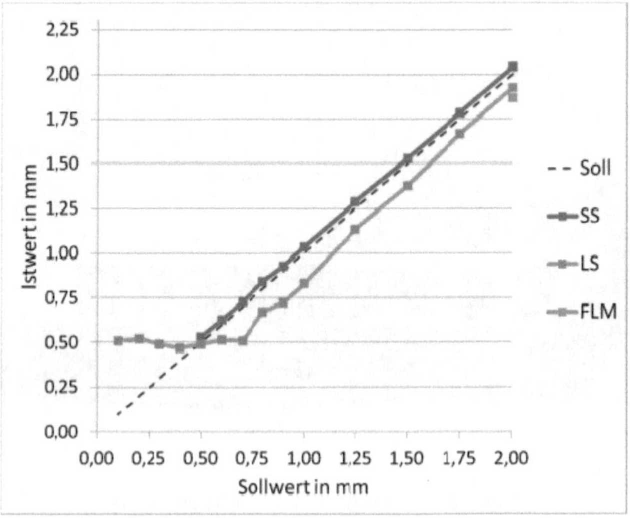

Fig. 10. Results of measuring minimal cylinder-diameters

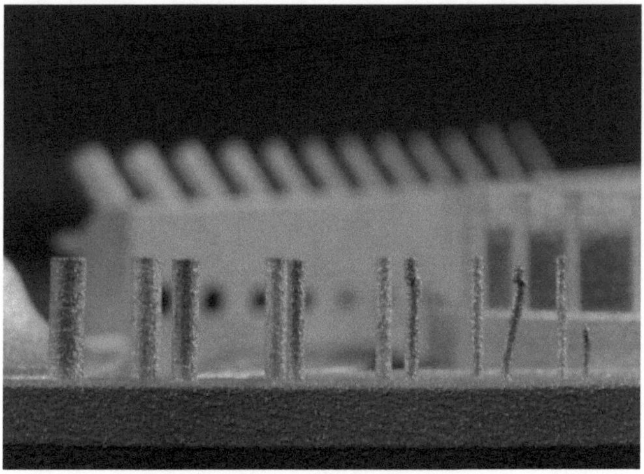

Fig. 11. Breaks in LBM-Cylinders

The results concerning cylinders with a polar angle of 45 and 90 degrees in LBM and LS are correlating with the results of cylinders with a polar angle of 0 degees regarding their accuracy. In FLM it is striking that smaller diameters can be built with increasing polar angles (see Figure 12). At a polar angle of 90 degrees, even cylinders with 0.1 mm diameter can be built. However, with increasing polar angles the form deviation in FLM becomes more visible. Due to the coarse resolution in FLM, caused by thick layers and a thick ABSplus wire, the deviation in Form and diameter for small cylinders becomes so large that inspecting their diameter is not possible any-more from 0.9 mm downwards (see Figure 13).

Fig. 12. Form deviation of FLM cylinders

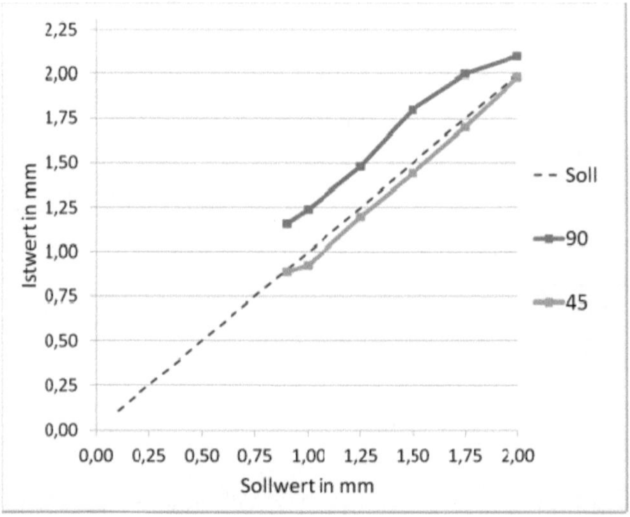

Fig. 13. Results of measuring cylinders with a polar angle of 45 and 90 degrees manufactured by FLM

3.3 Gaps and Bores

The evaluation of gaps and bores is reduced to a visual inspection. This is due to the fact that the accuracy of such filligreee bores cant be usefully inspected with a coordinate measuring machine, since the diameter of the measurement-end would be on the same scale as the diameter of the bores and the irregularities that are to be inspected. The results of the visual inspection are summarised in Table 1.

Table 1. Smallest depictable bores and gaps determined by visual inspection

	LBM	LS	FLM
bore along x-axis	0.5 mm	1.25 mm	0.7 mm
bores along y-axis	0.8 mm	1.5 mm	0.6 mm
bores along z-axis	0.6 mm	1.75 mm	0.8 mm
gaps along x-axis	0.4 mm	0.5 mm	0.3 mm
gaps alon y-axis	0.4 mm	0.4 mm	0.2 mm

One striking concerning bores in LS is their quality, which is worse compared to the other mannufacturing technologies. This becomes clear either by inspecting the smallest depictable diameters, but also by taking a look at the huge form deviation of bores in LS (see Figure 14). The eplanation for both, form deviation and resolution is found in the way of energy contribution in LS. As described above, in LS, less energy is necessary to melt the powder, compared tho LBM. Thus the threshold between melting the powder and not melting the powder is much smaller. Consequently, if excess energy is led into the part, surrounding powder ist melted and form deviations will occur.

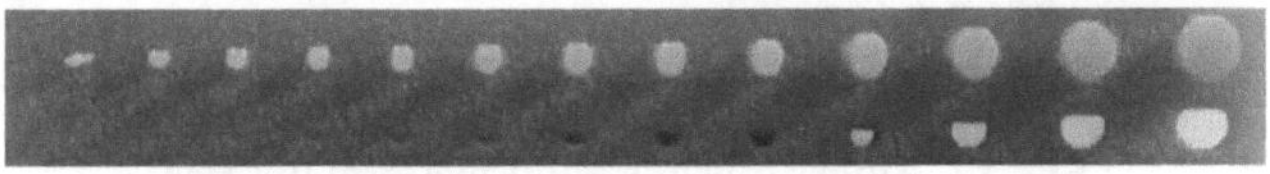

Fig. 14. Form deviation of bores along the y-axis in LS

3.4 Angles toward Build-Platform

The test-specimen contains five walls, inclined from 80 to 40 degrees towards the build-platform in steps of 10 degrees (see Figures 15-17). This walls serve as a visualisation of the decreasing surface quality with decreasing angles towards the build-platform. Again the walls are inclined to the negative x-direction in order to raise process stability and avoid process aborts. If possible, this walls should be built without support-structures, so deviations in form and surface quality can be displayed within the critical area.

Fig. 15. Angles towards build-platform in LS

In LS, the surface quality appears hardly affected by different angles towards the build-platform (see Figure 15). Even with an angle of 40 degrees, the stair-effect (visibility of layers on stongly inclined walls) is not visible.

Fig. 16. Angles towards build-platform in FLM

Taking a look at the walls built by FLM, it becomes clear that the stair-effect in FLM is visible right from the beginning (see Figure 16). This is due to the coarse resolution of FLM. Additionally, the wall inclined by 40 degrees even has a worse surface quality then the other walls. In FLM, supports are created automaticly. Therefore users are not able to erase supports before starting a FLM process. The wall, inclined by 40 degrees, was built up with supports. Thus the lack of surface-quality results from the connection between supports and part.

Fig. 17. Anlges towards build-platform in LBM

The walls in LBM convey a strong influence of angles between parts and build-platform and the surface-quality of downskin surfaces (see Figure 17). A first discoloration of the surface can be seen on the wall inclined by 60 degrees. This discoloration is a result of process-warmth, not beeing able to leave the part due to the fact that these walls dont have support-structures. At an inclination of 50 degrees, a serious deterioration of the surface quality becomes visible. This deteroration becomes even stronger with an inclination by 40 degrees. Additionally, the edge of the wall inclined by 40 degrees appears frayed. The reason for this can be found in the fact that with decreasing angle towrad the build-platform and increasing jam of heat inside the part, the curl-effect becomes stronger. In this case, the recoater-unit starts scantching the curled edge. This is a first sign, that at this angle of inclination, process aborts may occur depending on the orientation of the part towards the recoating-unit.

3.5 Stair-Effect

The bell-shaped feature on the test-specimen serves as a visualisation of the stair-effect. Comparing the built up test-specimens, a clear difference in surface quality can be recognised.

In LBM, steps are just slightly visible at an angle of 10 to 15 degrees towards the build-platform (see Figure 18). Due to the thin layer-thickness in LBM, the whole bell-profile apperas very fine and smooth. Taking a look at the LS-bell-profile, it becomes clear that the surfaces are a bit more rough than in LBM. The stair-effect is already visible at an angle of 20 degrees. In FLM, as mentioned above, single layers are always visible, due to the coarse resolution of the technology. In spite of this, the bell-profile conveys that, using the FLM-technology, angles of less then about 20 degrees inevitably lead to a loss of shape.

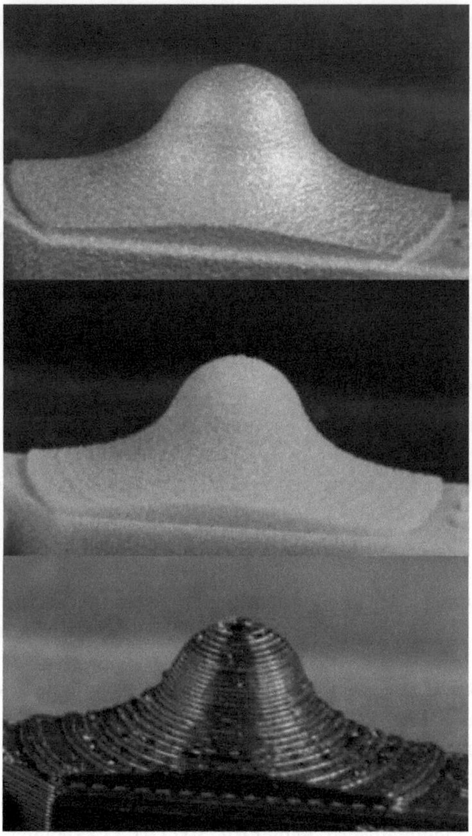

Fig. 18. Comparison of bell-shaped features on the test-specimen built by LBM, LS and FLM

4 Conclusions

Comparing the different test-specimen, built by LBM, LS and FLM, the first thing to be recognised is that in LBM the most filigree structures can be produced with the best accuracy. However, it becomes clear that the LBM-process is much more complex than for example the FLM-process. Both, designers and operators have to be aware of the typical constrains that are connect with the process-specific characteristics of LBM. This becomes particularly obvious, considering the huge influence that part-orientation and supports have on process stability and part-quality.

As mentioned above, LS is very similar to LBM concerning the course of procedure. This similarity can also be seen, comparing the test-specimen. In LS, most features are just slightly less filigree than in LBM. Due to the fact that support-structures are not needed for LS, a lot of time and money can be saved in pre- and, as a consequence, post-processing. In addition, the process-handling is easier and process aborts are a lot less likely.

Taking a look at the FLM-process, it is obvious that this technology is way less complex and filigree than LBM and LS. Fine features often can`t be displayed and deviations in form and dimension often can be recognised. However, the FLM-process is very easy to be handled. Supports are constructed automatically and when the part is built up, they can be removed by an alkaline-bath. Additionally, no precautions have to be taken and no cleaning effort has to be done handling any powder. The FLM-technology is much cleaner than LBM and LS and therefore much more suitable for an office-surrounding. The last thing to be taken into account for this comparison are process-costs. The FLM-technology is a lot cheaper than LBM (which is the most expensive technology) and LS.

References

1. Wohlers, T.: Wohlers Report 2011 - Annual Worldwide Progress Report, Fort Collins (2011)
2. Witt, G.: Taschenbuch der Fertigungs-technik. Fachbuchverlag Leipzig im Carl Hanser Verlag, München, Wien (2006)
3. Gebhardt, A.: Generative Fertigungsverfahren, 3. Auflage Hrsg. Carl Hanser Verlag, München (2007)
4. N. N., VDI-Guideline 3404: Additive fabrication - Rapid technologies (rapid prototyping) - Fundamentals, terms and definitions, quality parameter, supply agreements. Beuth Verlag, Berlin (2009)
5. Wiesner, A.: Selective Laser Melting - Eine Verfahrensvariante des Strahlschmelzens. Laser Technik Journal, 54–55 (2008)
6. Sehrt, J.T.: Möglichkeiten und Grenzen der generativen Herstellung metallischer Bauteile durch das Strahlschmelzverfahren. Shaker Verlag (2010)
7. Meixlsperger, M., Skrynecki, N., Wöllecke, F.: Vorrichtung und Verfahren zum Herstellen eines Bauteils in Schichtbauweise. Patent EP2502730 A1 (September 26, 2012)
8. Gibson, I., Rosen, D.W., Strucker, B.: Additive Manufacturing Technologies. Springer, New York (2010)
9. Rechtenwald, T., Roth, S., Pohle, D.: Funk-tionsprototypen aus Peak, vol. (11), pp. 62–68. Kunststoffe, Carl Hanser Verlag, München (2006)
10. Rietzel, D., Schmachtenberg, E.: Neue Kunststoffpulver für das selektive Laser Sintern, vol. (2), pp. 65–68. Kunststoffe, Carl Hanser Verlag, München (2008)
11. Wegner, A., Witt, G.: Design Rules For Small Geometric Feature in Laser Sintering, Duisburg (2011)
12. Reinhardt, T., Witt, G.: Ansätze zur Qualitäts-bewertung von generativen Fertigungsverfahren durch die Einführung eines Kennzahlen-systems. Rapid Tech Messe Erfurt (2012)

Mechanical Properties of Laser Beam Melting Components Depending on Various Process Errors

Stefan Kleszczynski[1], Joschka zur Jacobsmühlen[2], Jan T. Sehrt[1], and Gerd Witt[1]

[1] Institute for Product Engineering, University of Duisburg-Essen, Duisburg, Germany
{stefan.kleszczynski,gerd.witt}@uni-due.de
[2] Institute of Imaging and Computer Vision, RWTH Aachen University, Aachen, Germany

Abstract. Additive Manufacturing processes are constantly gaining more influence. The layer-wise creation of solid components by joining formless materials allows tool-free generation of parts with very complex geometries. Laser Beam Melting is one possible Additive Manufacturing process which allows the production of metal components with very good mechanical properties suitable for industrial applications. These are for example located in the field of medical technologies or aerospace. Despite this potential a breakthrough of the technology has not occurred yet. One of the main reasons for this issue is the lack of process stability and quality management. Due to the principle of this process, mechanical properties of the components are strongly depending on the process parameters being used for production. As a consequence incorrect parameters or process errors will influence part properties. For that reason possible process errors were identified and documented using high resolution imaging. In a next step tensile test specimens with pre-defined process errors were produced. The influence of these defects on mechanical properties were examined by determining the tensile strength and the elongation at break. The results from mechanical testing are validated with microscopy studies on error samples and tensile specimens. Finally this paper will give a summary of the impact of process errors on mechanical part quality. As an outlook the suitability of high resolution imaging for error detection is discussed. Based on these results a future contribution to quality management is aspired.

Keywords: Additive Manufacturing, Laser Beam Melting, process errors, mechanical properties, High Resolution Imaging.

1 Introduction

Additive Manufacturing (AM) offers many advantages for manufacturing of complex and individual parts. It provides a tool-free production, whereby physical parts are created from virtual solid models in a layer by layer fashion [1, 2]. In a first step the layer data is gained by slicing virtual 3D CAD models into layers of a certain thickness. Layer information could also be gained by slicing data from 3D scanning or CT scanning. In the following build process the layer information is converted into physical parts by creating and joining the respective layers. The principle of layer creation

G.L. Kovács and D. Kochan (Eds.): NEW PROLAMAT 2013, IFIP AICT 411, pp. 153–166, 2013.

classifies the AM process [3]. Laser Beam Melting (LBM) as an AM process offers the opportunity of small volume production of metal components. Here a thin layer of metal powder is deposited onto the build platform. In a next step the powder is molten into solid material by moving a laser beam (mostly Nd- or Yb-fibre laser source) across the current cross-section of the part. After this, the build platform is lowered and the two process stages are repeated iteratively until the solid metal part is fully produced (figure 1). As a result of the process the produced components show very good mechanical properties, which are widely comparable to conventionally processed materials [4, 5] or in some cases even better [5]. The density of components reaches approximately 100 % [4], [6 - 8]. Potential applications for LBM components are located in the domain of medical implants, FEM optimized lightweight components or the production of turbine blades with internal cooling channels [1, 2, 5]. There are about 158 factors of process influences [9] from which the parameters of laser power, scanning velocity, hatch distance (distance of melt traces) and layer thickness have been reported as the most influencing ones [4, 6, 9]. These main process parameters mentioned are often set into connection by means of the magnitude of volume energy density Ev [4, 6, 8] which is defined as:

$$E_v = \frac{P_l}{h \cdot v_s \cdot d} \tag{1}$$

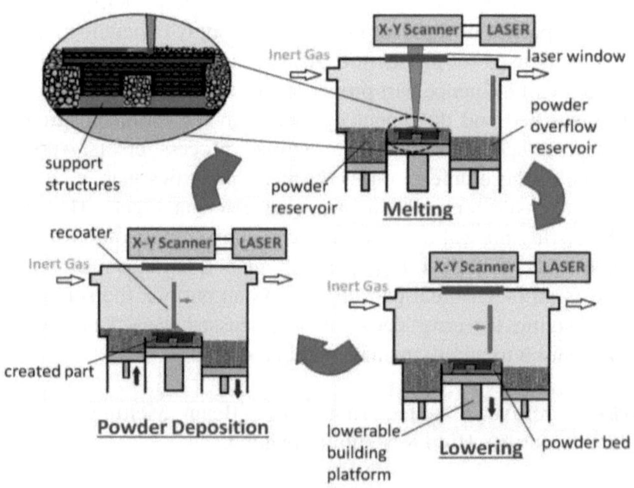

Fig. 1. Schematic process principle of LBM

where Pl stands for Laser Power, h stands for the hatch distance, v_s stands for the scanning velocity and d stands for the powder layer thickness. Since the process of layer creation determines the resulting part properties [1, 2, 4, 6] wrong process parameters or technical defects in certain machine components could also cause process errors which deteriorate mechanical properties.

Spierings et al. [4] show that the resulting part porosity mainly depends on the used process parameters and significantly affects the mechanical properties. In addition a correlation between volume energy density and the respective part properties is investigated with the result that volume energy density can be considered as the parameter determining part porosity. Yasa et al. [7] investigate the influence of double exposure strategies on resulting part properties. It is noted that the application of re-melting is able to improve surface quality and reduce part porosity.

Due to high security requirements of some potential domains of applications and actual standardisation efforts, a demand for suitable quality control for LBM technologies has been reported [8, 10, 11]. Thus far, some different approaches for process control and process monitoring have been given in literature. Kruth et al. monitor the current melt pool using a coaxial imaging system and control laser power to hold the size of the melt pool constant [11]. As the thermal conductivity of metal powder is about three orders of magnitude lower than those of solid metal [6] this system can improve the part quality for overhanging structures by lowering the laser power when the size of the melt pool shows fluctuations in these certain regions. Lott et al. [10] improve this approach by adding additional lighting to resolve melt pool dynamics at higher resolution. In [12] images of the deposited powder layers are taken additionally using a CCD camera. This enables the detection of coating errors due to a damaged coater blade. Doubenskaia et al. [13] use an optical system consisting of an infra-red camera and a pyrometer for visualisation of the build process and online temperature measurements. All approaches previously mentioned feature an implementation into the optical components or the machine housing of the respective LBM system. This makes it elaborate and expensive to equip existing LBM machines with these systems. Moreover the coaxial monitoring systems are limited to the inspection of the melt pool. The result of melting remains uninspected. The CCD camera used in [12] is restricted to the inspection of the powder layer. Possible errors within the compound of melt traces cannot be resolved.

In this work the influence of process errors on resulting part properties is investigated. First selected process errors are provoked and documented using a high resolution imaging system, which is able to detect errors at the scale of single melt traces. A further description of the imaging system is shown in paragraph 2.1 and in [14, 15]. In general, process errors can influence process stability and part quality [14]. Therefore error samples are built by manipulating the main exposure parameters and exposure strategies. Next, tensile specimens with selected errors are built and tested. The results are validated by microscopy studies on the tested tensile specimens. Finally a correlation between tensile strength, elongation at break, porosity and error type is discussed.

2 Method

2.1 LBM and High Resolution System

For the experiments in this work an EOSINT M 270 LBM system (EOS GmbH, Germany) is used. Hastelloy X powder is used as material, which is a nickel-base super alloy suitable for applications such as gas turbine blades. The documentation of process errors is carried out with an imaging systems consisting of a monochrome 29 megapixel. CCD camera (SVS29050 by SVS-VISTEK GmbH, Germany). A tilt and

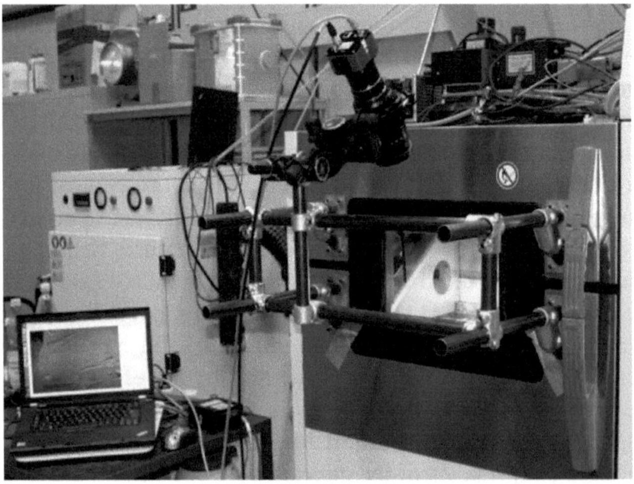

Fig. 2. Camera setup in front of LBM system EOSINT M 270

shift lens (Hartblei Macro 4/120 TS Superrotator by Hartblei, Germany) helps to reduce perspective distortion by shifting the camera back and allows placing the focal plane on the build platform using its tilt ability. A 20 mm extension tube reduces the minimum object distance of the lens. The imaging system is mounted in front of the LBM system using a tube construction which provides adjustable positioning in height and distance from the machine window (figure 2). Two orthogonally positioned LED line lights provide lighting for the build platform. Matt reflectors on the machine back and the recoater are used to obtain diffuse lighting from a close distance, which was found to yield the best surface images. The field of view is limited to a small substrate platform (10 cm x 10 cm) to enable best possible resolving power (25 µm/pixel to 35 µm/pixel) [15]. Image acquisition after powder deposition and laser exposure is triggered automatically using limit switches of the machine's coater blade and laser hourmeter.

2.2 Determination of Mechanical Properties

Test specimens for tensile testing are built as cylindrical raw part by LBM. The final specimen shape is produced by milling the raw parts into the standardised shape according to DIN 50125 - B 5x25 [16]. Tensile tests are performed according to the specifications of DIN 50125. A Galdabini Quasar 200 machine is used for the tests. The fragments of test specimens are used for further microscopy studies. For which unstressed material from the specimen's thread heads is prepared into grinding samples. The microscopy studies are carried out using Olympus and Novex microscopes. Porosity of error samples is determined using an optical method according to [7] and [8] where the acquired images are converted to black and white images using a constant threshold value. Finally the ratio of black pixels, representing the porosity, is measured.

3 Documentation of Process Errors

3.1 Process Errors

An overview of typical process errors has been given in previous work [14]. In this paper the main focus is on errors that influence part quality and in particularly on errors that affect mechanical properties. As mentioned in paragraph 1 mechanical properties strongly depend on process parameters, which define the energy input for the melting of the powder and in consequence the ratio of distance and width of the single melt traces. Technical defects of the laser source or the choice of wrong process parameter sets could therefore worsen the compound of layers and melt traces leading to porous regions. On the other hand too much energy input could lead to heat accumulation. In this case surface tensions of the melt induce the formation of superelevated regions, which could endanger process stability by causing collisions with the recoating mechanism. However higher energy inputs have been reported to increase mechanical properties due to a better compound of melt traces and lower porosity [7]. To provoke errors of these two kinds the main process parameters laser power, hatch distance and scanning velocity are changed by 20 % and 40 % arround the standard value, which was found by systematic qualification experiments (see [17]). Additionally the layer thickness is doubled from 20 μm to 40 μm for one sample keeping the process parameters constant. As illustrated in equation 1 these variations directly affect the energy input. Another sample is built using a double exposure strategy resulting in higher energy inputs. For the experiments the stripe exposure strategy is used. Hereby the cross sections of the current parts are separated in stripes of 5 mm length. The overlap value for these stripes is another process parameter which could affect the compound of melt traces. Therefore one sample with the lowest possible stripe overlap value of 0,01 mm is build.

3.2 Error Samples

Figure 3 shows an image, which was recorded during the build process of error samples using the high resolution imaging system. The samples are arranged in a matrix. The first three rows of the matrix represent the process parameters scanning velocity (vs), laser power (Pl) and hatch distance (h). These values are varied in the columns from - 40 % to + 40 % in steps of 20 %.The last row of the matrix contains a reference sample (left), a sample being built with double layer thickness (mid left), a sample being built with double exposure (mid right) and a sample being built with the lowest possible stripe overlap value (right).

As can be seen from figure 3, samples representing higher volume energy densities (reduced scanning velocity/hatch, increased laser power or double exposure) appear much brighter and smoother than those of samples representing low volume energy density (increased scanning velocity/hatch, reduced laser power or double layer thickness). A closer view at the documented error samples (figure 4) shows that surface irregularities and gross particles are visible at the sample of double layer thickness.

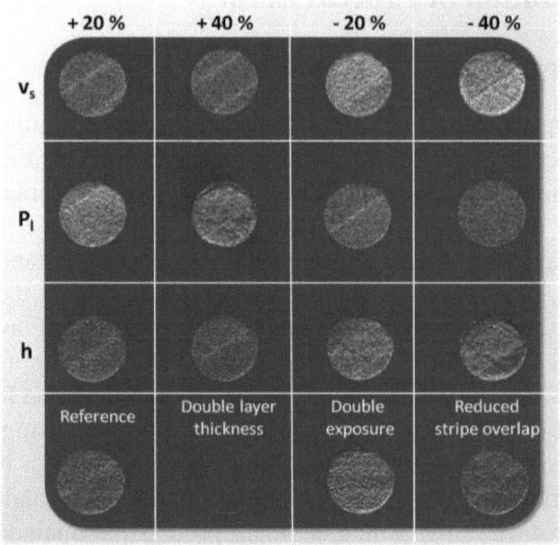

Fig. 3. Documentation of error samples using high resolution imaging

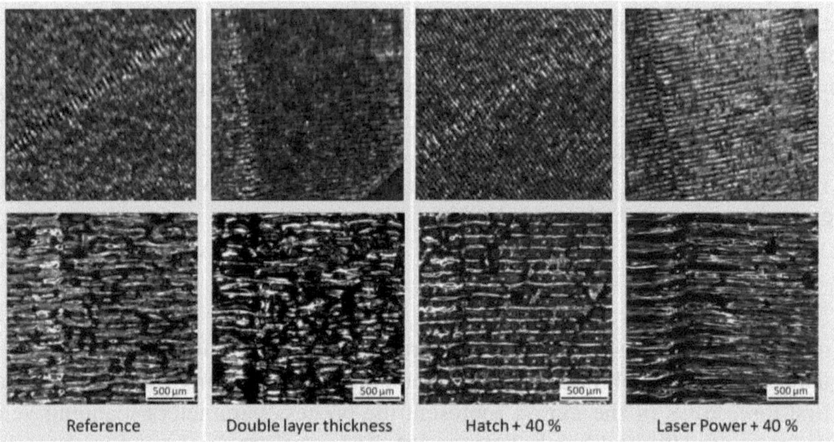

Fig. 4. Comparison of selected high resolution (top) images vs. images taken from microscopy (bottom)

The sample with 40 % enlarged hatch distance indicates a poor connection of melt traces, which could be a signal for increased porosity. The sample with 40 % increased laser power shows a strong connection of melt traces, although there are some superelevated regions visible at the edges. A comparison of the high resolution images with images taken from microscopy confirms these impressions. Here it is clearly visible that there is no connection between the melt traces at the sample build with enlarged hatch distances. The error samples with 40 % respectively increased or decreased parameters show the strongest deviation from the reference sample. These parameter sets are used for production of tensile test specimens. Additionally tensile test specimens representing standard, double layer, double exposure and reduced stripe overlap parameter sets are built.

4 Mechanical Properties

4.1 Tensile Strength

For each type of error six specimens are produced as described in paragraph 2.2 to ensure the level of statistical certainty. The determined values for tensile strength and the associated standard deviations are presented in figure 5. Additionally the calculated values for volume energy density are added into the chart. As can be seen the respective bars representing tensile strength and volume energy density show a similar trend for almost all specimens. In the case of the specimen of "double layer" this trend is not applicable. Higher values for the mean tensile strength are determined (comparing "double layer" to "H + 40 %" and "Vs + 40 %"), although this specimen has the lowest value for the volume energy density. Here it is remarkable that specimen "P - 40 &" shows a tensile strength which is about 14 % (117 MPa) lower than those of specimen "double layer" while the values for volume energy density are at the same level. Specimens produced using higher energy input parameter sets.

(VS - 40 %, H - 40 %, P + 40 %, double exposure) exhibit higher values for tensile strength compared to those of the reference specimen. On the other hand specimens produced with low energy input parameters (double layer, H + 40 %, Vs + 40 %,

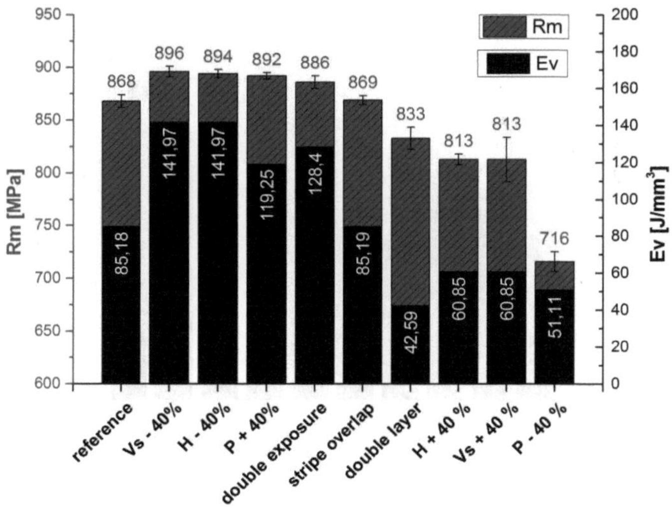

Fig. 5. Results from determination of tensile strength compared to calculated values of volume energy density

P - 40 %) show lower mean values than the reference specimen. Here it is remarkable that the lowest mean value for tensile strength (716 MPa) is determined for specimens built with 40 % decreased laser power. The highest mean value for tensile strength (896 MPa) is achieved at specimens produced with scanning velocities decreased by 40 %. The specimens with reduced stripe overlap value showed no significant deviation from reference specimens. Comparing the mean derivations, it is worth mentioning that "high energy" samples show lower values than those of "low energy". Specimens from the type of "Vs + 40 %" show the highest value for mean derivation. The determined mean value for the reference specimens (868 MPa) is comparable to the value determined in previous work (885 MPa) and values from literature (715 MPa - 1110 MPa) [17]. At this point it has to be stated that the maximum value (1110 MPa) is achieved after heat treatment.

4.2 Elongation at Break

Figure 6 shows the determined values for the elongation at break compared to the calculated values of volume energy density. Unlike the results for tensile strength, there seems to be no connection between the elongation at break and volume energy density. Furthermore, there are no significant divergent trends recognizable between high energy input and low energy input parameter sets.

It is remarkable that there are three different levels of values recognizable in the chart. First there is the level of about 30 % elongation at break which is determined for most of the specimens (reference, double exposure, stripe overlap, double layer, Vs + 40 %). Second there is the level of about 25 % to 28 % elongation at break which is detected at four specimens (Vs - 40 %, H - 40 %, P + 40 %, H + 40 %). Here it is remarkable that the three high energy input parameter sets (Vs - 40 %, H - 40 %,

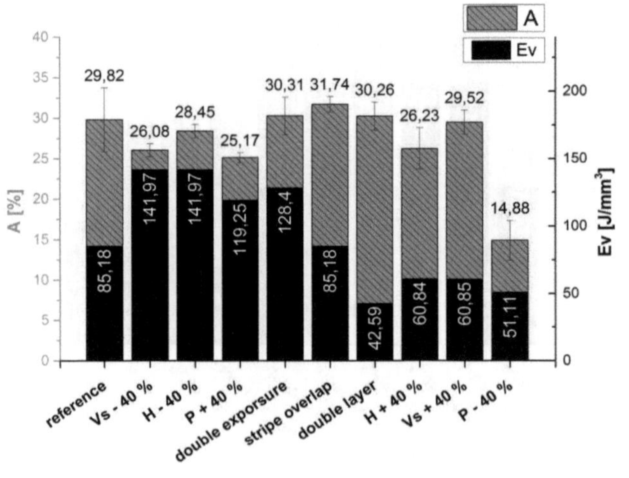

Fig. 6. Results from determination of elongation at break compared to calculated values of volume energy density

P + 40 %) show the lowest standard deviation compared to all other specimens. Finally the lowest value for elongation at break is measured for specimen "P - 40 %" representing the parameter set with lowest energy input. In literature the values for elongation at break for Hastelloy X are located in the range of 22 - 60 % [17] depending on the respective heat treatment. With exception of specimen "P - 40 %" all determined values are within this range. However the determined values are at least 50 % lower than the maximal values reported.

4.3 Porosity

After mechanical testing, selected specimens are used for determination of porosity using microscopy according to the procedure described in paragraph 2.2. For the reference specimen the porosity is determined to 0,04 %, which is comparable to results from previous studies [4 - 7] emphasising that LBM components achieve up to 99 % density. Specimen "Pl - 40 %", which has the lowest value for volume energy density, shows the highest porosity. The determined value is 3,94 %. The results from porosity analysis (as presented in figure 7) underline previously published statements, which say that porosity grows with sinking energy input. It has to be stated that in general "high energy input" specimens show very similar porosity values (0,020 % to 0,027 %, see figure 7). The determined porosity values are higher for low energy input specimens (0,227 % to 3,938 %), which confirms the assumption that porosity is strongly dependent on energy input. The porosity values of the "reference" and the "reduced stripe overlap" specimen differ from each other by 0,02 %. Thus the reduced stripe overlap specimens show a lower porosity value. This is remarkable due to the fact that the "reduced stripe overlap" was suspected to increase part porosity. One explanation for this result might be found in the exposure strategy. As mentioned in paragraph 3.1 the cross section of parts are subdivided in stripes of a certain width. After exposure these stripes are rotated and gaps in the compound of melt traces could be closed during exposure of the next layer. On the other hand it has to be stated that the details from photomicrographs used for the analysis show only one certain area of the whole cross section. Moreover pores are distributed stochastically, which makes it difficult to make a statement with an accuracy level of a hundredth of a percent.

Specimen	Porosity [%]
Reference	0.04
Vs - 40 %	0.025
H - 40 %	0.027
Pl + 40 %	0.024
Double exposure	0.025
Reduced stripe overlap	0.020
Double layer	0.227
H + 40 %	1.633
Vs + 40 %	1.176
P - 40 %	3.938

Fig. 7. Summary of determined porosity values

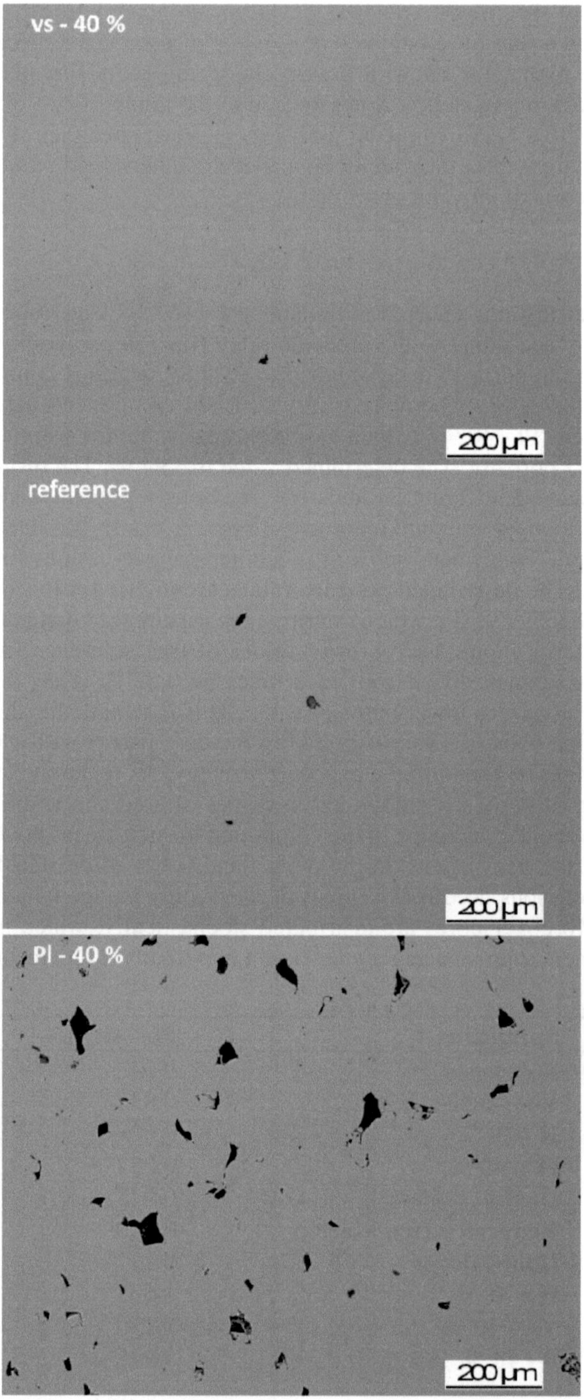

Fig. 8. Photomicrograph from selected test specimens

Figure 8 shows photomicrographs from reference specimen (middle), reduced scanning speed specimen (top, highest tensile strength) and reduced laser power specimen (at the bottom, lowest tensile strength). Specimen “vs - 40 %” shows little and small pores. The value of porosity is 0,025 %. The same appearance is visible at the photomicrographs from the reference sample, which shows slightly more but still small pores. Specimen “Pl - 40 %”, in contrast, shows clearly more and bigger pores, which seem to be distributed stochastically (figure 8, at the bottom).

5 Discussion

The results presented in the previous sections of this paper prove that mechanical properties strongly depend on process parameters. In general it can be stated that increasing energy input improves tensile strength and reduces porosity. It is to be expected that porosity affects tensile strength, due to the fact that irregularities like pores induce crack formation at a certain mechanical load. The elongation at break on the other hand is not systematically affected by different energy input parameter sets. Here, there are some groups of parameter sets which show values at similar levels. However there is no general connection between energy input and the elongation at break for the investigated material. It seems like the exposure strategies have more influence in this case. As can be seen from figure 6 the three high energy input parameter specimens “Vs - 40 %”, “H - 40 %”, “P + 40 %” show similar values for elongation at break. The “double exposure” specimen has a calculated volume energy density which is comparable to those of the other high energy specimens. Nevertheless the elongation at break of this sample lies in the same region as the “reference” sample and some “low energy input” samples. One possible explanation for this appearance could be that the “double exposure” sample was built using two different energy input parameter sets. One for melting the powder and another parameter set for re-melting the produced layer. Thus the heat flow has been different to those of the “high energy input parameter” samples, which has obviously induced different mechanical properties. The “high energy input” specimens show improved values for tensile strength but lower values for elongation at break compared to the reference sample. In contrast the “double exposure” sample shows an improved value for tensile strength at constant ductility.

Figure 9 compares the results of tensile and porosity studies depending on the volume energy density. For this purpose the respective numbers of specimens are plotted into the chart. For identification see the explanation in the chart. Comparing the two plots of logarithmic interpolations shows that both are working contrarily. Both of the magnitudes seem to run asymptotically to parallels of the x-axis for high values of the volume energy density. The tendencies at the tensile tests underline the results from porosity determination (specimen 7: double layer, Rm = 833 MPa, porosity 0.227 %, specimen 8: hatch distance plus 40 %, Rm = 813 MPa, porosity 1.633 %). In this case specimen 8 shows a higher value of porosity and lower tensile strength. Comparing these results with the images from figure 4 allows the conclusion that a poor connection of melt traces causes higher tensile strength values than no connection of melt traces. Specimens 7 shows that the previously mentioned correlation between tensile strength, volume energy density and porosity is not applicable to every kind of an error. Here the low value for the volume energy density does not correlate with the interpolation for tensile strength and porosity of other specimens.

This shows that volume energy density is more suitable for estimating tendencies concerning the magnitudes of tensile strength and porosity. A more significant influence is spotted at the type of error, respectively to the kind of energy input or exposure strategy.

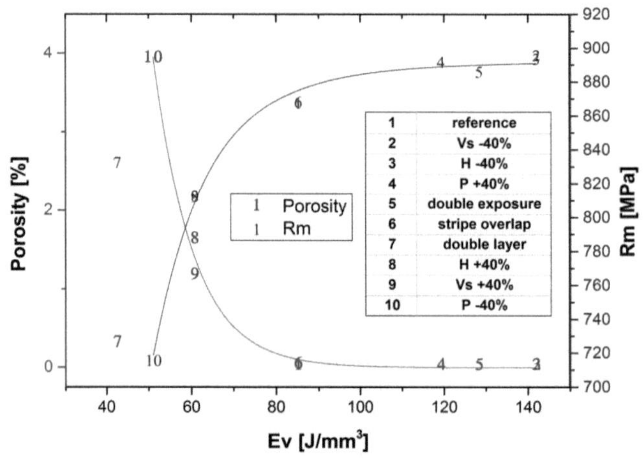

Fig. 9. Connection between porosity, tensile strength and volume energy density

6 Conclusions

In this paper a brief demonstration for documenting possible process errors in the area of LBM by using a high resolution imaging system was given. The results and validations via microscopy show a good correlation between the recorded images. High resolution imaging might be an alternative and more pragmatic approach for process monitoring and quality management in the area of LBM due to the fact that the system is easy to implement and compatible to every LBM systems that features a window for the inspection of the process.

In a second step the impact of process errors on tensile strength, porosity and elongation at break was investigated. It could be shown that a higher energy input mostly induces higher values for tensile strength and lower porosities. On the other hand it was found, that the lower the volume energy density is, the lower the determined tensile strength and the higher the porosity are. For some error samples it could be found that the measure of volume energy density is not in a direct correlation with the resulting part properties. This was noticed in detail by comparing the tensile strength of samples with similar values for volume energy density, which were varying for about 117 MPa. Here the nature of melt trace connection seems to have the bigger influence. The mentioned disagreement of volume energy density with resulting part properties was especially noticed at the determination of the elongation at break. Here some samples that were built with "high energy parameter sets" showed a reduced elongation at break, which induced that the higher energy input seems to embrittle the material compared to the values of the reference specimen. At the same time another specimen, with a comparable higher level of volume energy density resulting in an higher tensile strength, showed higher values for the elongation at break, which were at the same level as specimens produced with low energy input parameters.

Nevertheless all determined mean values for tensile strength and elongation at break were in the range of known values from conventionally produced samples. Only the sample with the lowest tensile strength, lowest elongation at break and highest porosity, which was produced by reducing laser power by 40 %, showed values which were at the lower end of the known range. The elongation at break, which is a measure for ductility of materials, did not reach more than 50 % of the known maximum value from literature. This means that for some applications, where high elongation at break values are required, heat treatments are still necessary to improve this certain part property.

For future work the further investigation of the influences of varying process parameters is necessary for different materials and different machine systems, which might use other laser sources or inert gases for flooding of the process chamber. Especially in case of elongation at break it would be interesting to analyse the influence of different exposure strategies. Using high resolution imaging systems for collecting data of different error types and materials could be a useful tool to create a knowledge database, which links process parameters, resulting surface images and resulting mechanical part properties. In a next step an automated image analysis could detect significant differences in the structure of melt traces and might therefore also be applicable to quality management and production documentation.

Acknowledgment. The IGF project 17042 N initiated by the GFaI (Society for the Promotion of Applied Computer Science, Berlin, Germany), has been funded by the Federal Ministry of Economics and Technology (BMWi) via the German Federation of Industrial Research Associations (AiF) following a decision of the German Bundestag.

References

1. Gibson, I., Rosen, D.W., Stucker, B.: Additive Manufacturing Technologies - Rapid Prototyping to Direct Digital Manufacturing. Springer, Heidelberg, ISBN 978-1-4419-1119-3
2. Gebhardt, A.: Generative Fertigungsverfahren - Rapid Prototyping - Rapid Tooling - Rapid Manufacturing. Carl Hanser Verlag, München (2007) ISBN: 978-3-446-22666-1
3. N. N.: VDI-Guideline 3404: Additive fabrication - Rapid technologies (rapid prototyping) - Fundamentals, terms and definitions, quality parameter, supply agreements. Beuth Verlag, Berlin (2009)

4. Spierings, A.B., Wegener, K., Levy, G.: Designing material properties locally with Additive Manufacturing technology SLM. In: Proc. of the Solid Freeform Fabrication Symposium, Austin, TX, USA (2012)
5. Wohlers, T.: Wohlers Report 2011 - Annual Worldwide Progress Report, Fort Collins (2011)
6. Meiners, W.: Direktes Selektives Laser Sintern einkomponentiger metallischer Werkstoffe. Ph.D. dissertation, RWTH Aachen University, Aachen, Germany (1999) ISBN: 3-8265-6571-1
7. Yasa, E., Kruth, J.: Application of Laser Re-Melting on Selective Laser Melting parts. Advances in Production Engineering & Management 6(4), 259–270 (2011)
8. N. N.: VDI-Guideline 3405 - 2. Entwurf. Additive manufacturing processes, Rapid Manufacturing – Beam melting of metallic parts – Qualification, quality assurance and post processing. Beuth Verlag, Berlin (2012)
9. Sehrt, J.T.: Möglichkeiten und Grenzen bei der generativen Herstellung metallischer Bauteile durch das Strahlschmelzverfahren. Ph.D. dissertation, University of Duisburg-Essen, Duisburg, Germany (2010) ISBN: 978-3-8322-9229-4
10. Lott, P., Schleifenbaum, H., Meiners, W., Wissenbach, K., Hinke, C., Bultmann, J.: Design of an Optical system for the in Situ Process Monitoring of Selective Laser Melting (SLM). Physics Procedia 2(pt. A), 683–690 (2011)
11. Kruth, J.-P., Mercelis, P., Van Vaerenbergh, J., Craeghs, T.: Feedback control of selective laser melting. In: Proc. 3rd Int. Conf. on Adv. Research in Virtual and Rapid Prototyping, pp. 521–527 (2007)
12. Craeghs, T., Clijsters, S., Yasa, E., Kruth, J.-P.: Online Quality Control of Selective Laser Melting. In: Proc. of the Solid Freeform Fabrication Symposium, Austin, TX, USA (2011)
13. Doubenskaia, M., Pavlov, M., Chivel, Y.: Optical System for On-Line Monitoring and Temperature Control in Selective Laser Melting Technology. Key Engineering Materials 437, 458–461 (2010)
14. Kleszczynski, S., zur Jacobsmühlen, J., Sehrt, J.T., Witt, G.: Error Detection in Laser Beam Melting Systems by High Resolution Imaging. In: Proc. of the Solid Freeform Fabrication Symposium, Austin, TX, USA (2012)
15. zur Jacobsmühlen, J., Kleszczynski, S., Schneider, D., Sehrt, J.T., Witt, G.: High Resolution Imaging for Inspection of Laser Beam Melting systems. In: Proc. of I2MTC 2013, Minneapolis, MN, USA (2013)
16. N. N.: DIN 50125 - Testing of metallic materials - Tensile test pieces. Beuth Verlag, Berlin (2009)
17. Sehrt, J.T., Witt, G.: Anforderungen an die Qualifizierung neuer Werkstoffe für das Strahlschmelzen. In: Proc. of Seminar: Rapid Manufacturing. IWB Anwenderzentrum Augsburg (July 08, 2010)

New Fiber Matrix Process with 3D Fiber Printer – A Strategic In-process Integration of Endless Fibers Using Fused Deposition Modeling (FDM)

Andreas Fischer, Steve Rommel, and Thomas Bauernhansl

Fraunhofer Institute for Manufacturing Engineering and Automation, Nobelstrasse 12. 70569 Stuttgart, Germany
{andreas.fischer,steve.rommel}@ipa.fraunhofer.de

Abstract. Product manufacturers are faced with a constant decrease in product development time, short product life cycles and an increase of complexity of the products. At the same time, it is becoming more and more difficult to justify the expected product features while satisfying the requirements of the trends like eco-design, resource efficiency, light weight design and individualization. These developments are key characteristics which favor the growing role and importance of additive manufacturing (AM). Additive Manufacturing techniques such as Fused Deposition Modeling (FDM), Selective Laser Sintering (SLS) or Selective Laser Melting (SLM) are transitions from building prototypes to manufacturing series products within short time and with highest requirements regarding material characteristics as well as product performance. Product performance is one of the main drivers of the creation of multi-material composites. These composites, existing of a matrix and embedded fibers, are more and more used in various industries not just the aerospace and automotive industry. They offer a solution to the light-weight requirements amongst others. Yet, so far the manufacturing of composites is only partially automated; it requires a lot of manual labor and is typically requiring tools resulting in added costs. A combination or integration of additive and composite manufacturing is very limited to not existing, being an indicator for the difficulties to bring these two technologies together.

To meet this challenge the 3D Fiber Print has been developed. It is an AM method using Fused Deposition Modeling (FDM) and composites in process simultaneously while manufacturing a product. The products are still manufactured layer-wise, with each layer containing a composite on-demand. The newly developed and patented print head is designed in the way that composite and matrix material are applied in the right setup and ration using a nozzle style print head to completely embed the composite. This paper presents the setup and possibilities of this new application.

Keywords: Additive Manufacturing, Fused Deposition Modeling, Fibers.

G.L. Kovács and D. Kochan (Eds.): NEW PROLAMAT 2013, IFIP AICT 411, pp. 167–175, 2013.

1 Introduction to Additive Manufacturing

Additive manufacturing is a technology that is considered to have the potential to start a new industrial revolution. Mostly known as rapid prototyping a variation of terms is used for "additive manufacturing" depending on the purpose of the product:

- Rapid Prototyping (RP) – building sample parts and prototypes
- Rapid Tooling (RT) – building tools for other manufacturing methods
- "Generative Manufacturing" or Additive Manufacturing (AM) – end-products with small- to medium-sized quantities.

The various technologies within additive manufacturing differentiate from conventional manufacturing technologies in the strategy the parts are produced. Contrary to CNC milling for example, being a subtracting manufacturing method and therefore manufacturing a product by removing material from a block, an additive manufactured product is built by placing material layer by layer onto each other until the final part shape is completed. Each layer is solidified individually, but also adhered to the ones next to it to form a continuous shape. Adhering these layers to each other is achieved by introducing an energy force, typically in the form of heat or light. This principle of building a product and its advantages is also the biggest disadvantage. The „created border" between these layers is a potential weak point within the product itself when it comes to mechanical or thermal stresses during use. One way to overcome this issue is taking precautions in the design and construction of a product following the stress analysis and strengthening the part where needed. This approach by itself is common to conventional design methods. Additionally the placement of the product to be built in the build space of the machine and therefore its orientation is also one way to reduce the risk of failure.

It can be argued that the advantages as well as the disadvantages of additive manufacturing result from the reduction of a three-dimensional manufacturing issue to a two-dimensional layerwise manufacturing issue. The selection of a manufacturing strategy or method and its corresponding planning process of conventional manufacturing are no longer applicable when it comes to additive manufacturing. The planning process can be considered lean. Benefits of this circumstance are saving in time and cost in the case of an increase in product complexity. Considering time and cost it is also true that an increase in detail by reducing the layer thickness or an increase in product size are typically resulting in a significant increase in production or building time.

Besides the classification of the additive manufacturing following DIN 8580 the following criteria can be used to classify additive manufacturing:

- Physical state of material
- Material the product is made of
- Usage of products within the product development process
- Manufacturing principle

One of the best describing classifica-tions is probably the manufacturing principle as shown in the table below.

Table 1. Classification of Additive Manufacturing according to the manufacturing principle

Principle group	AM	Abreviation	principle
Sintering	Selective Laser Sintering	SLS	localized powder melting
	Selective Laser Melting	SLM	
	Electron Beam Melting	EBM	
Extrusion	Fused Deposition Modeling	FDM	Application of liquified polymers using nozzles
	Multi-Jet-Modeling	MJM	
UV-Hardening	Stereolithographie	SLA	Localized co-polymerisation
Binder Technology	3-Dimensional Printing	3DP	Localized Application of binder
Laminating	Laminated Object Manufacturing	LOM	Contouring and applying of sheet material

Considering the production of serial end-use products Laminated Object Manufacturing (LOM) as well as 3D-Printing (3DP) are less well-suited. The strength of 3DP for instance lies within the possibility to produce colored samples and prototypes by coloring the binding material. The material of choice is modified gypsum for full-colored prototypes or silica sand for sand forming tools. The gypsum products have to be infiltrated with an epoxy or wax in order to gain mechanical strength. Building fine detailed products is, due to this low mechanical strength of the binding process, not possible. There are also some systems using metal powder for the application in tooling. These systems, however, are very limited and harder to find.

The so-called Layer Oriented Manufacturing (LOM) uses foil, paper or metal sheets to manufactured product. Each sheet represents one layer. The principle uses laser cutting or stamping tools to contour the sheets, which are then placed on top of each other and adhered. The current significance of this principle is limited. On the other hand, combining this principle with other additive manufacturing principles like Fused Deposition Modeling (FDM) could result in new applications, due to the advantage of LOM in generating relatively big surfaces in a short amount of time. The biggest disadvantage of LOM is the amount of scrap.

All the other principles mentioned in the table currently play a more important role with regards to additive manufactured products. The main reason is the combination of fine detail, handling and the mechanical strength of their products. Most significant for the quality of these products are the following criteria:

- Surface quality (in the direction of the layer as well as alongside the layer set up)
- Mechanical strength within a layer
- Mechanical strength between two layers
- Density of the additive applied material (e.g. air pockets)

This paper will in particular focus on Fused Deposition Modeling (FDM) as the principle of choice for the application.

2 State-of-the-Art of Fused Deposition Modeling (FDM)

Fused Deposition Modeling (FDM) uses liquefied or melted thermoplastic material which is applied in layers using minimum one nozzle. The fast cooling of the thermo-plastic material itself in the contour applied is responsible for the solidified product. De-pending on the part shape and complexity, a support structure using support material may be required. This is typically the case when the angle of the overhanging contour is bigger than 45°. This support structure is then in a post-processing step removed using either mechanical or thermal force or a bath in which the support material dissolves. The support material is applied by a nozzle just like the product material. This can be either the same nozzle or a second nozzle. All material applied is supplied in a string form run of from a spool. In general, every thermoplastic material can be used for FDM. The current material selection is, however, limited to a few. These include ABS, ABSi, PLA, PC and PPSU. Professional FDM equipment is currently capable of handling a maximum of ten (10) different materials which are only layer-wise or in various mixtures combined. These base materials are also available in a variety of colors increasing the application field. The possibility of layer-wise building with different materials and different colors within the same build job or product is responsible for colored single-material or multi-material products.

The size limit of products produced as a single part with no post-gluing of segments is determined by the building volume of the equipment. The largest build volume is currently 914mm x 610mm x 914mm.

The surface quality or roughness is de-fined by the extruded layers. Individual layers vary in thickness between 0,178 mm and 0,254 mm and are also visible. If this is an issued from an esthetics standpoint the surface can be smoothed using acetone-steaming afterwards. The new surface is close to that of an injection molded part. Alternatively, FDM products can also be surface coated in order to achieve higher resistance rates or to include other functions of the product.

Tests have shown that the mechanical strength of a FDM product is approx. 80-85% of a comparable injection molded product.

FDM has one advantage compared to other additive manufacturing technologies. It is easy to stop the process at any giving point in order to place other products or by-products into the product and then continue and finish the building process. Successfully tested examples include metal components shown in figure 1 and even electronics as shown in figure 2. This expands the functionality of additive manufactured products e.g. a lighting function.

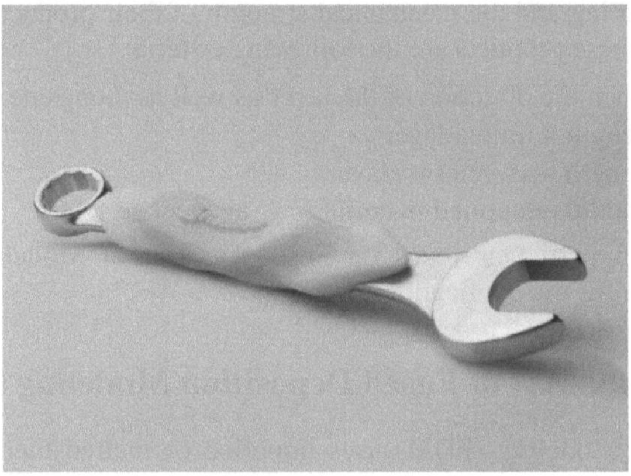

Fig. 1. Embedded wrench | Fraunhofer IPA

Fig. 2. AM part with integrated EL-Foil and Converter I Fraunhofer IPA

3 3D Fiber Printer

3.1 Basic Mechanical Challenges

3.1.1 Fiber Feeding

Trying to feed a fiber together with a thermoplastic material through an extruder the basic difference between the two be-comes evident immediately. The thermo-plastic filament usually supplied in thick-nesses either 1,75 mm or 3,0 mm is able to handle pull and push forces. This circumstance allows the feeding of the filament without bigger issues by using either two gears or using one gear and a ball bearing to restrain the filament. By turning the gear a continuous feed rate is applied, which in turn enables the material to be transported all the way through the heating area and the nozzle to the tip of the nozzle. It is important that during this process of feeding the filament through the nozzle is happening with as little friction as possible. It is also beneficial for the filament to move as straight as possible through the print head unit.

Contrary to the filament, the fibers are unable to handle push forces. In order to move a fiber forward, pull forces have to be applied. Fibers are acting from a mechanical point of view like a true rope, only being able to handle pull forces. Under these circumstances it became clear that the mechanism of feeding the filament forward will not work for the fibers alike.

3.1.2 Weight and Mass Ratio of Fiber to Thermoplastic Material

The diameter of the fiber, e.g. a carbon fiber or a carbon fiber bundle, is much less than the diameter of the thermoplastic filament. The thickness of a 1k carbon fiber cable for instance is only 190 µm. This is also the reason that the carbon fiber cannot be transported by the molten thermoplastic material, simply by pulling it along. Assuming a round cross section of the fiber its cross sectional area calculates to approximately A_{Fiber} = 28.300 µm². The cross sectional area of a 0,5 mm diameter nozzle is in comparison A_{nozzle} = 196.000 µm². The cross sectional area of a 1,75 mm thick filament finally is $A_{Filament}$ = 2.405.000 µm².

In order to completely fill the nozzle with fiber and thermoplastic material, the later has to fill the remaining cross sectional area of $A_{nozzle} - A_{Fiber} = 167.700\ \mu m^2$.

Assuming the same feed rate for the fiber as well as the filament the nozzle will be receiving 2.405.000 μm^2, which is more than 14 times the amount of the remaining cross sectional area. The effect would be excess material resulting in a clogged up nozzle.

This basic and simple calculation brought up the idea of embedding the fibers prior to it entering the extruder. This idea would have been a much easier task to complete. The fibers could have been equally applied with the thermoplastic material in an external apparatus.

In order to solve the problem stated above, a filament string with the exact diameter of the nozzle would need to be manufactured. Considering the mentioned example a 0,5 mm thick filament string of thermo-plastic material including the fiber already would be needed. Using such a theoretically and practically possible filament would result in the same problems as the fibers itself. The filament would be too weak to handle the push forces required for the FDM process and the nozzle to work.

3.2 Solution

The problems discussed under point 3.1.1 and 3.1.2 can be reduced and simplified to the problem of the fiber feeding process. The fiber(s) and the filament have to be run at different feed rates and therefore it is impossible to feed the fibers with the filament using the same transportation unit. As mentioned above, transporting the filament is not an issue and can be considered state-of-the-art. To feed the fiber(s) at the matching ration of the feed rate of the filament is the challenge. To reduce the complexity of finding and running the matching feed rate of the fiber, one idea was to fix the endless fiber directly on the building platform prior to starting the building process. This procedure would then automatically assure that the fiber is applied depending on the feed rate of the extruder head. Proceeding this way, however, will cause other problems. One being the fact that the fiber would need to be manually attached, the other being that the fiber and with it the matrix already produced will be under constant force and therefore additional stress would be put on the part during the build process. This issue comes especially into effect when the head has to move in a curvature or moving into a corner (depending on the geometry of the product) instead of a straight line. Describing a curvature or running into a corner applies strong shear forces to the fiber-matrix so that the risks of delamination or pull-off are increasing. Additionally the strong bending of the fiber at the nozzle tip poses the risk of dam-aging the fibers as well as an increased wear of the nozzle itself. These two risks are especially high due to the dynamic pull forces with which the fibers are pulled through the nozzle.

With these issues and risks in mind the search for a solution ended in the modified application of the water-jet pump principle.

This principle is characterized through the fact that a flowing medium is guided through a pipe and within that pipe along an opening. The flow of the medium is creating suction and thereby pulling the medium from the opening. This principle cannot be copied one to one to fiber and filament but requires adjustments. On the one hand, a very slow moving medium like the molten thermo-plastic material does not create suction. On the other hand, such a thin fiber would not be moved by suction. These issues are solved by the design of the 3D Fiber Printer nozzle. The fiber must be fed through a side channel into the main channel in which the filament is fed. The position of the meeting point of the two channels has to be exactly at the point where the diameter of the nozzle has reached the 0,5 mm. The extruder nozzle is designed in the way that the molten filament is fed through several mm of length through the channel with the diameter of 0,5 mm. Within these few millimeters the fiber is introduced to the molten filament and "adheres" to it forming the mechanical bond. This bond will now pull the fiber along with the molten material achieving the right mix ratio of thermoplastic material to fiber to create the best embedding of the fiber as well as the correct feed rate of both.

3.3 Design of the 3D Fibre Print Nozzle

The basis for the design of this special nozzle is the standard design of conventional brass nozzle in order to realize the same simple exchange mechanism of the nozzles within the print head. An additional side channel is designed meeting the main channel at a specific point. The fibers are introduced through this channel. The design and size of this channel are too small to be manufactured with conventional technologies. The solution is manufacturing this nozzle by the additive manufacturing technology of Selective Laser Melting (SLM). The side channel is also forming a curve see figure 3. This curve is needed to smooth the entrance of the fiber into the main channel and by that reducing the wear of the nozzle at the meeting point during the use of the nozzle. Car-bon fibers are known for acting like sanding paper to any burr or hard break resulting in the destruction of the nozzle in the long run.

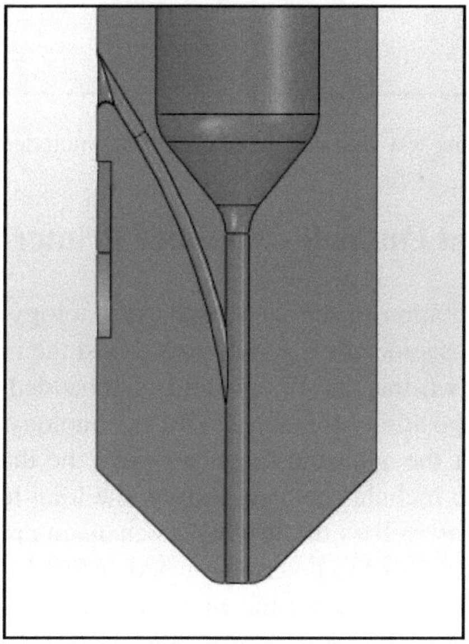

Fig. 3. FDM 3D Fiber Printer nozzle design proposal including a curved fiber channel Fraunhofer IPA

4 First Test Results

The first test runs were performed using the designed and built nozzle together with a 3K carbon fiber and a thermoplastic elastomer from the urethane group from Bayer Material Science. Starting point was the application of a straight line, to test the complete embedding of the fiber with the molten filament see figure 4. The base plate for the test was a heated aluminum plate with the temperature set to 110°C. For adhesion purposes a kapton adhesive tape was applied to the aluminum plate. The kapton adhesive tape is a high heat resistance adhesive tape which also shows good adhesion to the urethane material.

The first test results showed success at the application and feeding of both materials in the right setup. The visual inspection showed a good embedding of the fiber within the product.

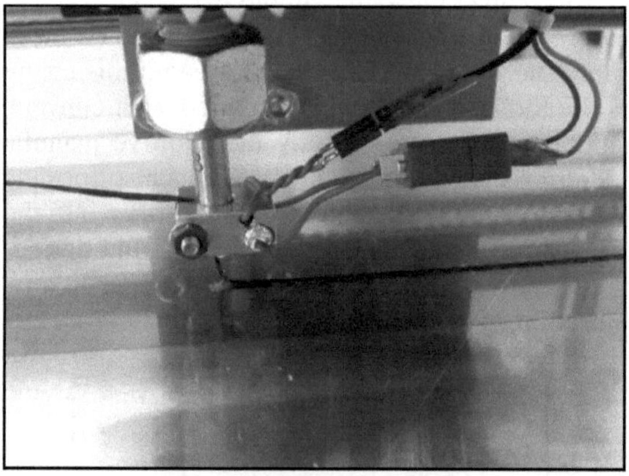

Fig. 4. First printing test with new nozzle and fiber included I Fraunhofer IPA

5 Next Steps and Outlook - 3D Fiber Printer

The industrial implementation of the de-veloped technology requires further optimization and tests. These tests include e.g. pull tests to test the matrix.

It was, however, shown that the fibers can be embedded into products manufactured with the Fused Deposition Modeling (FDM) technology.

Besides the testing of the adhesion forces between the thermoplastic material and the fibers, which have to include chemical testing, the long-term testing of the nozzle and its design for wear as well as the feeding mechanism are the next steps planned. The later tests are also intended to determine whether SLM is the right technology to manufacture the nozzles and whether the SLM manufactured nozzles are suited for industrial use.

References

1. Jäger, et al.: Carbonfasern und ihre Verbund-werkstoffe: Herstellungsprozesse, Anwendungen und Marktentwicklung. Die Bibliothek der Technik, Band 326. Verlag Moderne Industrie (2010)
2. Flemming, et al.: Faserverbundbauweisen, Fasern und Matrices. Springer-Verlag (1995)
3. Flemming, et al.: Faserverbundbauweisen, Halbzeuge und Bauweisen. Springer (1996)
4. Flemming, et al.: Faserverbundbauweisen, Fertigungsverfahren mit duroplastischer Matrix. Springer (1999)
5. Grießbach: Praxis Rapid Technologien, Handbuch für Produktentwickler, Techniker, Kaufleute. Vonroth & Bode KG Verlag (2009)
6. VDI-Richtlinie: Generative Fertigungsverfahren – Rapid-Technologien (Rapid Prototyping) - Grundlagen, Begriffe, Qualitätskenngrößen, Liefervereinbarungen. VDI, Düsseldorf (2009)
7. Zäh, M.: Wirtschaftliche Fertigung mit Rapid-Technologien: Anwender-Leitfaden zur Auswahl geeigneter Verfahren. Carl Hanser Verlag (2006)

From Rapid Prototyping to eManufacturing

Nikolai Zaepernick

Electo Optical Systems GmbH, Krailling, Germany
nikolai.zaepernick@eos.info

Abstract. Additive Manufacturing technologies have been a revolution to R&D process efficiency and are in place for many years now. The technologies have very much matured ever since and thus seem to pave their way to direct parts business.

Looking back to traditional manufacturing methods, the industrial market has developed highly productive and flexible Machine Tools over time to comply with the requirements of a wide spectrum of industrial applications. These machines have set standards and thus set the expectation benchmark of industry customers.

Additive Manufacturing opens up an enormous additional potential for industrial applications, however competing with traditional production methods. Electro Optical Systems (EOS) is gearing its portfolio and market activities clearly towards manufacturing applications. We will show, which activities EOS are taking to go this road. The example of handling and robotics industry illustrates which additional value Laser Sintering can achieve.

In this context, the gripper technology is a front runner for innovative solutions – lightweight and integrated functionality are key value drivers for handling & robotics. Existing applications are intensively profiting from this new technology: From gripper weighing less than 20g and being able to can handle 10kg parts to high-speed pick and place solutions being fully integrated into today's automation environment.

Additive Manufacturing is opening up economically valuable production up to lot size one – metal implant parts for human bodies are another perfect example for this mass individualization approach.

Keywords: Digital process chain, rapid prototyping, additive manufacturing, design-driven manufacturing, serial production, plastic Laser Sintering, handling & robotics, medical implants.

1 Introduction

EOS was founded in 1989 and is today the world-leading manufacturer of Laser Sintering systems.

Laser Sintering is the key technology for e-Manufacturing, the fast, flexible and cost-effective production of products, patterns and tools. The technology manufactures parts for every phase of the product life cycle, directly from electronic data. Laser Sintering accelerates product development and optimizes production processes.

G.L. Kovács and D. Kochan (Eds.): NEW PROLAMAT 2013, IFIP AICT 411, pp. 176–180, 2013.

EOS serves customers in more than 50 countries operating sales, application and service offices in 11 countries. With its focus on innovation, EOS reinvests a large portion of its revenues into R&D, resulting in a strong patent portfolio with more than 700 active patents in nearly 100 patent families.

EOS completed its business year 2012/2013 with revenues of more than 110 million with a worldwide staff of almost 500 employees, thereof 320 in Germany.

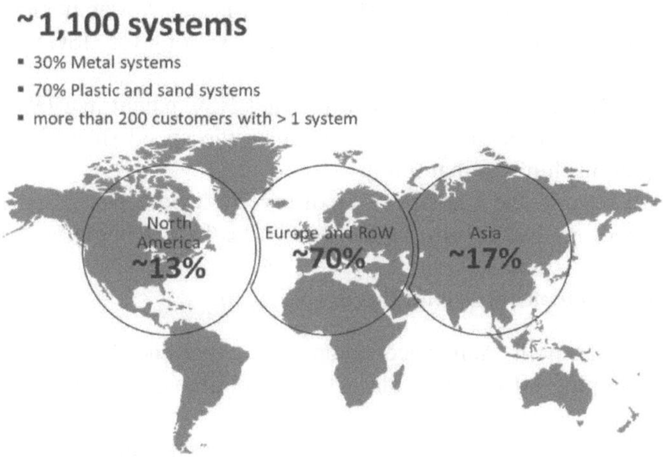

Fig. 1. EOS installed machine base 2012

2 The History of High Speed Cutting

High speed cutting today is probably the most exciting example for a production technology in terms of productivity. However this has not always been the case. It took quite a while for milling to get where it is today.

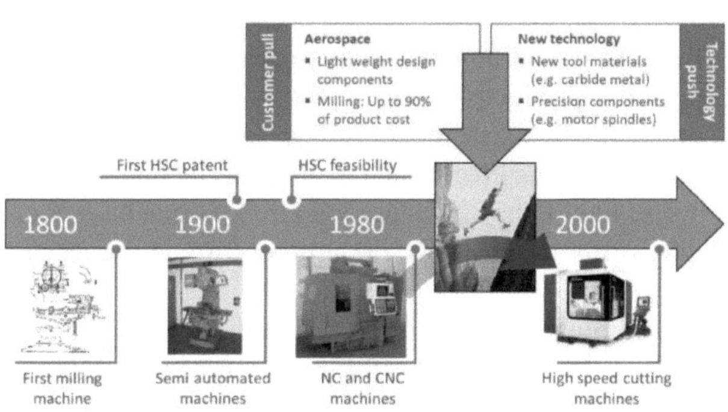

Fig. 2. Schematic development of High Speed Cutting

It all began in 1818 when Eli Whitney (USA) designed and built the first milling machine for metals. Industrialization has then generated a huge pull for machines tools – manual and semi-automated machines were about to rise. In 1952 J. Parsons constructed the first NC machine using punch tape for machine operation. From the 1970s onwards, CNC machines developed.

Today the High Speed Cutting technology (HSC) is available providing cutting speed over 5.000 m/min, feed rates factor 10 higher than normal milling and thus 30% higher material removal rates. Even surface quality is superior at much higher productivity levels. The core question is: 'Why did it take so long for HSC to be established in production'?

Interestingly Dr. Salomon was granted the first HSC patent in 1931 and Lockheed proved HSC feasibility in the 1950s at 10 times the speed (material removal rates >200 than actual machines at that time).

It's about 'Crossing the Chasm' as Geoffrey A. Moore describes in his famous book. The technology had to surpass a critical mass before being widely accepted in the market.

Two catalysts were necessary for this development:

- Strong customer pull: Aerospace companies experienced plane ribs to require enormous chipping effort, more than 90% of total cost related to the part production were related to milling.
- Technology push: New carbide metal tool materials and motor spindles were developed

3 Readiness of Additive Manufacturing Technologies

Additive Manufacturing technologies provide high value compared to traditional manufacturing methods. The freedom of design almost makes traditional design obsolete. Engineers can freely design their parts almost without restriction of the production process. Industry experts talk about a 'design-driven manufacturing' process instead. This freedom of design allow for lightweight design as well as very complex geometries.

Furthermore, integrated functionality help to save assembly effort after manufacturing and integrate this production step in the Laser Sintering process itself. This way, hinges and other functional elements can be 'printed in one shot'.

Additive Manufacturing machines make economies of scale obsolete. An economic production of 'lot size one' is no issue any longer. The machine does not care to produce 100 completely different parts or 100 similar parts at the same time. This is the start of mass customization.

Finally, Additive Manufacturing has its roots in the Rapid Prototyping field. A major value proposition is the fact, that machines can produce parts overnight. Time to market has a major influence on R&D processes and thus open up doors for much more flexible and short term focused product developments.

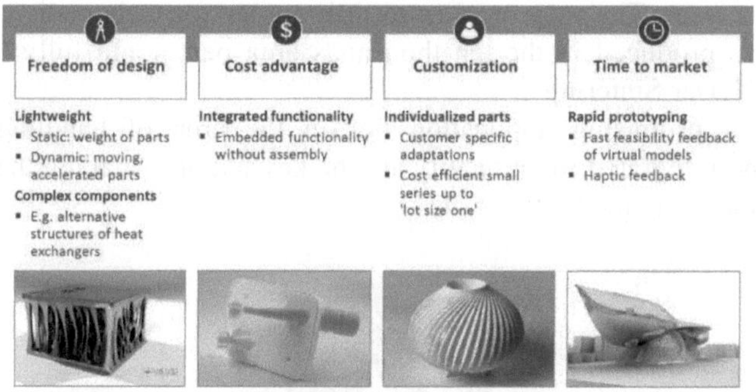

Fig. 3. Key differentiation criteria for Laser Sintering

Are Additive Manufacturing methods yet ready to replace existing production methods like casting or injection molding on the plastic side? At this point in time certainly not – there are some hurdles to be overcome for the broad field of all applications.

Firstly part quality is key and needs to fully comply with customer requirements. This comprises mechanical properties as well as dimensional accuracy and surface quality. With this as prerequisite, AM processes need to be robust – within the build platform, between several jobs, machines and even suppliers. Finally different industries require specific certifications or quality assurance methods and thus also need to be respected.

In many cases, Additive Manufacturing already fulfills these requirements. There are already very interesting applications out there where Additive Manufacturing is a fully accepted mass production method.

And looking ahead on the product roadmaps of AM machine OEMs, the major imitating factors like productivity or build size are being addressed in the near future.

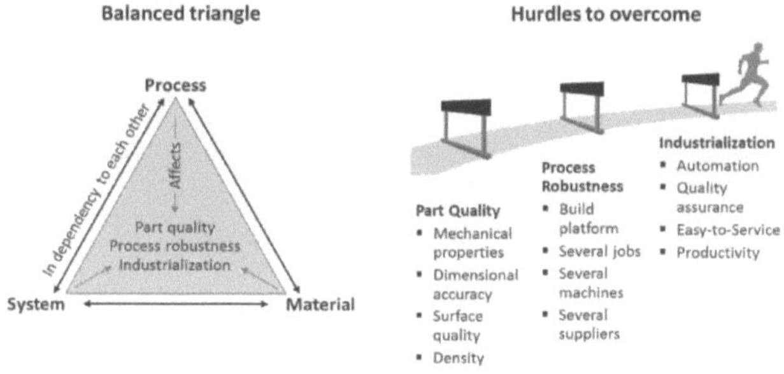

Fig. 4. Implementation hurdles for Additive Manufacturing

4 Application Examples

For example dental copings and bridges are a 'lot size one' application. Today, there are multiple machines installed to produce 4-5 million individualized metal bridges and copings per year. The same applies to drill guides for orthopedic surgery. These plastic parts are produced in the ten thousands area per month fully leveraging the possibilities of Laser Sintering.

Another very promising application lies in the area of handling & robotics. Lightweight and integrated functionality are the key features making the difference to existing manufacturing methods.

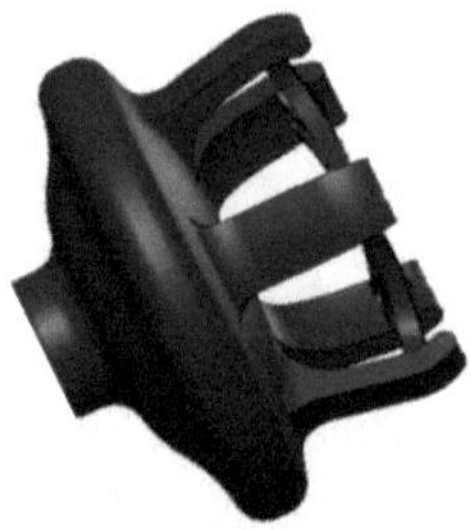

Fig. 5. Example lightweight gripper developed by Kuhn-Stoff

The example of a hole-gripper for part handling purposes recently developed by the company Kuhn-Stoff (www.kuhn-stoff.de) is a perfect one: the gripper weights less than 20g and can handle parts up to 10kg. It operates an integrated pneumatic membrane to apply gripping force.

The advantages of this gripper compared to traditionally manufactured alternatives are

- Weight reduction of about 80%
- Printed in one shot - no final assembly
- Geometry fully flexible and scalable
- Tested to >5 mio. Cycles

Energy Turnaround: Printing of Thermoelectric Generators

Ines Dani[1], Aljoscha Roch[1], Lukas Stepien[1], Christoph Leyens[1,2], Moritz Greifzu[2], and Marian von Lukowicz[2]

[1] Fraunhofer Institute Material and Beam Technology IWS, Dresden, Germany
ines.dani@iws.fraunhofer.de
[2] Technical University Dresden, Germany

Abstract. Waste heat in the temperature range up to 100 °C can be found in virtually all areas of industry and building and is often the hardest to recover cost-efficiently. It accounts for more than 50 % of the total heat generated in industry [1].

Therefore, increasing interest is noticed in thermoelectric harvesters in the operation temperature range from room temperature up to 100 °C. The aim of this work is the development of materials and processes for the material and cost efficient scalable production of TEGs.

PEDOT:PSS is an interesting p-type conducting polymer suitable for printing of TEGs. A printed unileg generator, i.e. by using just a p-conductive material, was realised by dispenser printing of DMSO-doped PEDOT:PSS on polyimide. Contacts were made from silver. A TEG with 60 legs was characterized. The internal resistance is increasing linearly with the number of legs, confirming the good reproducibility of the printing process. By using a multilayer design the resistance can be decreased by more than 50 %. A temperature difference of 90 K results in a voltage of 37 mV (at 20 °C at the cold side).

1 Introduction

In developed countries, ca. 40% of the total fuel consumption is used for heating. Of this, about one-third is wasted due to insufficient transformation technologies into electricity. This wasted heat may be used by exploiting the Seebeck effect, enabling special materials to generate electrical energy if exposed to a temperature gradient. Thermoelectric generators (TEGs) built up from these materials are a durable way for energy supply without any moving components. Drawbacks of state-of-the-art materials are their low efficiency, limited availability or toxicity of the raw materials, and high costs.

Polymer materials in combination with printing techniques offer the possibility to manufacture flexible generators from non-toxic and easily available raw materials. The operation temperature of these materials ranges from room temperature up to 100 °C; small temperature gradients of 1 K can be exploited.

G.L. Kovács and D. Kochan (Eds.): NEW PROLAMAT 2013, IFIP AICT 411, pp. 181–184, 2013.

2 Material Development

PEDOT:PSS (poly(3,4-ethylenedioxythiophene) poly(styrenesulfonate) is an intrinsically p-type conductive polymer and possesses a high conductivity, good stability and flexibility. All printing experiments presented here are based on the printing paste SV3 (Heraeus) with a viscosity of 1 Pa s (at a shear rate of 250 s^{-1}). The solids content is about 2.3 wt%.

Due to the high viscosity of the PEDOT:PSS printing paste a dispenser printer (Asymtec) was used (figure 1).

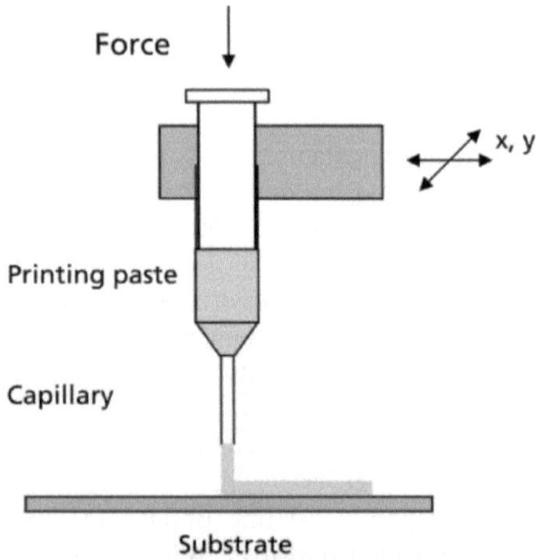

Fig. 1. Scheme of the dispenser printing process

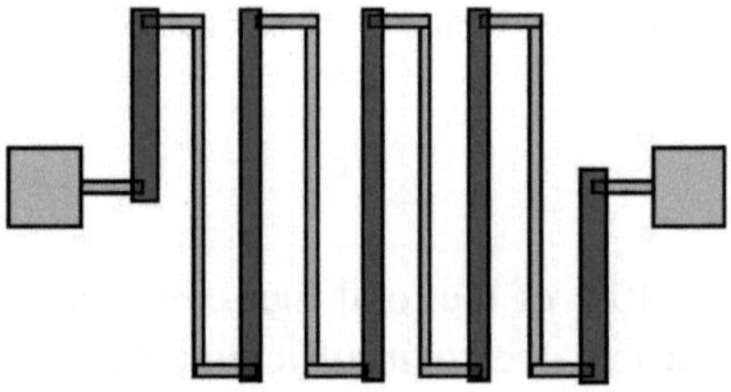

Fig. 2. Scheme of an unileg TEG – p-type conductive thermoelectric material (blue) and silver contacts (grey)

To increase the charge carrier concentration of the polymer, PEDOT:PSS was doped with dimethyl sulfoxide (DMSO). For characterisation of the thermoelectric properties the pastes were printed on glass substrates and dried at 50 °C for 2 days. The resulting films with a thickness of 22 μm were tested with a modified van der Pauw method [2]. The electrical conductivity increased from 8 S/cm without DMSO addition to 84 S/cm with 6 wt% DMSO. The Seebeck coefficient of about 15 μV/K was not influenced by DMSO doping.

3 Printing Process and Results

Until now, no n-type conductive polymer for thermoelectric applications is commercially available. Therefore only p-type conductive PEDOT:PSS was used to print an unileg thermoelectric generator (figure 2).

Polyimide film was chosen as substrate due to its temperature stability up to 400 °C and its flexibility. In a first step, silver paste (Heraeus) was printed as contact material and sintered at 200 °C. PEDOT:PSS with 6 wt% DMSO serves as the active thermoelectric material. It was printed on top of the silver contact and dried at 60 °C for 24 h. A multilayer structure of the active material reduces the internal resistance by half, enabling a higher number of thermocouple legs with the same internal resistance.

Using a dispenser printing process enables a fast adaption of the unileg geometry; standard parameters are a length of 1 mm and a height of 10 mm for one unileg. A TEG with 60 unilegs corresponds to a total length of about 30 cm (figure 3).

Fig. 3. Dispenser printing process of a TEG with 60 unilegs

For analyzing the thermoelectric performance of the device, the printed unileg film was wrapped around an adapter, made from two aluminum parts and a PEEK connector with low heat conductivity (figure 4).

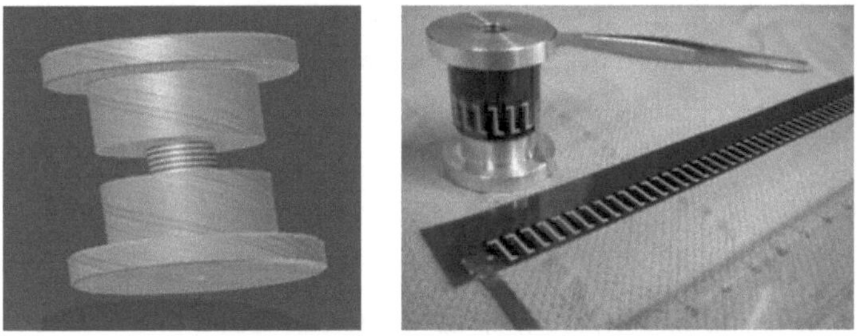

Fig. 4. CAD model of the adapter for winding-up the unileg film (left) and complete TEG with contacts (right)

The cold side was held constant at 20 °C, the temperature of the hot side was stepwise increased. Increasing the temperature gradient leads to a linear increase in the measured voltage (figure 5). For a temperature difference of 90 K a resulting voltage of 37 mV was determined.

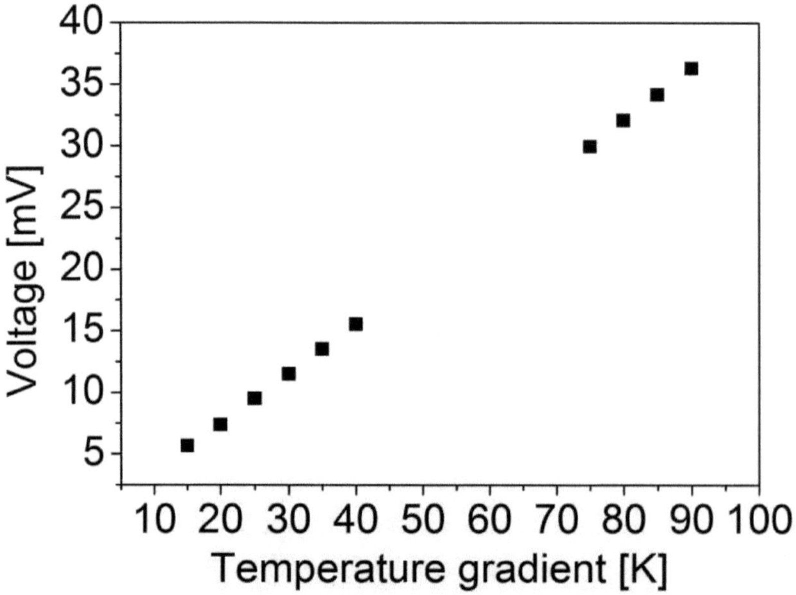

Fig. 5. Voltage in dependence of the temperature gradient in the TEG, cold side at 20 °C

4 Summary

A printed unileg generator consisting of only a p-conductive material was realised by dispenser printing of DMSO-doped PEDOT:PSS on polyimide to demonstrate the manufacturing of flexible thermoelectric generators from non-toxic and easily available raw materials. A TEG with 60 legs and silver contacts was characterized. By using a multilayer design the internal resistance was decreased by more than 50 %. With a cold side temperature of 20 °C and a temperature difference of 90 K a voltage of 37 mV was generated.

References

1. Quoilon, S.: Appl. Thermal Engineering 31, 2885–2893 (2011)
2. De Boor, J., Schmidt, V.: Adv. Materials 22, 4303–4307 (2010)

Generative Manufacturing and Repair of Metal Parts through Direct Laser Deposition Using Wire Material

Steffen Nowotny[1], Sebastian Thieme[1], David Albert[1], Frank Kubisch[1], Robert Kager[1], and Christoph Leyens[2]

[1] Fraunhofer Institute Material and Beam Technology IWS, Dresden, Germany
[2] Technical University Dresden, Germany
steffen.nowotny@iws.fraunhofer.de

Abstract. In the field of Laser Additive Manufacturing, modern wire-based laser deposition techniques offer advantageous solutions for combining the high quality level of layer-by-layer fabrication of high value parts with the industry's economical requirements regarding productivity and energy efficiency. A newly developed coaxial wire head allows for omni-directional welding operation and, thus, the use of wire even for complex surface claddings as well as the generation of three-dimensional structures. Currently, several metallic alloys as steel, titanium, aluminium, and nickel are available for the generation of defect-free structures. Even cored wires containing carbide hardmetals can be used for the production of extra wear-resistant parts. Simultaneous heating of the wire using efficient electric energy increases significantly the deposition rate and energy efficiency. Examples of application are light-weight automotive parts, turbine blades of Nickel super alloys, and complex inserts of injection moulds.

1 Introduction

Laser buildup welding is a well-established technique in industrial applications of surface cladding and direct fabrication of metallic parts. As construction materials, powders are widely used because of the large number of available alloys and the simple matching with the shaped laser beam. However, the powder utilization is always less than 90 %, characteristic values are in the range of 60 %, and so a serious part of the material is lost. Additionally, there is a risk for machine, operators and environment due to dangerous metal dust. The alternative use of wires as buildup material offers a number of advantages: The material utilization is always 100 %, and rightly independent from the part's shape and size. The process is clean and safe, and so the effort for protecting personnel and environment is much less. Also the wire feed is completely independent from the gravity which is of great advantage especially in applications of three-dimensional material deposition.

The main challenge compared to powder is the realization of an omni-directional welding operation with stable conditions of the wire supply. The only possible solution therefore is to feed the wire coaxially in the centre axis of the laser beam. This technology requires a complex optical system which permits the integration of the

G.L. Kovács and D. Kochan (Eds.): NEW PROLAMAT 2013, IFIP AICT 411, pp. 185–189, 2013.

wire material into the beam axis without any shadowing effects of the laser beam itself. Accordingly, the work presented here was focused of the development of a new optics system for the practical realization of the centric wire supply as well as the related process development for the defect-free manufacturing of real metallic parts.

2 Laser Wire Deposition Head

Based on test results of previous multi beam optics [1], a new optical system suitable for solid-state lasers (slab, disk, fiber) has been developed. The optical design of the head shown in Figure 1 is based on reflective optical elements and accommodates a power range of up to 4 kW. The laser beam is symmetrically split into three parts so that the wire can be fed along the centre axis without blocking the beam. The partial beams are then focused into a circular spot with a diameter ranging from 1.8 to 3 mm. The setup enables a coaxial arrangement for beam and wire, which makes the welding process completely independent of weld direction. The coaxial alignment is even stable in positions that deviate from a horizontal welding setup.

The wire is fed to the processing head via a hose package that contains wire feeder, coolant supply and protection gas delivery. Wire feeders from all major manufactures can be easily adapted to this setup and are selected based on wire type, feed rate, and operation mode. Typical wire diameters range from 0.8 to 1.2 mm. However, the new technology is principally also suitable for finer wires of about 300 μm in diameter. The wires can be used in cold- and hot-wire setups to implement energy source combinations. The new laser wire processing head is useful for large-area claddings as well as additive multilayer depositions to build three-dimensional metallic structures.

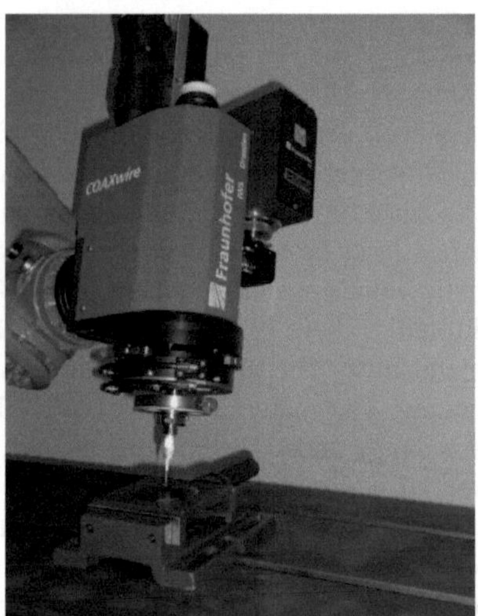

Fig. 1. Laser processing optic with coaxial wire supply

For process monitoring, the wire deposition head may be equipped with a camera-based system which measures dimensions and temperature of the melt bath simultaneously during the running laser process. Optionally, an optical scanning system controls the shape and dimension of the generated material volume in order to correct the build-up strategy if necessary [2].

3 Deposition Process and Results

Fig. 2 shows a typical laser wire cladding process during a multiple-track deposition. The process shows a stable behaviour with extremely low emissions of splashes and dust, compared to powder-based processes. The integrated on-line temperature regulation keeps the temperature of the material on a constant level during the whole manufacturing duration.

Fig. 2. Process of laser wire deposition of a metal volume

The part is built from a large number of single tracks, which are placed according to a special build-up strategy. This strategy is designed by computer calculation prior to the laser generative process. Normally, an intermediate machining between the tracks and layers is not necessary.

The primary process parameters laser power, wire feeding rate and welding speed have to be adapted to each other to enable a continuous melt flow of the wire into to the laser induced melt pool. For certain primary parameters, the process stability depends on the heat flow regime during the build-up process. Besides the temperature regulation mentioned above, also interruptions between selected layers may be useful to cool down the material. If necessary, also an active gas cooling of the material can be applied [3].

In Fig. 3 the cross section of a generated structure of the Nickel super alloy INCONEL718 is shown. The solidified structure is defect free and each layer is metallurgically bonded to the other. Through optimization of the process parameters, even the crack-sensitive IN718 structure is crack-free. An optimized build-up strategy allows minimal surface roughness of RZ = 63 μm. A layer thickness of 1.4 mm and a build-up rate of 100 cm³/h can be achieved with 3.0 kW laser power and 2.0 m/min welding speed. Simultaneous heating of the wire using efficient electric energy (hot-wire deposition) increases significantly the deposition rate up to about 160 cm³/h [4].

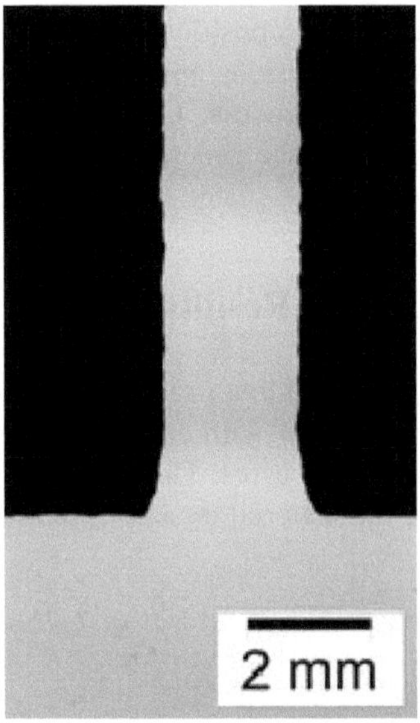

Fig. 3. Cross-section of a laser generated wall of INCONEL718

Standard welding feedstock wires can be used as deposition material. The suitable wire diameters range from about 1.2 mm down to 0.3 mm. Using the fine wires, the best lateral resolution of the generated structures lie in the range of 0.6 mm. Here the deposition rate amounts to 30 cm^3/h.

Figure 4 illustrate two examples of layer-by-layer generated parts. Figure 4a shows a turbine blade out of INCONEL718. The blade's height is 100 mm, and inside it has a hollow structure. The height of the inlet tube of Figure 4b is 85 mm, and it consists of light-weight alloy AlMg5. The height of the single layers is 0.4 mm for the Ni alloy and 0.7 mm for the Al alloy.

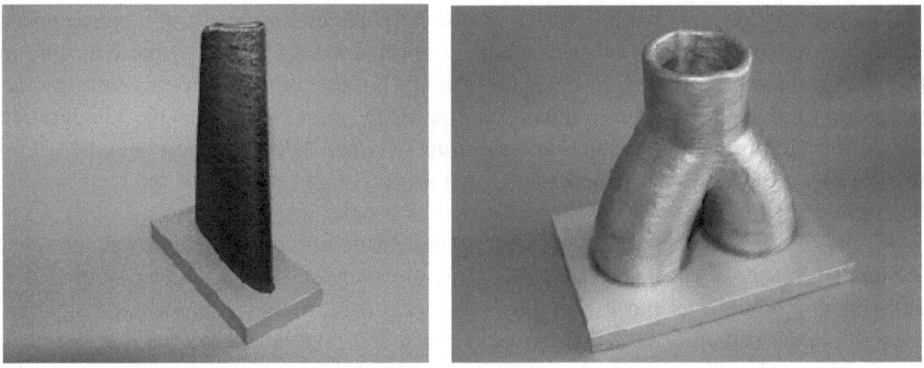

Fig. 4. Turbine blade out of INCONEL718 Inlet tube out of AlMg5

4 Summary

The current state of laser wire deposition shows the wide range of potential applications of this new technique. In addition to the well-established powder welding and powder-bed melting techniques, wires represent an advantageous alternative for high-quality laser deposition. A specially developed laser head with coaxial wire supply permits omni-directional welding operation and thus new dimensions in additive manufacturing. Also, equipment for the on-line process regulation is available and can be used for quality management. In particular, the regulation concerns the melt bath's dimensions and temperature on its surface.

As construction material, commercially available welding feedstock wires can be used. The material utilization is always 100 %, the welding process is clean, and the variant of hot-wire cladding advantageously increases productivity and energy efficiency. The generated metal structures are completely dense, as important precondition for a high mechanical strength of the final parts. The surface roughness is typically lower than RZ100 μm, and the model-to-part-accuracy lies in the range of some tenth of a millimetre.

Examples of application are corrosion protection coatings on cylinders, turbine parts of Nickel and Titanium [5] alloys as well as light-weight parts for automobile use.

References

1. Nowotny, S. (ed.): Laser Cladding with Centric Wire Supply. Fraunhofer Verlag (2009) ISBN 978-3-8396-0020-7
2. Hautmann, D.: Adaptive Laser Welding, REPORT, ed. by MTU AeroEngines, vol. 2, pp. 30–33 (2010) (in German)
3. Beyer, E.: High-Power Laser Materials Processing. In: Proceedings of the 31st International Congress on Applications of Lasers and Electro-Optics, Anaheim, CA, September 23-27, paper OP2 (2012)
4. Pajukoski, H., Näkki, J., Thieme, S., Tuominen, J., Nowotny, S., Vuoristo, P.: Laser Cladding with Coaxial Wire Feeding. In: Proceedings of the 31st International Congress on Applications of Lasers and Electro-Optics, Anaheim, CA, September 23-27, paper P124 (2012)
5. Brandl, E., Michailov, V., Viehweger, B., Leyens, C.: Deposition of Ti–6Al–4V using laser and wire. Surface & Coatings Technology 206, 1120–1129 (2011)

Capp Model for Prismatic Parts in Digital Manufacturing

Vidosav D. Majstorović[1], Tatjana V. Šibalija[2], Bojan M. Erčević, and Marko M. Erčević[3]

[1] University of Belgrade, Faculty of Mechanical Engineering, Belgrade, Serbia
vidosav.majstorovic@sbb.rs
[2] Faculty of Engineering International Management, European University, Belgrade, Serbia
[3] IVA 28 d.o.o, Novi Beograd, Serbia.

Abstract. The paper discusses the feature-based process planning in a digital manufacturing concept, for prismatic parts manufacturing. STEP standard as an enabler of feature-based manufacturing is presented in detail, along with belonging application protocols. The case study on the feature-based CAPP/CAM model in Serbian SME is shown to depict the advantages, followed by the research recommendations for the improvement of the presented concept.

Keywords: Digital manufacturing, Feature-based manufacturing, CAPP, STEP.

1 Introduction

Digital manufacturing, within digital factory, has become the main manufacturing concept in the twenty-first century and highly important for companies to enhance their competitiveness. One of the research priorities in Strategic Research Roadmap 2010-2013 of European public-private partnership for Factories of Future (PPP FoF), is ICT-enabled intelligent manufacturing [1]. It was stressed here that ICT is a key enabler for improving manufacturing systems at three levels: smart factories, virtual factories and digital factories. In this context, digital factories are defined with the ultimate goal to improve design productivity using software for the digital representation and test of products and processes prior to their manufacture and use. This also implies better understanding and design of production and manufacturing systems for better product life cycle management involving simulation, modelling and knowledge management from the product conception to manufacturing, maintenance and disassembly/recycling. The major sub-topics are: knowledge and analysis; enhanced, interoperable models for products and processes; design environments; and lifecycle management [2].

Digital manufacturing is considered a part of a digital factory. From the industrial application aspect, digital manufacturing implies IT support for planning and engineering processes and 3D computer visualization. The digital manufacturing could be defined as a methodology that uses deeper and extensive knowledge and IT. In general, there are three main elements that define the digital manufacturing: (i) IT system

G.L. Kovács and D. Kochan (Eds.): NEW PROLAMAT 2013, IFIP AICT 411, pp. 190–204, 2013.

and its application; (ii) theoretical model of a digital manufacturing – the knowledge level used as a methodology for digital manufacturing: and (iii) usage of specific advanced methods and techniques, such as e.g. web-based multi-agents systems. From the informational point of view, the fundamental characteristics of information/data in digital manufacturing are: digital presentation/format; multiple usage; and, independence with respect to the distance, time and place of utilization [3].

IT integrated with traditional manufacturing technologies improved automation and digitization of the planning, design, production, inspection, management, and other related activities. Hence, the overall digital output presents a mirror of the actual manufacturing using limited amount of data. The digital presentation of a product is the very essence of the digital manufacturing in terms of numerous activities performed in the factory. The digital model of a product could be used to simulate and analyse the related production activities flow or to verify different production planning, machining/tool path, inspection and resource utilization scenarios.

It could be summarised that the main features of digital manufacturing are [4]:

- Ambiguity of any knowledge must be eliminated, and knowledge must be presented in forms that can be processes/analysed/utilised by various IT tools and transformed into digital data. This is a basis for factory-wide knowledge integration in terms of widespread application of Computer Aided Design (CAD)/Computer Aided Process Planning (CAPP)/Computer Aided Manufacturing (CAM)/Computer Aided Inspection (CAI)/CAx, and for the cooperation: factory system – process - numerical control machining. Therefore, one of the main problems is a synthesis of different formal presentations of different aspects/elements of digital manufacturing system.
- Simulation and prediction of product development is of essential importance, since it allows analysis and evaluation of manufacturability, product's performance, and other related characteristics.
- All manufacturing-related activities supposed to be independent of distance, time and location (i.e. in a networked environment).

Enabling technologies for digital manufacturing could be divided into following groups [4]:

- for product design/development and simulation: CAD, VM (Virtual Manufacturing), CAPP, CAE (Computer Aided Engineering), CAM, etc.;
- for material processing (machining) and material flow control, such as NC (Numerical Control), CNC (Computer Numerical Control), DNC (Direct Numerical Control), FMS (Flexible Manufacturing System), control systems for material flow, etc.;
- for production management: IGES (Initial Graphics Exchange Specification), STEP (Standard for the exchange of product data), ERP (Enterprise Resource Planning), MES (Manufacturing Execution System), PDM (Product Data Management), database, Internet, Intranet, ...;
- for collaboration among enterprises: networked manufacturing, client relation management and supply chain management, etc.

The objective of this research is to analyse process planning for prismatic parts in a digital manufacturing concept. After the introduction, the second section presents the process planning model based on product features for prismatic parts. As an example, the existing feature based CAPP/CAM model for prismatic parts used in Serbian SME IVA-28 is discussed in the third section. Finally, concluding remarks are given with respect to the advantages and limitations of the feature based process planning model in a digital manufacturing, as well as recommendations for our future research.

2 Feature Based Process Planning

The objective of process planning in a manufacturing enterprise is to generate the optimal sequence of necessary manufacturing processes in order to produce the observed part in the technologically feasible and economically effective way. Today, this is performed by using computer-aided process planning (CAPP) systems. Process planning is based on the relevant data of a part design model, such as geometrical features, material, tolerances (accuracy), surface roughness, etc. The set of all these data could be considered as manufacturing features, which present a link between computer-aided design (CAD) and computer-aided manufacturing (CAM). The main issue here is how to translate the manufacturing features (derived from a part design model) into the manufacturing operations sequence [5]. Beside the integration with CAD and CAM, it is very important to design CAPP systems at such a way to provide a real-time link with the other existing IT systems in the enterprise such as ERP, MES, etc.

Probably the most important advantage of the feature technology is the ability to associate not only geometric and topological information, but also form features, tolerances, material properties and other information which may be used during the process planning. Today, Standard for Exchange of Product Model (STEP) is used for the implementation of feature technology. STEP provides the neutral, interoperable, computer interpretable file format of product data, that is independent of any particular system and suitable for transfer, processing and communication between different systems such as CAD and CAM. STEP file applies object oriented database method to model the relationship within the file data, and uses a formal specification language EXPRESS to define product data (entities, attributes and relationships). Therefore, STEP has been accepted as an international standard for feature based technologies. The official designation of STEP standard is ISO10303: Industrial automation systems and integration - Product data representation and exchange. STEP covers a wide range of applications. To specify the part data representation for a specific application, STEP uses Application Protocols (AP).

Application Protocol 224 (AP224) gives manufacturing feature definitions based on prismatic and turning parts. AP224 is defined in the part ISO 10303-224: Application protocol: Mechanical product definition for process planning using machining features. The latest version of this standard was issued in 2006. The standards includes the application protocol for the representation and exchange of information for manufacturing mechanical parts and shape representation of machining features, addressing manufacturing part properties, process control documentation, manufacturing specifications, administration data, and requisitions [6].

The extension of STEP standard is STEP-NC standard, that defines independent tool paths and volume removal features based on geometric features (constructs), and adds geometric dimension and tolerance data for inspection. Basically, it combines and interconnects STEP (ISO10303) standard with ISO 14649: Data model for computerized numerical controllers that defines the requirements for machining. This is formally defined in Application Protocol 238 (AP238) of ISO 10303 standard. The main idea behind it was to replace ISO 6983 G code standard with a modern format that connects the CAD part design data used to determine the machining requirements with the CAM process data that solves those requirements (CNC process data).

It is also important to mention AP203: Application Protocol for Configuration Controlled Design, that is used to exchange design data (geometry, product structure, tolerances, configuration management) among different CAD systems [6].

For feature based CAPP, AP240 is of essential importance (AP240: Application Protocol For Process Plans For Machined Products). It defines the information for macro process planning, and provides process plans, machine tool resources (i.e. fixtures, cutting tools), process planning activities and their sequencing, setups, materials, properties, process requirement documents, and part shape with features and tolerances [6].

According to the above, integration of feature-based product knowledge directly in CNC programs in STEP-NC manufacturing is presented in Figure 1.

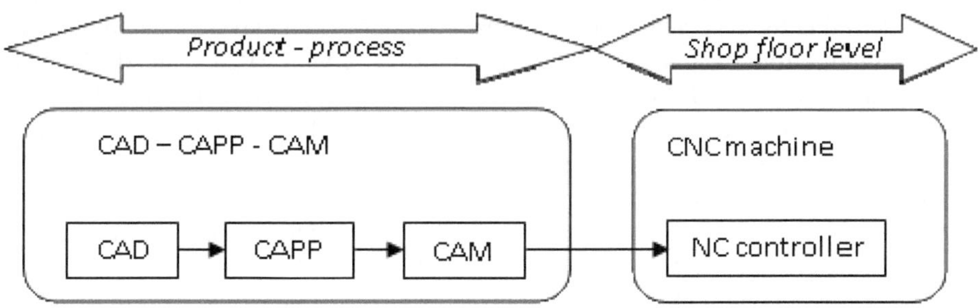

Fig. 1. STEP-NC manufacturing concept

The above commonly accepted standards and protocols enable further advances in feature based CAD/CAM/CNC collaborative systems. Several recent researches used different approaches to solve problems in feature based process planning. Yongtao and Jingying [5] constructed a feature based process planning model for the hole's machining that consists of three parts: the features framework, the semantic precedence-relations-net and the sequencing mathematical model. Ming and Mak [7] used the combination of neural networks (NNs) to solve setup planning problem in CAPP, in an intelligent manufacturing environment. Capponi et al. [8] presented feature based CAPP developed for the aircraft industry, characterised by complex geometries of parts. Deja and Siemiatkowski [9] presented the CAPP approach which focuses on the identification of process alternatives and sequencing adequate working steps, based on part design data in terms of machining features. Álvares and Ferreira [10] developed a system for web-machining, aiming at integrating CAD/CAPP/CAM for the remote manufacturing of cylindrical parts, adherent to STEP-NC. Xu and He [11] analyse a total integration of CAD, CAPP, CAM and CNC based on STEP-NC. Campos and Miguez [12] developed the collaborative STEP-based CAD/CAM/CNC chain, that integrates both types of information - process data and standard machining (STEP-NC) program, enabling full monitoring and traceability of activities in the system.

There are few recent papers from a literature that discuss feature based process planning for prismatic parts in an integrated CAD/CAPP/CAM system. Sharma and Gao [13] presented a feature based CAPP for prismatic parts using STEP AP224, where process plans and related documentation is automatically generated from AP224 files. The decision logic is stored in an external database, making the system compatible with various manufacturing applications. Amaitik and Kilic [14] developed an intelligent CAPP system for prismatic parts using STEP features, which generates process plan (including operations selection, tool selection, machining parameters determination, machine tools selection and setup planning) and corresponding STEP-NC file. In this system, the inference engine uses hybrid approach consisted of NNs, fuzzy logic and rule-based system. Collazos and Álvares [15] proposed a methodology for the integration of CAD/CAPP/CAM for prismatic parts, in accordance to STEP-NC standard. It starts with a product design using DSG (Destructive Solid Geometry) where the features of machining are associated to machining operations (it can incorporate also the elements related to the process planning, such as selecting the type and diameter / length of a tool for removing material in a cavity). The output from the CAD module is in conformance with AP203 and AP224, and it presents an input for CAPP module. The CAPP module uses AP224 and STEP-NC to define the machining features of the part to be machined. This research considers the extended CAPP or an integrated CAPP/CAM system, hence one of the process planning outputs is the NC program in the format STEP-NC ARM ISO 14649 that contains information of machining features, cutting parameters, cutting tools, fixturing, working steps, workplans, among others. This further generates the tool path for each working step described in a NC program in the G and M codes (RS274) format to be used in traditional CNC machines without support for STEP-NC. In this approach, the data generated at each interface of the modules are stored in a database in XML format. Nassehi et al. [16] applied collaborative multi-agent systems in designing an STEP-NC compliant, object-oriented CAPP for prismatic parts. It is highlighted here that the data model, i.e. the ability to store complex geometries and multi-level class hierarchies, plays an essential role in CAPP. STEP-NC environment provides potentials for the above data model, due to clear entity definitions and expandable architecture, and also due to the fact that STEP-NC architecture stores full information about a product. Hou [17] presented a research on CAD/CAM integration, mainly focusing on the connection of CAPP and CAM utilising machining features for prismatic parts. In this approach, the existing CAPP system is integrated with the commercial CAD/CAM system by implementing the prototype integration layer. This provides the whole set of data available as an input for CAM system: product geometric features, process plan information together with machining features. Hence, the tool paths can be automatically generated from solid models and process plans. Sun et al. [18] proposed an integrated setup/fixture planning approach for prismatic parts machining, that consists of four steps: definition of the requirements and constraints for setup/fixture planning; description of the precedence constraints between setups; selection of the suitable location surfaces; evaluation of the candidate setup/fixture plans.

It is important to clarify how STEP-NC and the related APs impact process planning systems, considering a fact that process planning occurs between design and manufacturing. AP203 presents an interface for design, and AP238 - an interface for manufacturing. Hence, the popularity of AP238 will contribute to popularity of CAPP systems, where enterprises will define their inputs and outputs according to STEP APs. To facilitate this, AP224 (part features) and AP240 (process planning) were defined. In the long term it is expected that CAD system will write AP203 edition 2 or AP224; CAPP system will read AP224 and write AP240; CAM system will read AP-240 and write AP238; and, finally, CNC system will read AP238 to manufacture a part [19].

Figure 2 presents the primary data flow in STEP-NC manufacturing concept based on design features and using STEP APs, enabling a direct exchange and a better data flow between CAD-CAPP-CAM-CNC systems (for prismatic parts). The steps could be described as follows [12, 19]:

1. In the part design phase (CAD), the part is defined using the AP-203 (edition 2) that includes geometric and tolerance data, and other basic data about the part (i.e. product structure, configuration management, etc.). Hence, is suitable for CAD data exchange, but it does not include the feature information needed for macro process planning (CAPP) and for CAM.
2. Based on AP203 data detailed part design is performed including features modelling (CAD-f), and gives AP224 data. AP224 data contains AP-203 geometric information and new information - features to be machined and their tolerances.
3. Based on AP224 data CAPP system extract design features and performs macro process planning implying information about facilities, available CNCs, and the distribution of work. This also includes definition of fixtures and setups, process selection and operations definition, machining tools selection and machining parameters definition. All these data are added to AP224 to generate AP240 output.
4. CAM system generates data for CNC controllers programming using AP240 data. These data are added to AP240 to create AP238 output for feature based machining.
5. Based on the received AP238 data, CNC system performed manufacturing operations.

As presented at Figure 3, the basic architecture of CAPP system is based on the interaction with databases that contains information about features, as well as information about CNC machines, cutting tools, fixtures, types of material, etc. [15]. Data from these databases are needed for the optimization of processes/machines sequence and for CNC and cutting tools selection, to generate AP240 file that can be used by CAM system that create AP238 file. This file contains data about part design, manufacturing feature, setup and fixture requirements, and sequences of processes, setups and cutting tools, and can be directly used by CNC controllers to perform machining processes.

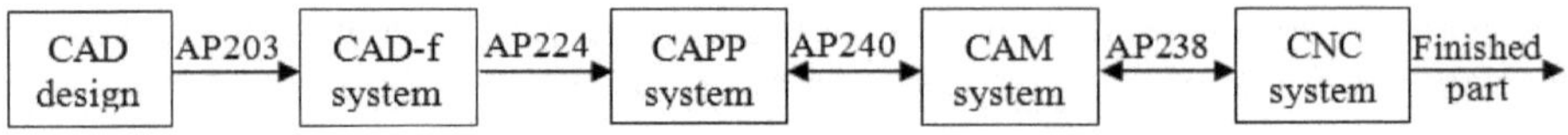

Fig. 2. The simplified scheme of CAD/CAM/CAPP/CNC integration enabled by STEP-NC for prismatic parts

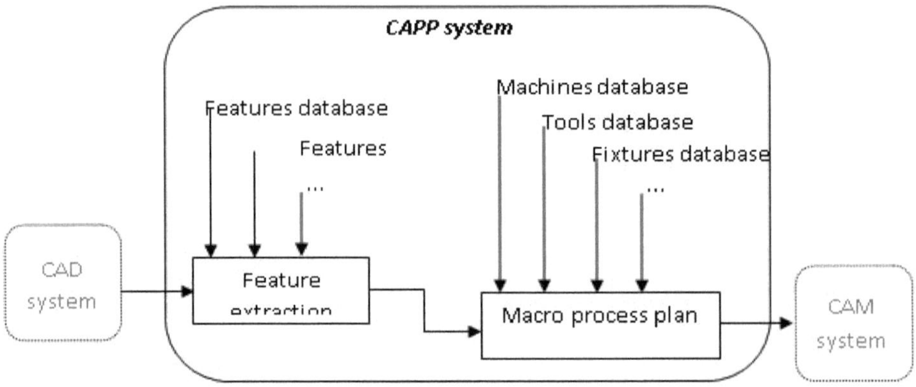

Fig. 3. The basic architecture of CAPP system in STEP-NC manufacturing concept

In the above STEP-NC manufacturing concept, CAPP system that performs macro process planning is supposed to carry out the following activities:

- features extraction and translation of design features into manufacturing features;
- selection of the machining processes based on manufacturing features, considering constraints related to dimensions, tolerances, material, etc.;
- selection of the machining sequence with alternatives and setups for fixture;
- selection of the cutting tools, considering the machine tool, material of the part, toll characteristics (material, dimensions, geometry, tool life, etc.);
- calculation of the time – cost model;
- creation of the process plan documents including alternative plans and time - cost estimates.

Besides the above activities, in some instances (e.g. [10]) the extended CAPP system or the integrated CAPP/CAM system performs also the following:

- selection of the strategies for tools paths creation based on STEP-NC;
- determination of the technological conditions of optimized machining;
- generation of the NC program (AP238), etc.

The AP224 of STEP standard specifies a set of features for use in process planning. Figure 4 presents a part of the feature classes defined in STEP AP224, for prismatic parts. A manufacturing feature identifies the types of features necessary to manufacture a machined part. In general, manufacturing form features are divided into three categories: (i) machining feature presents volume that should be removed by machining to obtain the final part geometry (e.g. hole, pocket, slot, step); (ii) transition feature presents an area between two surfaces, which include chamfer, edge round and fillet.; (iii) replicate feature is the arrangement of identical copies of the base shape, or a pattern of features, where the same features in terms of size and sort are arranged on the final product in circular, rectangular or general pattern [17, 19, 20]. The above features definition according to STEP AP224 is shown in Table 1.

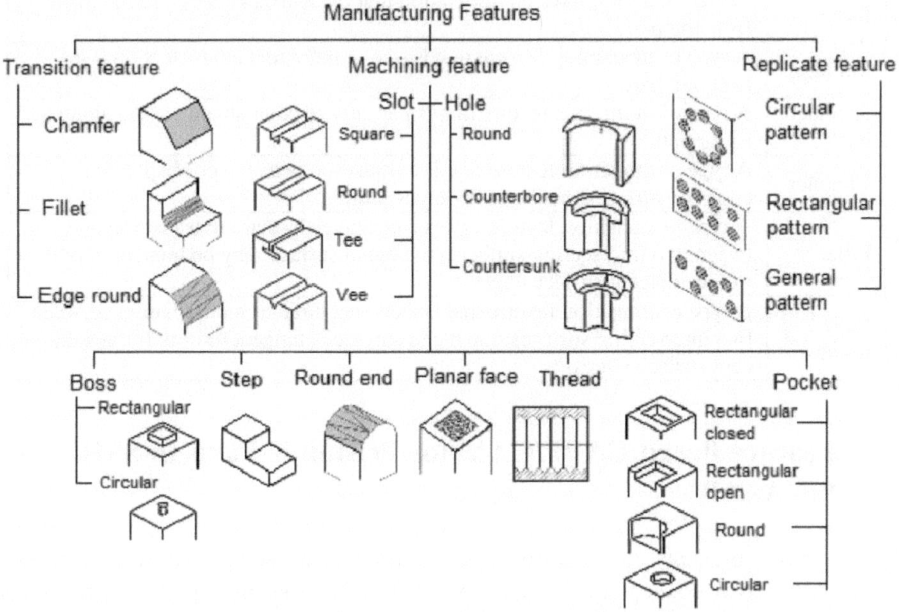

Fig. 4. A part of form feature categorisation in STEP AP224 for prismatic parts [20]

Table 1. STEP AP224 feature definitions for prismatic parts [19, 20]

Feature	Definition/explanation
Hole	A type of multi-axis feature that is the removal of a cylindrical volume from a part (a counterbore, countersunk, or a round hole).
Pocket	A type of multi-axis feature that is a volume with a specific shape, removed from the part; the sides of a pocket may be parallel to the pocket's orientation vector coming out of the pocket or the sides may be tapered (a general, rectangular closed, or a rectangular open pocket).
Slot	A type of multi-axis feature that is a channel or depression with continuous direction of travel (a square, round, tee, or a vee slot).
Boss	A type of multi-axis feature that is a closed shape that protrudes from the surface of the part; the intersection of the boss and the base surface may have a radius shaped blend between them (a circular boss or a rectangular boss).

Table 1. (*continued*)

Step	A type of multi-axis feature that is a linear sweep of a shape; the shape shall be specified by two lines that connect at a point and extend infinitely; the enclosed angle shall be smaller than 180 degrees; the intersection of the two lines need not be blended with a radius.
Thread	A type of machining feature that is a ridge of uniform section on the form of a helix on the external or internal surface of a cylinder.
Planar face	A type of multi-axis feature that is an unbounded planar cut of a part; the Planar face shall have an orientation that the Z-axis is the direction away from the part.
Marking	A type of machining feature that is one or more text characters on a surface of a part.
Rounded end	A type of multi-axis feature that is partially circular shape passed along a linear path.
Chamfer	A type of transition feature that is a transition between corresponding edges of surfaces, having a flat cross section.
Fillet	A type of transition feature that is a concave circular arc transition between two intersecting surfaces; the blend surface may be tangent to both of the adjacent surface edges.
Edge round	A type of transition feature that is a convex circular arc transition between two intersecting surfaces; the blend surface is tangent to both of the adjacent surface edges.

3 Feature Based CAPP/CAM for Prismatic Parts in SME "IVA-28"

This section presents a feature based production setup, or the integrated CAD/CAPP/CAM concept in Serbian SME "IVA-28" that produces precise, mainly prismatic parts for machine building industry. The STEP-based production scenario is as follows:

- The product specification is received from the customer in a form of 3D CAD model or STEP file. In case of 3D CAD model, the corresponding STEP file is generated from the CAD model.
- CAM technology is prepared base on the product design, including the following steps:
  - the product geometry is divided into the predefined feature;
  - all known features are imported from the features knowledge database in order to connect the product geometry with the existing features from the database;
  - if there are no existing features in the database that correspond the product geometry, new features are generated using pre-prepared tools and added to the database.
- Then the program post-processing is performed for different machine controllers included in the macro process planning, in order to generate the corresponding NC codes. The generated NC codes are stores on the server.

- Machines access the corresponding NC codes store on the server and perform machining operations.
- After machining, the automatic measurement is performed on the 3D measuring machine using the above STEP file.

The approach used in "IVA-28" is the features recognition where the part geometry/topology is examined and matched with the predefined feature from the knowledge base, in contrast to the design by feature approach where the part is formed from the predefined features with their attributes.

It could be notice that the whole system is organized as a direct numerical control (DNC), where CNC machines are networked and their operations are controlled from the central computer. Each machine has its own IP address which allows them to behave as a computer. Some of the direct benefits of this STEP-based manufacturing concept in "IVA-28" are: company-developed standards for machining particular forms/ features; independence off CNC machines; proven strategies for machining the most important features, assuring high quality of final products; very fast and risk free setup of new products in case of known features and known materials; price, cost and capacity predictability.

Figure 5 presents the feature based CAD model of a typical product – a slide used for various kind of machine tools (for x, y and z axis kinematics). The features base is organized as a knowledge base, with systemized groups of elements and corresponding solutions. The set of geometric features (shapes) forms one functional whole. The corresponding group of machining operations is applied over the observed set of geometric features, and this presents an independent part of a program. For each set of geometric features and product material, there is a predefined sequence of machining operations together with the optimal machining parameter values, corresponding tool geometry and material, etc.

The above shown product (slider) consists of two groups of features designated as Feature A and Feature B, which are stored in features knowledge based together with other features. The corresponding CAM technology is generated first by loading the set Feature B and set Feature A from the knowledge base, and then connecting Feature B and Feature A with the product geometry. The set Feature B contains machining operations sequence, process parameters, tool data and other machining-related information. The same applies for the set Feature A. Figures 6.a) and 6.b) show generation of CAPP/CAM program for the set Feature B and set Feature A, respectively.

For the existing program it is possible to change or reselect the machine, tool library (e.g. select the library for some other material of a product), or machining operations. Each machine is presented by its postprocessor. The machining operations library contains systematized group of products, where each group could contain almost infinite number of functional group of operations that correspond to the functional wholes of a product. Figure 7-left presents the selection of machine group properties, namely the selection of a postprocessor/machine and related tool and operation libraries. The selection of tool-related parameters is presented in Figure 7-right, with two options: to load tools from the library into the program with the pre-defined regime parameters, or to automatically re-calculated the tool-related parameters with respect to the selected material of a part.

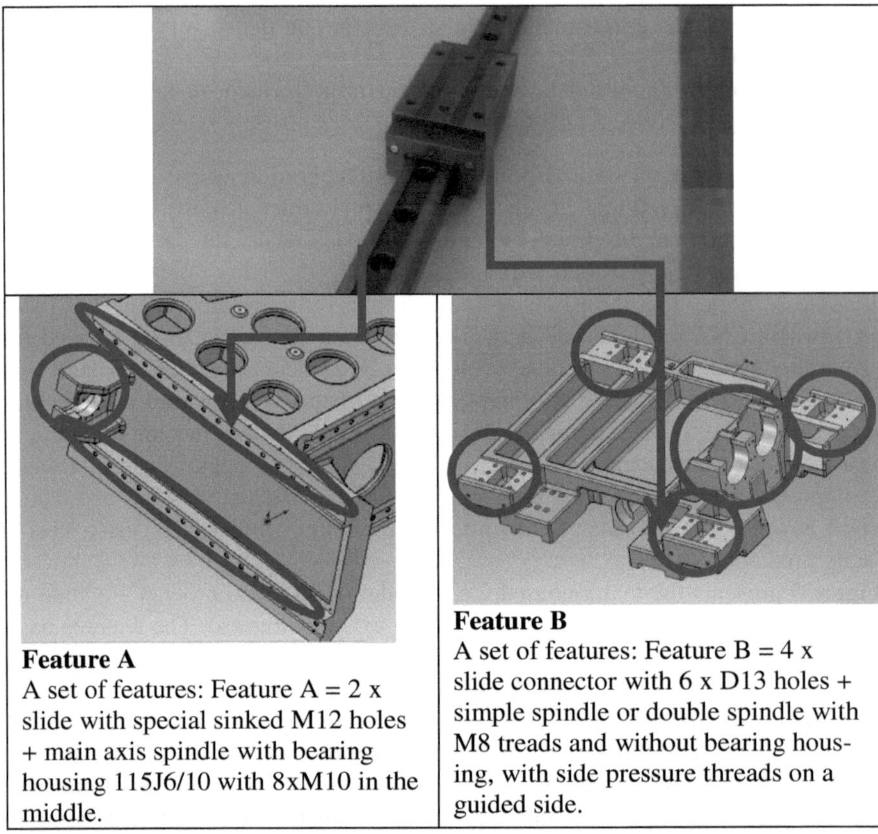

Fig. 5. A real product (slide) – top, and CAD model with typical features - down

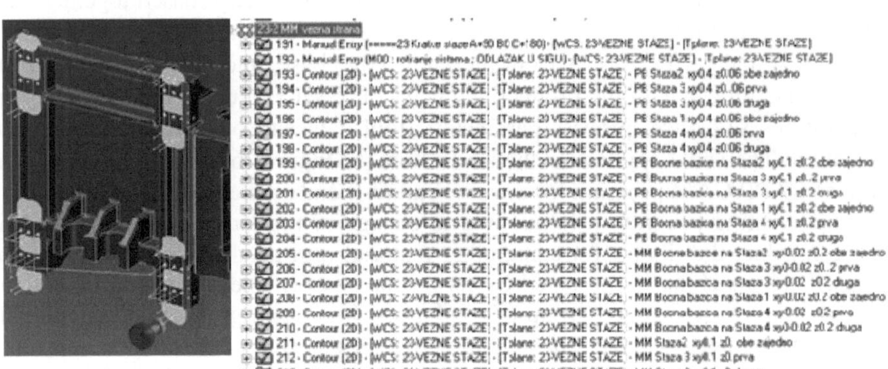

Fig. 6. a) Generation of CAPP/CAM program for the set Feature B

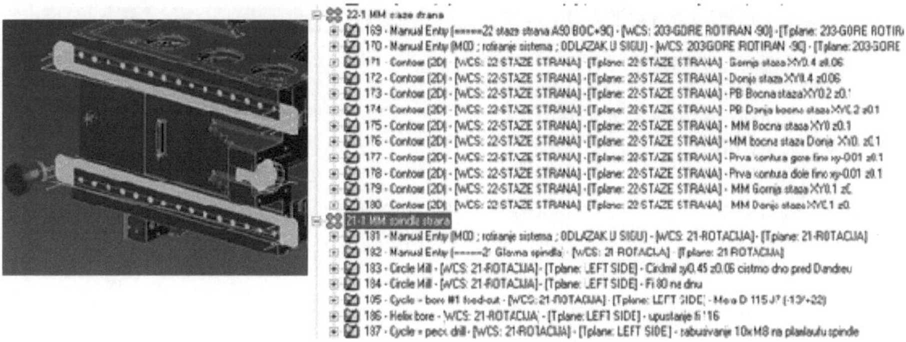

Fig. 7. b) Generation of CAPP/CAM program for the set Feature A

Hence, in case of a new material of a part, all regime parameters must be re-calculated according to the material selection. Prior to this, it is necessary to define how to calculate the regime parameters for each material. This is performed according to the table for material definition (Figure 8).

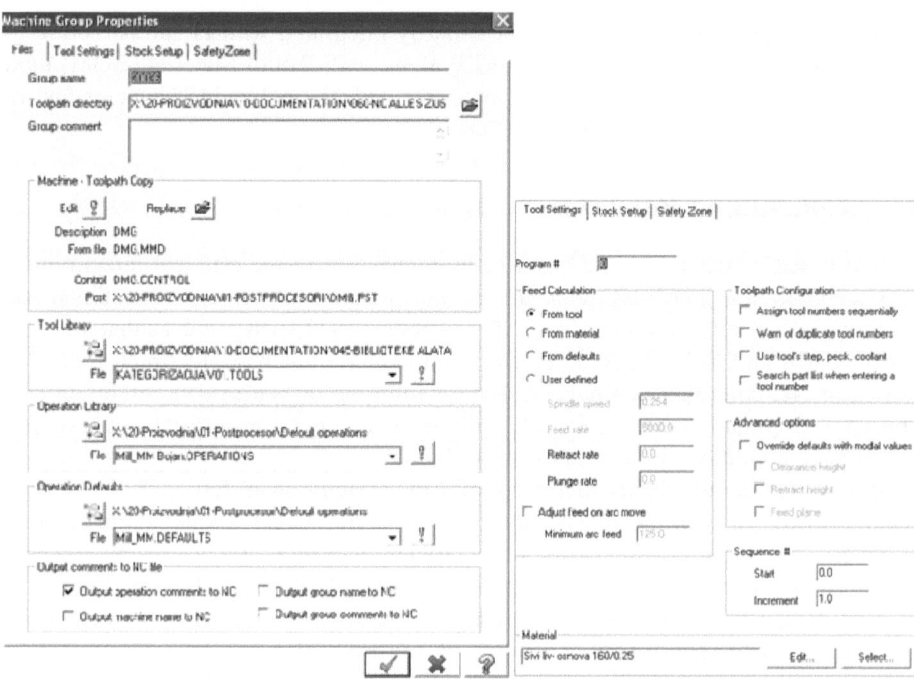

Fig. 8. The selection of machine group properties (machines, tools, and operations) – left, and tool settings – right

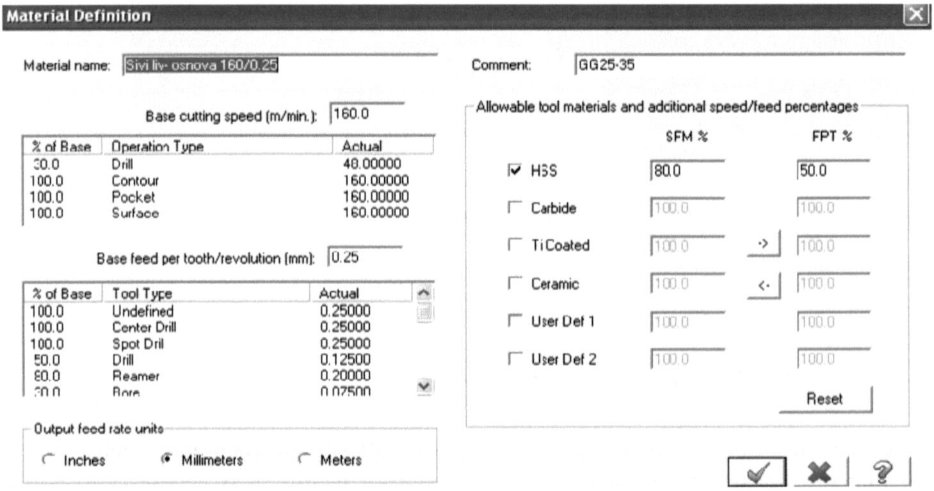

Fig. 9. The material definition: the adjustment coefficients for regime parameters (with respect to the operation and tool type) for the selected material of a part

The above process planning concept is partly integrated with IT infrastructure for manufacturing management: MES and ERP along with Bill of Material (BOM) module, where planning process considers products with the belonging elements and resources (e.g. machines with belonging elements).

4 Concluding Remarks

Feature based model is assumed as a main contributor to the efficient integration of CAx technologies. STEP-NC standard allows for more effective, bi-directional data exchange between CAx systems and NC controllers. STEP-NC enables that all CAD, CAPP, CAM and NC data could be stored in a single database, to which each CAx system or NC controller can have access for information retrieval or alteration [11]. This is of essential importance for full realisation of a digital manufacturing concept.

Dynamic conditions in a global supply chain, as well as in an enterprise itself, lay down complex requirements in front of CAPP systems in a digital, STEP - based manufacturing. In such dynamicity, the main tasks that CAPP should simultaneously respond to are: identification of the most efficient and effective machining sequence (including selection of machining operations, machines, tools, fixtures, and other relevant data) and alternatives, considering all possible constrains in the actual production; automated and timely generation of a process plan based on design features of a part/product (from CAD system); generation of the optimal process plan based on the time-cost criteria. The output of such CAPP system is a STEP-NC process plan, that could be directly used by CAM system to control the CNC system and perform the actual production operations. One of the main challenges is also the complexity of today's products, so future activities should include the enrichment of a feature database.

The future research direction for the improvement of the existing IVA-28 digital manufacturing concept implies the following: (i) the unambiguous definition and further development of STEP manufacturing features, and their full implementation in the practice; (ii) the design of a feature recognition model, the selection of machines, tolls, machining operations and optimal machining parameters, and the generation of process plans using artificial intelligence techniques.

References

1. Ad-hoc Industrial Advisory Group, Factories of Future PPP: Strategic Multi-annual Roadmap (2010), http://ec.europa.eu/research/industrial_technologies/pdf/ppp-factories-of-the-future-strategic-multiannual-roadmap-info-day_en.pdf (accessed February 2013)
2. Šibalija, T.: Factory of the Future PPP Strategic Multi-Annual Roadmap 2013. Key speaker, Invited paper, Factories of the Future for Thailand 2013 Conference, Bangkok (January 2013)
3. Majstorović, V.: Towards a Digital Factory - Research in the World and our Country. Journal of Applied Engineering Science 10(3), 161–165 (2012)
4. Xiong, Y.-L., Yin, Z.-P.: Digital manufacturing—the development direction of the manufacturing technology in the 21st century. Front. Mech. Eng. China 2, 125–130 (2006)
5. Yongtao, H., Jingying, M.: A knowledge-based auto-reasoning methodology in hole-machining process planning. Computers in Industry 57, 297–304 (2006)
6. http://www.iso.org (accessed February 2013)
7. Ming, X.G., Mak, K.L.: Intelligent setup planning in manufacturing by neural networks based approach. J. Intell. Manuf. 11, 311–331 (2000)
8. Capponi, V., Zirmi, O., Brissaud, D., Villeneuve, F.: Computer aided process planning, strategy, and models in the aircraft industry. Proc. IMechE 220(pt. B), 541 (2006)
9. Deja, M., Siemiatkowski, M.: Feature-based generation of machining process plans for optimised parts manufacture. J. Intell. Manuf. (2012), doi:10.1007/s10845-012-0633-x
10. Álvares, A.J., Ferreira, J.C.E.: WebMachining: Implementation of a Collaborative CAD/CAPP/CAM System for e-Manufacturing Throught the Internet. In: 38th CIRP – International Seminar on Manufacturing Systems, Florianopolis, SC (2005)
11. Xu, X.W., He, Q.: Striving for a total integration of CAD, CAPP, CAM and CNC. Robotics and Computer-Integrated Manufacturing 20, 101–109 (2004)
12. Campos, J.G., Miguez, L.R.: Standard process monitoring and traceability programming in collaborative CAD/CAM/CNC manufacturing scenarios. Computers in Industry 62, 311–322 (2011)
13. Sharma, R., Gao, J.X.: Implementation of STEP application protocol 224 in an automated manufacturing planning system. Proc. IMechE 216(9, pt. B), 1277 (2002)
14. Amaitik, S., Kiliç, E.: An intelligent process planning system for prismatic parts using STEP features. Int. J. Adv. Manuf. Technol. 31, 978–993 (2007)
15. Collazos, J.G., Álvares, A.J.: Methodology for integrating CAD/CAPP/CAM aiming for manufacturing of prismatic parts adherent to the STEP-NC. In: 21st Brazilian Congress of Mechanical Engineering, Natal, RN, Brazil (October 2011)

16. Nassehi, A., Newman, S.T., Allen, R.D.: The application of multi-agent systems for STEP-NC computer aided process planning of prismatic components. Int. J. Mach. Tools Manufact. 46, 559–574 (2006)
17. Hou, M.: PhD Dissertation: CAD/CAM Integration Based on Machining Features for Prismatic Parts, Department of Mechanical Engineering, University of Kansas (2008)
18. Sun, X., Chu, X., Xue, D., Su, Y., Tang, C.: An integrated setup/fixture planning approach for machining prismatic parts. Int. J. Prod. Res. 50, 1009–1027 (2012)
19. `http://www.steptools.com/library/stepnc/` (accessed February 2013)
20. Amaitik, S.: Development of a STEP Feature-based Intelligent Process Planning System for Prismatic Parts, PhD Dissertation, Middle East Technical University (2005)

Product Quality Inspection Combining with Structure Light System, Data Mining and RFID Technology

Kesheng Wang and Quan Yu

Knowledge Discovery Laboratory, Norwegian University of Science and Technology, Norway
{kesheng.wang,quan.yu}@ntnu.no

Abstract. 3D vision inspection has been rapidly developed and increasingly applied in product quality inspection. Structure Light System (SLS) is one of these technologies with good performance and lower cost. To achieve the automated quality inspection, a decision support system is required to provide with the SLS together to automatically perform the quality check. However, for extracting quality information, the quality inspection system has to be combined together with a data logging system, so that the quality information of the product is available for production query and quality checking. In this paper, Radio Frequency Identification (RFID) technique is used to accomplish the production tracing and tracking together with the quality inspection system. By assigning a RFID tag to each inspected part, it is possible to identify the type of the product and write the quality inspection result decided by the data mining classifier for the real-time quality query. The proposed approach will be an alternative for SMEs considering the fast product type update with respect to the fast changing market.

Keywords: Quality inspection, Structure Light System, Data mining, RFID.

1 Introduction

Product quality inspection is a nontrivial procedure during the manufacturing process both for the semi-finished and end products. 3D vision inspection has been rapidly developed and increasingly applied in product quality inspection, with the advantages of high precision and applicability comparing with commonly used 2D vision approaches. 3D vision is superior in inspection of multi-features parts due to providing the height information. 3D vision techniques comprise approaches on the basis of different working principles [1]. Among various approaches, the Structure Light System (SLS) is a kind of cost effective techniques for the industrial production [2, 3, 4]. By projecting specific patterns on the inspected product, the camera captures corresponding images. The 3D measurement information of the product is retrieved in the form of the point cloud on the basis of the images. With the generated 3D point cloud, automated quality inspection can be performed with less human interference, where data mining approach is commonly used [5, 6].

However, although the quality information is able to be decided on the basis of the SLS and data mining approaches, it will not become more valuable until it is used to

G.L. Kovács and D. Kochan (Eds.): NEW PROLAMAT 2013, IFIP AICT 411, pp. 205–220, 2013.

improve the production process and also achieve the real-time data access and the quality traceability.

In this paper, Radio Frequency Identification (RFID) technique is used to integrate the quality information together with the product. RFID uses a wireless non-contact radio system to identify objects and transfer data from tags attached to movable items to readers, which is fast, reliable, and does not require physical sight or contact between reader/scanner and the tagged objects. By assigning a RFID tag to each inspected product, it is possible to identify the product type and query the quality inspection history. An assembly quality inspection problem is selected as a case study to test the feasibility of proposed complex system. The proposed approach will be an alternative for SMEs considering the fast product type update with respect to the fast changing market.

The paper is organized as the following: Section 1 introduces the general applications of 3D vision in manufacturing and the importance of combining SLS, Data mining and RFID technology. Section 2 introduces the architecture of the complex system and the working process in each system level respectively. Section 3 shows a case to study the feasibility of the system combining the three techniques. Section 4 comes to the conclusion of the applicability of the complex system.

2 Quality Inspection System Architecture on the Basis of Structure Light System, Data Mining and RFID

The RFID 3D quality inspection system combines the function of product quality inspection and RFID tracing and tracking. By attaching the RFID tag to the product, the system takes the pictures of the inspected product as the inputs, generates the 3D point cloud and finally writes the quality related information of the product in the RFID tag. Thus, it is available to monitor the product quality along the production line and achieve the real-time quality control.

2.1 System Architecture

The quality inspection system comprises 4 levels as shown in Figure 1, which are 3D vision level, data processing level, computational intelligence level and RFID level respectively. Within each level of the system, the data is converted as the sequence of the point cloud, the feature vector, the quality information and the writable RFID data. Each level is introduced as following:

1. 3D vision level consists of the Structure Light System, which uses camera together with the projector to generate the point cloud of the inspected product.
2. Data processing level comprises quality related feature determination and extraction. The product quality is quantified on the basis of the point cloud according to the design requirements. The feature vector is generated after the processing as the input of the next level.

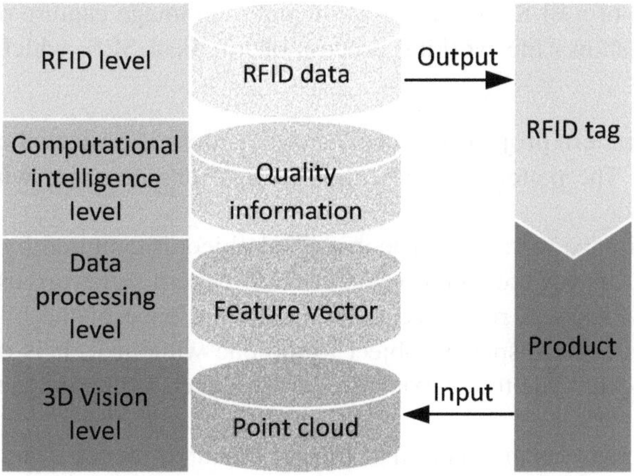

Fig. 1. System architecture

3. Computational intelligence level uses data mining approaches to achieve the automated quality classification on the basis of the feature vector.
4. RFID level comprises the RFID hardware and software, which achieves the product tracking and controlling by writing and reading RFID tag attached on the product.

2.2 Introduction of the Structure Light System

Structured Light System (SLS) is one of typical 3D vision techniques. SLS accomplishes to point cloud acquisition by projecting specific patterns onto the measured object and capturing the corresponding images. The point cloud of the object surface can be generated with the images analysis approach.

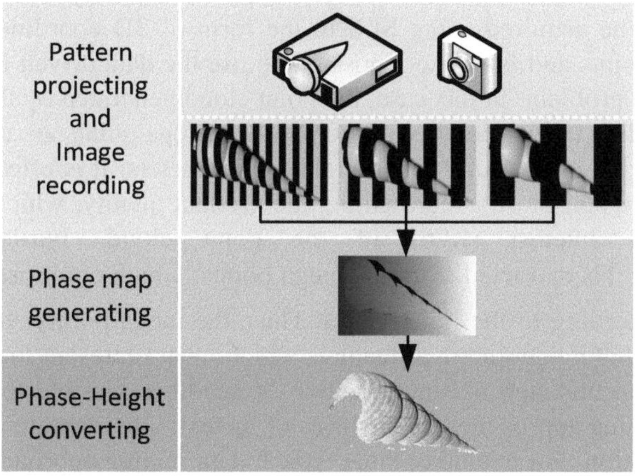

Fig. 2. General working process of a Structured Light System

The hardware of a SLS consists of a computer, an image capture device and a projector. Figure 2 shows the working process of a typical SLS, which can be divided into 4 steps.

1. Step 1 is the pattern projection. A coded light pattern is projected onto the scene by the projector. The pattern can be either a single one or series with respect to the type of the code.
2. Step 2 is the image recording. The inspected object is captured by the camera, and then captured images are stored in sequence if pattern series are used. The scene is captured previously as references without the presence of the object. Comparing the images with the inspected object to the one without it, it is observed that the pattern is distorted due to the existence of the object, which indicates the height information.
3. Step 3 is the phase map generating. Images captured in step 2 are analysed by the computer with fringe analysis techniques on the basis of the pattern encoding rule. The wrapped phase maps are obtained firstly, and to be unwrapped to obtain the map with continuous phase distribution.
4. Step 4 is the transformation from phase to height. The height value of each image pixel is derived from its phase by phase calibration or phase-height mapping comparing with the reference obtained from step 2. After the calibration, the pixels in the image are transformed to points in metric unit, and the height value in each pixel is calculated, so that the 3D point cloud of the inspected object is formed.

2.3 Decision Support Using Data Mining Approaches

To achieve the automated product quality inspection, it is of importance to use computational intelligence approaches to classify the product quality with less artificial interference. Data mining method is often used to analyze the massive data and find out the knowledge to ascertain the product quality [7]. Data mining based quality inspection requires input variables related to product quality. Although the point cloud is able to be acquired using SLS in the form of 3D coordinates, it generally includes mega points and is unlikely possible to use the data driven methods to solve the classification problem. In this case, the point cloud generated by the SLS has to be processed according to the specifics of the product. Mega points are converted to be a vector including the most useful quality related parameters. It is effecient to focus on the partial point cloud mostly representing the product quality, which can be seen as the Region of Interest (ROI) of the point cloud. Furtherly, a vector $X = \{x_1, x_2, \ldots, x_n\}$ is extracted from the mega points, which comprises feature values xi calculated according to the point cloud. Thus, the mega points are converted to a single vector including geometrical features which most represent the product quality information. With this step of simplification, it becomes feasible to select the most suitable data mining approaches on the basis of the extracted feature vectors to do the quality classification. For example, three typical data mining approaches are common used in the classification problem, which are Artificial Neural Networks (ANN), Decision tree and Support Vector Machines (SVM) respectively.

1. Decision Tree

A decision tree is one of data mining approaches applied in many real world applications as a solution to classification problems. A decision tree is a flowchart-like tree structure, where each internal node denotes a test on an attribute, each branch represents an outcome of the test, and each leaf node holds a class label. The construction of decision tree classifiers does not require any domain knowledge or parameter setting, and therefore is appropriate for exploratory knowledge discovery.

C4.5 is a classic algorithm for decision tree induction. The succedent algorithm C5.0 is available in IBM SPSS Modeler®. By using this software, it is easy to accomplish the decision tree induction and test.

2. Artificial Neural Networks

As another effective data mining approach, an Artificial Neural Network consists of layers and neurons on each of them. Parameters are adjustable in an ANN such as the number of the hidden layers and neurons, the transfer functions between layers and the train method etc. A powerful ANN toolbox is available in Matlab® and can be highly customized to get the best result by the user.

3. Support Vector Machines (SVM)

A Support Vector Machine (SVM) is a supervised learning method for data analysis and pattern recognition. The standard SVM is designed for binary classification. Given a set of training examples, each marked as belonging to one of two categories, several SVM training algorithms are available to build a model that assigns examples into corresponding category. New examples are then predicted to belong to a category based on the constructed model. For multi-classification problem, a commonly used approach is to construct K separate SVMs, in which the kth model yk(x) is trained using the data from class Ck as the positive examples and the data from the remaining K − 1 classes as the negative examples, which is known as the one-versus-the-rest approach.

2.4 RFID Level

Radio Frequency Identification (RFID) is one of numerous technologies grouped under the term of Automatic Identification (Auto ID), such as bar code, magnetic inks, optical character recognition, voice recognition, touch memory, smart cards, biometrics etc. Auto ID technologies are a new way of controlling information and material flow, especially suitable for large production networks [8]. RFID is the use of a wireless non-contact radio system to transfer data from a tag attached to an object, for the purposes of identification and tracking. In general terms, it is a means of identifying a person or object using a radio frequency transmission. The technology can be used to identify, track, sort or detect a wide variety of objects [9]. RFID system can be classified by the working frequency, i.e. Low Frequency (LF), High Frequency (HF), Ultra High Frequency (UHF) and Microwave. Different frequency works for various media, e.g. UHF is not applicable to metal but HF is metal friendly. Thus, the working frequency has to be used on the basis of tracked objects.

Hardware of RFID system includes RFID tag, RFID reader and RFID antenna. RFID tag is an electronic device that can store and transmit data to a reader in a contactless manner using radio waves, which can be read-only or read-write. Tag memory can be factory or field programmed, partitionable, and optionally permanently locked, which enables the users save the customized information in the tag and read it everywhere, or kill the tag when it will not be used anymore. Bytes left unlocked can be rewritten over more than 100,000 times, which achieves a long useful life. Moreover, the tags can be classified by power methods i.e. passive tags without power, semi-passive tags with battery and active tags with battery, processor and i/o ports. The power supply increases the cost of the tag but enhance the readable performance. Furthermore, a middleware is required as a platform for managing acquired RFID data and routing it between tag readers and other enterprise systems. Recently, RFID become more and more interesting technology in many fields such as agriculture, manufacturing and supply chain management.

3 Case Study

In this paper, a wheel assembly problem is proposed as the case study of the implementaion for combining the SLS, data mining approaches and RFID technology. In the first step, the assembly quality classification is introduced. Secondly, the point cloud of the object is acuqired using SLS and converted to be the feature vector, which is defined according to the assembly requirments and provided to the data mining classifier as the input. At last, the quality is decided by the classifier and converted to RFID data, which is saved in the RFID tag attached on the object.

3.1 Problem Description

To certify the feasibility of proposed 3D vision based quality inspection, LEGO® wheel assembly inspection is taken as the example in this paper. In the supposed wheel assembly inspection, the object is to check the assembly quality. A wheel is constituted of 2 components, the rim and the tire. Possible errors occur during the assembly process as shown in Figure 3, which are supposed to be divided into 5 classes, according to the relative position of the rim and tire:

1. Wheel assembly without errors
2. Tire is compressed
3. There exists an offset for one side
4. Rim is detached from the tire
5. Rim is tilted

Fig. 3. Wheel assembly classification

Each class has a corresponding inner layout, which is shown respectively in Figure 4. The section views show the differences among the classes.

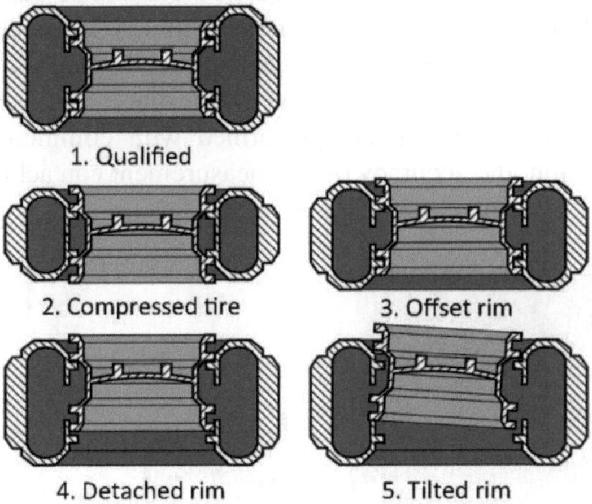

Fig. 4. Inner layout of the wheel

2D vision is not applicable to distinguish some cases because of the similarity of pictures, as shown in Figure 5.

It is noticed that there is not so much difference between the two wheels from the top view in the image. However, the height is different if seen from the side view. Structured Light System (SLS) is an effective solution which can get 3D point cloud of the inspected part, so that the real height value of the parts are obtained while errors can be recognized with the metric information directly.

Fig. 5. Similarity from the top of the view

3.2 Feature Extraction for the Classification

The hardware configuration is shown in Figure 6. The image capture device is used with a SONY XCG-U100E industrial camera, with UXGA resolution (1600×1200 pixels) and Gigabit Ethernet (GigE) interface, and together with a Fujinon 9mm lens. A BenQ MP525 digital projector is employed to project the patterns. The hardware control and the image processing are performed with commercial software Scorpion®. After calibration, the accuracy of the measurement can achieve 0.01mm in this case study.

Fig. 6. SLS hardware

In proposed assembly inspection, geometry features are extracted as the input attributes on the basis of the 3D point cloud and regression analysis, while the corresponding class of the inspected wheel is predefined. 200 examples have been recorded for each class. 180 of them are used as the train set while the left examples are used as the test set.

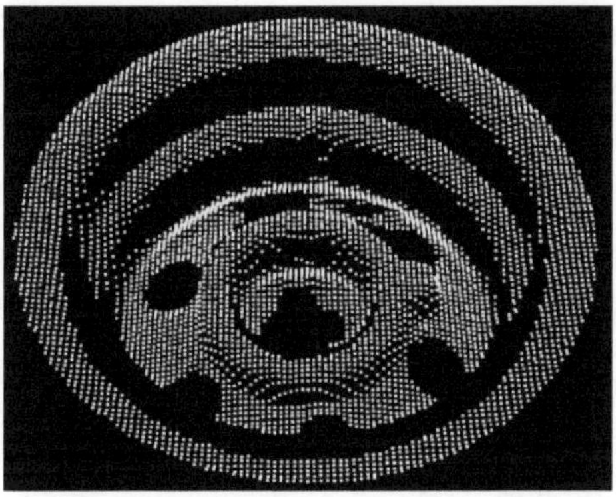

Fig. 7. Point cloud acquired with SLS

Because the tire has much lower reflectivity comparing with the rim, more noises will be generated when acquiring the point cloud of both. To get a cleaner point cloud and simplify the feature extraction, the tire part is disregarded by lowering the contrast value. Thus, only the rim part is kept for the further analysis as shown in Figure 7. 6 line profiles are extracted from the point cloud of the rim part for feature extraction in the next step, with 3 horizontals and 3 verticals. Only the end sections of each line profile are kept for the regression analysis

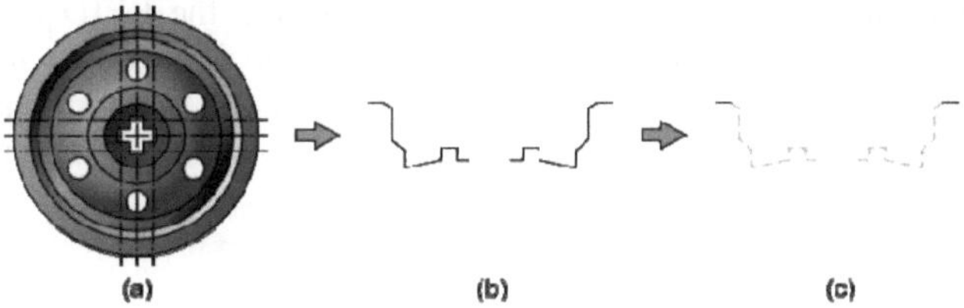

Fig. 8. Section profile processing

As shown in Figure 8, two sections of profile are used to define the attributes of the rim. Only the beginning and ending section of the profile are kept while other sections are ignored. The horizontal data of the points are considered as the predictors, which correspond to the x-axis and y-axis in terms of the horizontal and vertical profiles respectively, while the vertical data are considered as the responses, which means the height value of the points. It is possible to use statistics to estimate the regression line of the values corresponding to the profile section [10]. Taking into account the slope of the regression line, it is possible to estimate the height and the pose of the rim part.

The least squares linear regression is used to acquire the regression line. With the numerical data of a profile and the regression line corresponding to a wheel, 5 values are obtained, which are respectively

x1: slope of the regression line
x2: bias of the regression line
x3: mean height value of points in the beginning section
x4: mean height value of points in the ending section
x5: width between the starting point and the ending point

where x1 and x2 are acquired by linear regression analysis while x3, x4 and x5 are measured from the points of the profile sections directly shown in Figure 9. With all the extracted values, a data vector is constructed for each of the six profiles of the rim:

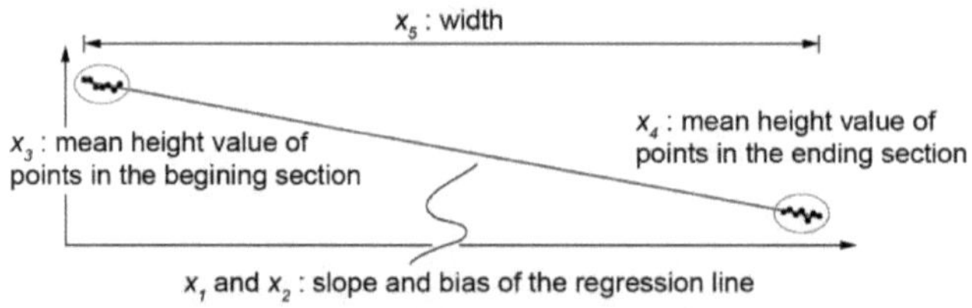

Fig. 9. Feature extraction

Regarding the 5 classes in the wheel assembly problem, it is of importance to extract features most related to the assembly status from the point cloud. These 5 features extracted above are able to describe the pose of a wheel, which indicates the height and inclination. For each profile, 5 feature values are extracted. Thus, a vector XS is used to denote a wheel where $X_S = \{X_1, X_2, \ldots, X_6\}$ as the inputs of data mining approaches.

3.3 Quality Information Embeded Using RFID Tag

After the assembly quality has been approved using SLS and data mining decision support system. The quality information will be written into the RFID tag placed in the wheel. Thus, the quality information is combined with the product together for the check in the future. In this wheel assembly quality inspection problem, the tag accessing is to be completed with the Reader Test Tool of the RFID reader as shown in Figure 10.

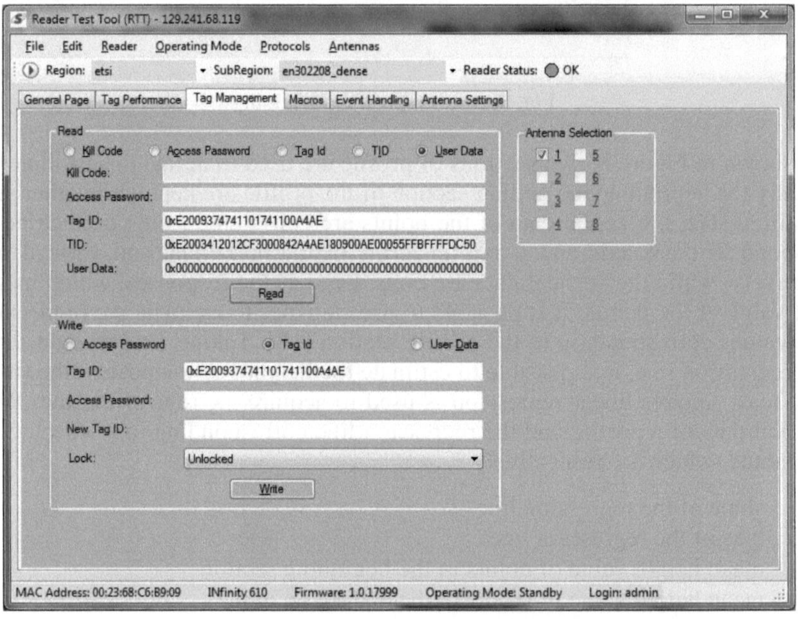

Fig. 10. RFID reader test tool

In this case study, the OMNI-ID RFID tag, the SIRIT RFID reader and the IMPINJ near-field antenna are used to construct the RFID system, as shown in Figure 11.

Fig. 11. RFID system hardware

During the test, a RFID tag is placed in the tire in the first step as shown in Figure 12.

Fig. 12. Attach a RFID tag to a tire

Then the tagged tire and a rim are assembled. Because the EPC code of each RFID tag is set to be unique, each wheel is given by a unique . Finally the assembly quality of the wheel is to be inspected using SLS and data mining based decision support system. After the classification, the quality information will be written into the tag and kept with the product as shown in Figure 13.

Fig. 13. Quality inspection and information writing

The memory of the OMNI-ID RFID tag is divided into three parts, which are allocated for the EPC code of 96bits, the user data of 512bits and the TID data of 64bits. The 512bits user data are reserved for the customized information, which can be converted to 64 ASCII characters. In this case, the quality inspection related information is written in the user memory of the RFID tag. The information comprises the classified assembly quality, the inspection time and the inspection date, in the form of “QUALITY=X TIME=HH:MM DATE=DD.MM.YY”. Because the tag memory is written in the form of hexadecimal, the texts have to be converted to HEX before writting. Supposing the quality is classified to be 1 and the inspection is taken at 14:05 on 30.04.13, the information is converted from the ASCII characters to hexadecimal as shown in Figure 14.

Text	QUALITY=1 TIME=14:05 DATE=30.04.13
HEX	0x5155414C4954593D312054494D453D3134 3A303520444154453D33302E30342E3133

Fig. 14. Conversion from the ASCII characters to the Hexadecimal

The information writing is completed using reader test tool as shown in Figure 15.

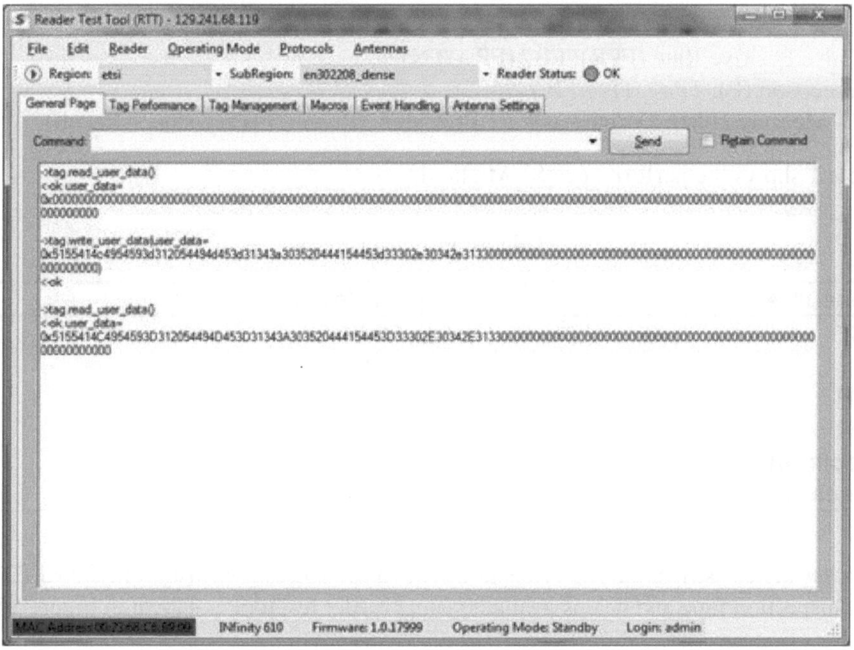

Fig. 15. RFID tag information control

After the quality information is written in the user memory of the tag, it is able to be read out with any other RFID reader. Using the same decoding approach, the hexadecimal can be restored to characters again for later check.

3.4 Result and Analysis

During the quality inspection test. Six measures are obtained for the classifier on the basis of 4 outcomes: True Positive (TP), True Negative (TN), False Positive (FP) and False Negative (FN), which constructs a confusion matrix

TP (correctly accepted)	FN (incorrectly refused)
FP (incorrectly accepted)	TN (correctly refused)

For each validation, the confusion matrixes of each class are constructed respectively, and the final confusion matrix for each validation would contain the average values for all classes combined.

The 6 measures are defined as following

1. Correctly Classified Instances (CCI): Percentage of samples correctly classified.
2. Incorrectly Classified Instances (ICI): 100% -CCI.

3. True Positive Rate (TPR): TP / (TP + FN). Also called Recall measure.
4. False Positive Rate (FPR): FP / (FP + TN)
5. Precision (P): TP / (TP + FP)
6. F- Measure (FM): Defined as (2 * Precision *Recall) / (Precision + Recall)

Table 1 shows the performance SVM classifiers.

Table 1. Results of SVM

Measures	Support Vector Machine
CCI	95.80 %
ICI	4.20 %
TPR	0.853
FPR	0.011
Precision	0.958
F-measure	0.902

After the classification, the inspection related information is written in the RFID tag. The inspection time and date is available using reader test tool as shown in Figure 16.

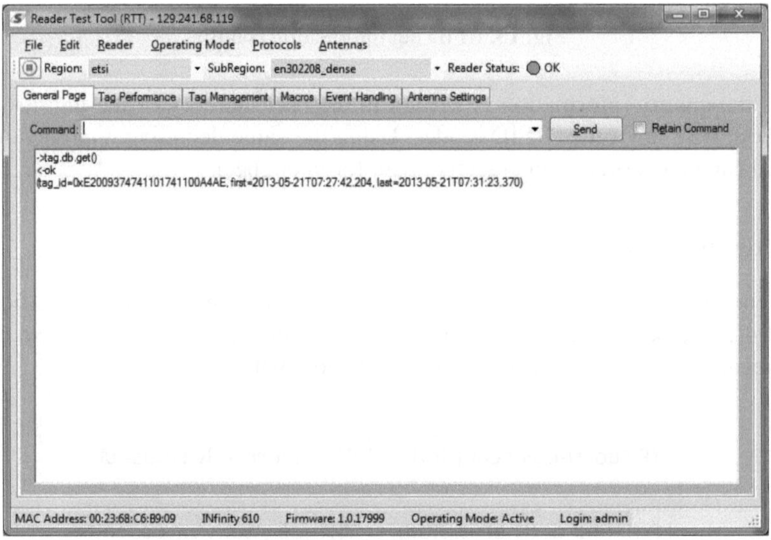

Fig. 16. Acquiring the RFID tag information

Combining with the quality classfication result, the text to be written is

QUALITY=1 TIME=07:27 DATE=21.05.13

Which is converted to be HEX as following using compiler

0x5155414C4954593D312054494D453D30373A323720444154453D32312E3035 2E3133

The converted HEX is written in the user memory of the tag in command line of the reader test tool as shown in Figure 17.

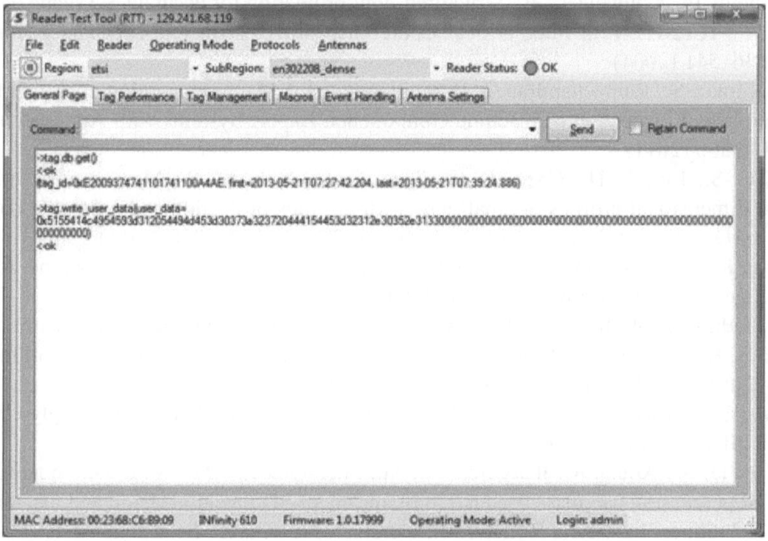

Fig. 17. Information writting using RFID system

4 Conclusions

The combination of Structured Light System, the data mining approach and RFID technology is tested in this paper. SLS is applicable in this proposed wheel assembly quality classfication problem. According to the assembly requirements, the feature definition on the basis of the point cloud is suitable for similar product type. The feature vector extraction provides the SVM classifier availalbe inputs and ahecieves 95.8% Correctly Classified Instances. Meanwhile, the RIFD system successfully converts the quality inspection result to acceptable data format for the tag and writes the information in it. This step is able to improve the traceablity of the product quality. Suppose multiple SLS inspection stations are assigned along the assembly line, the quality inspection results are saved in the RFID tag respectively. The earlier inspection result is available for the system before the product enters into the following processing station. The tag embeded information does not require the remote database access. The interuption due to the product quality is able to be avoided. In the following work, the middleware for integrating the three system will be developed.

References

1. Barbero, B.R., Ureta, E.S.: Comparative study of different digitization techniques and their accuracy. Computer-Aided Design 43(2), 188–206 (2011)
2. Pernkopf, F.: 3D surface acquisition and reconstruction for inspection of raw steel products. Computers in Industry 56(8-9), 876–885 (2005)

3. Xu, J., Xi, N., Zhang, C., Shi, Q., Gregory, J.: Real-time 3D shape inspection system of automotive parts based on structured light pattern. Optics & Laser Technology 43(1), 1–8 (2011)
4. Skotheim, Ø., Couweleers, F.: Structured light projection for accurate 3D shape determination. In: ICEM12-12th International Conference on Experimental Mechanics, Bari, Italy, pp. 536–541 (2004)
5. Ravikumar, S., Ramachandran, K.I., Sugumaran, V.: Machine learning approach for automated visual inspection of machine components. Expert Systems with Applications 38(4), 3260–3266 (2011)
6. Lin, C.-S., Lin, C.-H., Chen, D.-C., Tien, C.-L., Yeh, M.-S.: Measurement method of three-dimensional profiles of small lens with gratings projection and a flexible compensation system. Expert Systems with Applications 38(5), 6232–6238 (2011)
7. Wang, K.: Applying data mining to manufacturing: the nature and implications. Journal of Intelligent Manufacturing 18(4), 487–495 (2007)
8. Elisabeth, I., Zsolt, K., Péter, E., László, M.: The RFID Technology and Its Current Applications. In: Proceeding of the Modern Information Technology in the Innovation Process of the Industrial Enterprise-MITIP, pp. 29–36 (2006)
9. Lewis, S.: A Basic Introduction to RFID technology and its Use in the Supply Chain. Laran RFID, White Paper (2004)
10. Tellaeche, A., Arana, R., Ibarguren, A., Martínez-Otzeta, J.M.: Automatic quality inspection of percussion cap mass production by means of 3D machine vision and machine learning techniques. In: Graña Romay, M., Corchado, E., Garcia Sebastian, M.T. (eds.) HAIS 2010, Part I. LNCS, vol. 6076, pp. 270–277. Springer, Heidelberg (2010)

Additional Methods to Analyze Computer Tomography Data for Medical Purposes and Generatively Produced Technical Components

Philipp Sembdner, Stefan Holtzhausen, Christine Schöne, and Ralph Stelzer

TU Dresden, Faculty of Mechanical Science and Engineering
christine.schoene@tu-dresden.de

Abstract. The result of a computer tomography (CT) record is an image layer stack, which can be applied for medical diagnostic purposes, as well for virtual 3D object representation (e.g. for a skull bone) depending on the used threshold values. Based on these representations, medical 3D objects could be easily produced by generative manufacturing processes. The additive manufacturing of individual implants is also based on the use of CT data for the representation of the remaining bone. This knowledge was further used for the evaluation of industrial CT data.

For technical applications, the industrial computer tomography becomes more and more important. Compared with devices from the medical area, industrial CT devices have a significantly higher radiation power and a much higher resolution. Thus, these data are not only suitable for the reconstruction of a 3D model in the sense of reverse engineering, but also for the inspection of components of metal workpieces (density differences, cavities, joint connection). The known CT evaluation procedures from medical use have been qualified and further modified in the research group. The developed software solution allows the generation of free-form sections through a layer image stack e.g. to evaluate the quality of a soldered connection in a tube. This procedure is also suitable to detect and determine density differences and the detachment of individual layers of a generative produced component. The topic is illustrated by selected practical examples.

Keywords: Computer tomography, Reverse Engineering, 3D-Inspection.

1 Introduction

In the context of industrial manufacturing and assembly, continuous quality analysis is absolutely necessary to guarantee the production guidelines predefined in an operation. Optical 3D measuring systems for contactless measurement of component geometries and surfaces are increasingly being applied in the production process. These measuring techniques are also being used more and more often in combination with automated manufacturing supervision processes to maintain consistently high standards of quality [1].

G.L. Kovács and D. Kochan (Eds.): NEW PROLAMAT 2013, IFIP AICT 411, pp. 221–229, 2013.

One disadvantage of such systems is that it is only possible to inspect visible regions of the manufactured object. It is impossible to check inner areas of components or joints, such as welded, soldered or adhesive joints, by means of these nondestructive measuring techniques. Here, it makes sense to inspect the formation of blowholes or inclusions in pre-series manufacturing to optimise the production process or to safeguard the quality standards during series manufacturing [2].

Computer tomography (CT) is an imaging technology that provides a proven solution to this problem. State of the art in the medical environment, this technology has become more and more established in other technological fields. However, in mechanical engineering, we are faced with other requirements that must be fulfilled by the procedure, both in terms of the definition of the measuring task and strategy and with consideration of the issue of measuring uncertainty.

Because high accuracy is needed, micro CT systems are frequently used, resulting in huge data volumes in the form of high-resolution slice images. However, increases in the capacities of computer systems in recent years make image analysis, as well as 3D modelling, based on these slices images a promising technology. Consequently, it is necessary to develop efficient analysis strategies for data gathered by means of imaging techniques to find new strategies for quality assurance and process optimisation.

a) Model reconstruction and – processing | b) Execution of operation planning | c) Reverse engineering for design

Fig. 1. Application of medical CT for operation planning [4]

The Reverse Engineering team at the Chair of Engineering Design and CAD of the Dresden University of Technology has been studying the analysis and screening of CT data, at first mainly from medicine [3, 4], for several years. An example is given in Fig. 1, in which a discrete 3D model is generated from CT data. In this process, the calculation of the iso-surfaces is performed by means of the Marching Cubes Algorithm [5]. In the next step, the segmented model of the lower jaw bone is used for operation planning and the design of an individual implant for the patient.

Due to industry demand, processing of CT image data in the technical realm is becoming more and more important. The paper elucidates opportunities for component investigation from CT data by means of efficient image processing strategies and methods using the example of a soldered tube joint.

2 Fundamentals

Computer tomography (CT) is an imaging technique. As a result of its application, we obtain stacks of slice images. As a rule, these data are available in the DICOM format, which has been established in medical applications. Apart from the intrinsic image data, it includes a file head, incorporating the most essential information about the generated image. Relevant information here includes patient data, image position and size, pixel or voxel distance and colour intensity. In the industrial realm, the images are frequently saved as raw data (RAW) or in a standardised image format, such as TIFF. For ongoing processing of the image data in the context of the three-dimensional object, additional geometric data (pixel distance, image position etc.) must be available separately. The colour intensity values of the image data, which depend on density, are saved in various colour intensities (8 bit and higher). In medical applications, a 12 bit scale is often used.

Table 1. Comparison of a medical and an industrial CT data record

Characteristics	Medical CT of a skull	Industrial CT of a tube joint
Image format	DICOM	RAW, TIFF
Image size	512 x 512 pixels	991 x 991 pixels
Pixel size	0.44 x 0.44 mm	0.08 x 0.08 mm
Distance between images	1 mm	0.08 mm
Image number	238	851
Data volume	121 MB	1550 MB
Measuring volume	about 225 x 225 x 238 mm	about 80 x 80 x 68 mm

Table 1 offers an example outlining the difference between a medical CT and an industrial micro-CT based on two typical data records. In this representation, the differences in accuracy can be seen. It is possible for high-resolution industrial CTs to achieve measuring inaccuracy values of less than 80 μm. This results in a clearly higher data volume, which is frequently many times that of a medical CT (in the example shown in Table 1, it is approximately 13 times greater). It is very difficult to handle such a flood of data. Consequently, either we have to have available powerful computer systems capable of handling huge data volumes, or data stock has to be reduced, which, in turn, leads to losses in accuracy. For the latter option, one solution is to remove individual layers off the slice stack, thereby reducing image resolution or colour intensity.

Evaluation of CT data is often made more challenging by measuring noise and the formation of artefacts due to outshining. It is impossible to solve these problems simply by using individual image filters. For this reason, noise and artefact reduction are discussed in many publications [6].

In the following, the authors elucidate the CT data analysis methods implemented as program modules to read, process and display CT data developed at the Chair.

3 Inspection of a Soldered Tube Joint

The task was to inspect two soldered flange joints on a pipe elbow in co-operation with a manufacturer and supplier of hydraulic hose pipes (see Fig. 2). The goal was to dimension the tube joint to withstand higher pressure values. It was first necessary to demonstrate impermeability. What this means, in practical terms, is that the quantity and size of blowholes or inclusions of air (area per image, volume in the slice stack) in the soldering joints have to be inspected in order to guarantee that a closed soldering circle of about 3…4 mm can be maintained. It is possible to execute the measurements using pre-series parts, sample parts from production, or parts returned due to complaints.

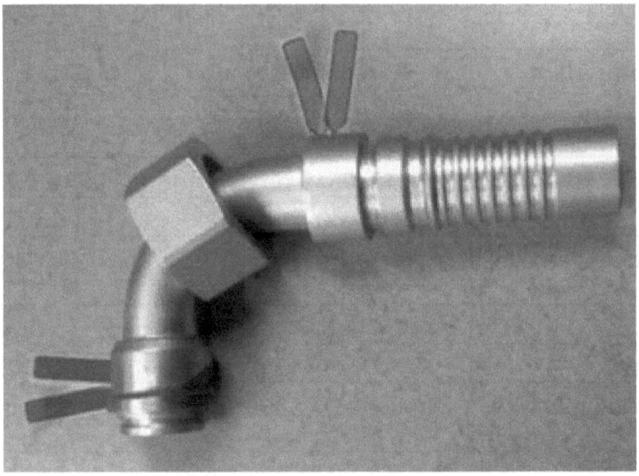

Fig. 2. Tube joint with two soldered flanges

It was impossible to solve the problem by means of the optical measuring instruments available at the chair, so the measurement was performed by the measuring firm Werth Messtechnik GmbH by means of the WerthTomoScope 200. As a result, we obtained 2D slice images in the RAW format with a resolution of approximately 80 µm and a slice image distance of about 80 µm. Data volume per measuring position is about 1.5 GB assuming approximately 850 images [7].

3.1 Methodes for Blowholds

A slice image resulting from the CT record is shown in Fig. 3. On the right side, one may clearly see inclusions in the region of the soldering joint. It is necessary to detect these positions and to quantify their area. If this is done using several images in the slice stack, we can draw conclusions regarding the blowholes' volume.

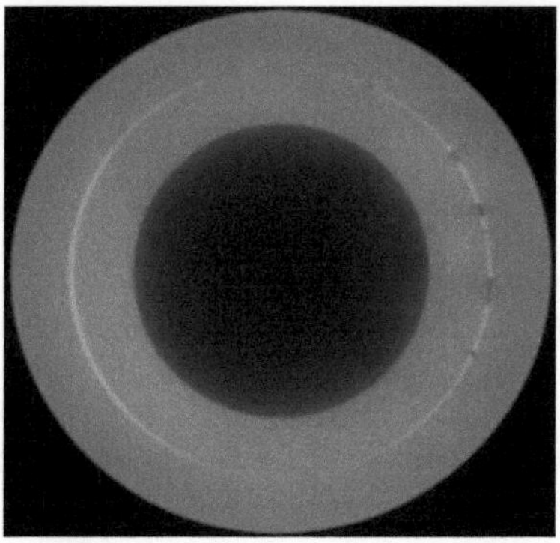

Fig. 3. Slice image with blowholes in the soldering region

To guarantee that only the zone of the soldering joint was considered for the detection of air inclusions, instead of erroneously detecting inclusions in the tubes themselves, we first determined the tube's centre point in the cross section at first. This approach only works if the slices are perpendicular to the tube centre line, so that the internal contour of the tube forms a circle. The centre point is determined as follows (Fig. 4):

a) A threshold filter is applied. For our example, we chose a threshold of 1600 in the 12 bit colour range. As a result, we obtain a binary image.
b) Next, the outer and inner contours of the tube joint are segmented by a cut off filter.
c) Afterwards, the centre of the inner contour is determined by means of a Hough transformation based on the specification of the known tube inner diameter.

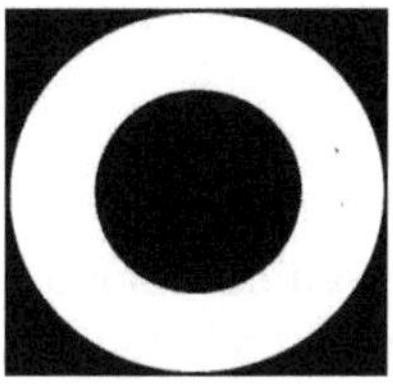

Application of a threshold filter

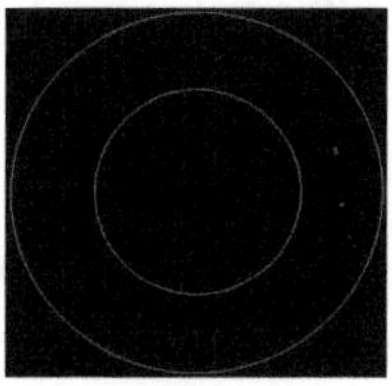

Application of a cut off filter

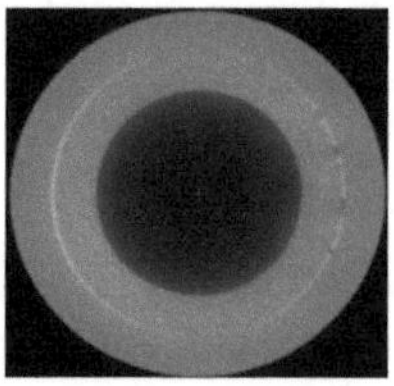

Centre point obtained by means of Hough transformation

Fig. 4. Determination of the tube centre point in the slice image

Having determined the centre point, the blowholes are identified by defining a threshold. The sequence of the individual procedural steps is shown in Fig. 5.

Inverting of imgage and smoothing | Calculation of a threshold and “BLOBS” filter | Image with identified blowholes

Fig. 5. Sequence of image filtering to identify blowholes

a) First, the image is inverted and smoothed.
b) Afterwards, the threshold is calculated. To do this, the image is subdivided into two segments based on a predefined value (in the example: 2240 at a colour intensity of 12 bits). Next, bright ranges in the image are localised by means of a “BLOBS“ filter [8]. This filter detects each separate object in an image. The objects are marked with polygons (in our example, rectangles).
c) Then the identified objects lying in the soldering region are found as a function of the calculated centre point and the given soldering circle diameter (= outer diameter of the inner tube). In this search, a tolerance is added to the soldering circle diameter (in our example: ±10%). Our goal is to record only the soldering joint rather than to detect inclusions in the tube itself. Now, the bright pixels (in our example with gray scale 4095) are identified inside the rectangle. They represent the air inclusions, for which we are searching. It is possible to quantify the area of the blowholes in the image using the known pixel width in both image directions.

The result of blowhole detection is especially dependent on the choice of an adequate threshold. This threshold has to be predefined by the user. If this threshold must be used over several slice images, the value has to be adjusted again if necessary.

3.2 Generation of 3D Freeform Cross Sections

Analysis of planar (2D) cross sections using the CT data record does not allow for a complete analysis of cylindrical or freeform inner structures in one view. Especially when evaluating a soldered joint, alternative slice images provide a way to represent the area to be soldered in a manner rolled out on a plane. To do this, slice images are generated, which cope with Spline surfaces according to their mathematical representation. The basis for this approach is that all slice images with their local, two-dimensional co-ordinate systems <st> are transformed into a global co-ordinate system <xyz> (Fig. 6).

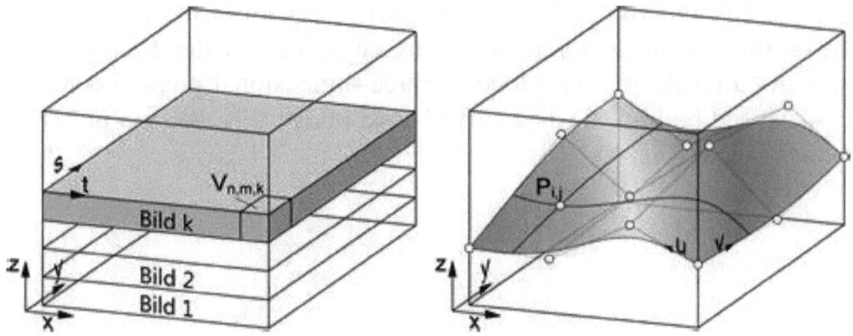

Fig. 6. Representation of the slice images in a compound of slice images Left: In a global reference co-ordinate system, all slice images are unambiguously defined. This way, indexing of a voxel V is possible (feasible) by indexing n, m, k. Right: In this global reference system, one may define an arbitrary Spline surface, whose discrete surface points P may be unambiguously transformed into the reference system.

Consequently, each image pixel of a slice image k can be represented as a three-dimensional voxel V by its co-ordinates [n,m,k]. This voxel is also defined in three-dimensional space. In this reference system, one may define any Spline surface of the type F(u,v) (Bezier, Hermite, etc.). In the case investigated here, we used Hermite surface patches, which are described by defining a mass point matrix G_{ab}. It is possible to calculate discrete points P_{ij} on the patch surface. These points are also defined in the global reference system. The quantity of points on the Spline surface in u- or in v directions determines the resolution of the desired slice image. The gray intensity values for one patch point can be determined by trilinear or tricubic interpolation of the gray intensity values of the adjacent voxel.

The procedure to create a freeform slice through the slice image stack can be described as follows:

1. In one or more CT slice images, the soldered joint is marked by a Spline curve (shown in red colour in Fig. 7). In this process, the quantity of defined curves is arbitrary. The quantity of supporting points per layer has to be the same in order to establish the mass point matrix G_{ab}.

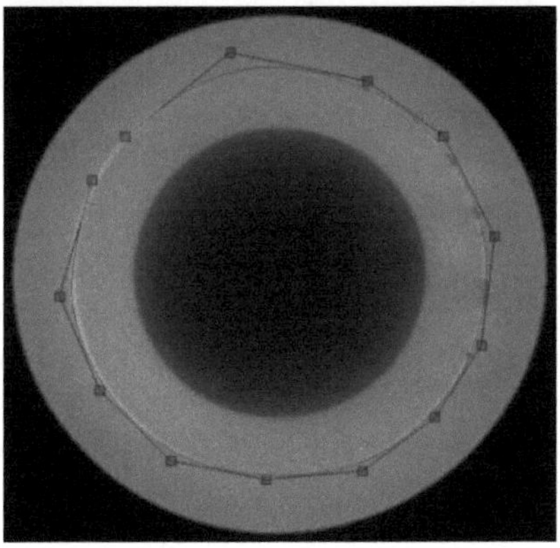

Fig. 7. Marking of the soldered joint in the CT image

2. Now we can calculate the Spline surface with the help of the mass points. Subsequently, on this surface, discrete points are calculated by iteration of the <uv> coordinates. As a result, a point cloud of three-dimensional points is created, which can be visualised both in 2D as a curved slice image (Fig. 9), and in 3D space as a triangulised object (Fig. 8).

Fig. 8. 3D cross section through a tube joint in the region of the soldering position

Fig. 9. Rolled out cross section of a tube joint

Thus, it is now possible to make qualitative statements about the joint's impermeability. Furthermore, one may perform an analysis to measure, for example, the size of the blowhole on the generated slice image. However, if this cross section is executed repeatedly in the region of the soldering position, concentric to the originally defined cross section, then we obtain a number of these three-dimensional panorama views of the soldered joint. Visual inspection of these views in their entirety, without taking into account further evaluation strategies, may provide an initial estimate of impermeability.

4 Summary

The use of computer tomography in industry offers a great potential for contactless and nondestructive recording of non-visible component regions. Since it is possible in this context to apply a significantly higher radiation level than for medical CTs, measuring uncertainty may be clearly reduced and data volumes concomitantly increased. Additionally, since the test objects are mostly stationary, as a rule, movement blurs can also be avoided [2].

The results provided by computer tomography can be used equally effectively for various tasks. In the narrower sense of Reverse Engineering, it is possible to use the data for modelling, for example. However, the most common applications come from measuring analyses, such as wall thickness analyses and test methods within the context of quality assurance. The example discussed in the paper shows that the implementation of efficient analysis strategies is essential for process monitoring and automation. The option of generating arbitrary freeform cross sections by means of a slice image stack particularly opens up new strategies for component investigation.

References

1. Bauer, N.: Handbuch zur industriellen Bildverarbeitung – Qualitätssicherung in der Praxis, 2. Auflage, Stuttgart (2008)
2. Zabler, E., Rosenberger, M., Bergmann, R.: Röntgen-Computertomographie in der industriellen Fertigung (Kraftfahrzeug-Zulieferer) – Anwendungen und Entwicklungsziele, DGZfP-Jahrestagung, Mainz (Mai 2003)
3. Schöne, C., Stelzer, R., Sembdner, P., Markwardt, J., Reitemeier, B., Engel, G.: Individual Contour Adapted Functional Implant Structures in Titanium. In: Proceedings of the 21st CIRP Design Conference, Daejeon, Südkorea, pp. 149–153 (2011)
4. Sembdner, P., Schöne, C., Stelzer, R.: Forming the interface between doctor and designing engineer – an efficient software tool to define auxiliary geometries for the design of individualized lower jaw implants. International Journal of Computer Assisted Radiology and Surgery 7(suppl. 1), 418–420 (2012)
5. Seibt, T.: Umsetzung eines geeigneten Marching Cubes Algorithmus zur Generierung facettierter Grenzflächen, Diplomarbeit, TU Dresden (2011)
6. Hahn, M.: Verfahren zur Metallartefaktreduktion und Segmentierung in der medizinischen Computertomographie, Dissertation, Universität Karlsruhe (2005)
7. Schöne, C., Stelzer, R.: Reverse Engineering in der Produktentwicklung, Aktuelle Herausforderungen. 10. Gemeinsames Kolloquium Konstruktionstechnik, Dresden, pp. 351–363 (Juni 2012)
8. `http://www.aforgenet.com/framework/features/blobs_processing.html` (State of October 29, 2012)

Correction of Highlight Line Structures

György Gyurecz[1] and Gábor Renner[2]

[1] Department of Machine Design and Safety Engineering, Óbuda University, Hungary
gyurecz.gyorgy@bgk.uni-obuda.hu
[2] Computer and Automation Research Institute, Hungarian Academy of Sciences, Hungary
renner@vision.sztaki.hu

Abstract. Class A surfaces are external surfaces of industrial objects, to which aesthetic appeal is attributed. Their design involves beyond functional criteria also subjective ones related to style and appearance. The design criteria include appropriate highlight line structures of the surfaces. Correction of highlight lines is usually performed interactively by the designer, which is time consuming, tedious and inaccurate.

In the current paper an algorithmic method is suggested to automatically improve the quality of the highlight line structure. Application of the method is demonstrated by the correction of highlight line structure of industrial surfaces.

Keywords: Highlight lines, highlight line structure.

1 Introduction

Most important class A surfaces can be found on cars, airplanes, ship hulls, household appliances, etc. Beyond functional criteria, the design of class A surfaces involves aspects concerning style and appearance. Creating tools supporting the work of a stylist is a challenging task in CAD and CAGD.

A highlight line structure is a series of highlight lines, representing visually the reflection and the shape error characteristics of the surface. They are calculated as the surface imprint of the linear light source array placed above the surface [1].

The structures are evaluated by the pattern and the individual shape of the highlight lines. A comprehensive quality inspection can be carried out by the comparison of the highlight line structures of different light source and surface position settings. The uniform or smoothly changing highlight line pattern is essential for the high quality highlight line structures.

Following the inspection, the defective highlight curve segments are selected and corrected. Based on the corrected highlight curves, the parameters of the surface producing the new highlight line structure can be calculated [4].

In our method the correction of highlight line structure is carried out in two steps. First, sequences of evaluation points are defined to measure the error in terms of distance and angle functions. Next, these functions are smoothed and based on the new function values, new highlight line points are calculated. New highlight curve curves are constructed using these points. The outline of the method is summarized in Figure 1.

G.L. Kovács and D. Kochan (Eds.): NEW PROLAMAT 2013, IFIP AICT 411, pp. 230–238, 2013.

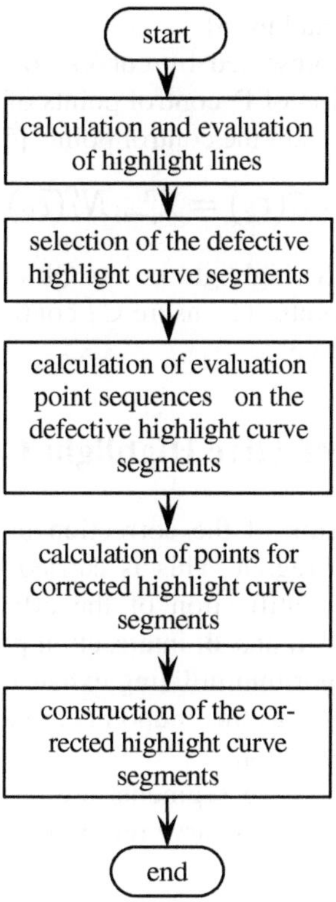

Fig. 1. Block diagram of the highlight line structure improvement method

2 Surface Representation and Calculation of Highlight Curves

The surfaces are represented in Bézier, B-spline or NURBS form that are widely used in CAD applications. The shape of **S**(u,v) surface is defined by an array of control points $\mathbf{P}_{i,j}$ and the Bézier, B-spline or NURBS basis functions [5]. The highlight lines are reflections of linear light sources on the surface, calculated as set of discrete highlight points. The highlight points are points on the surface where the corresponding surface normal and the light source intersect each other. Let $\mathbf{L}(\lambda)=\mathbf{A}+\mathbf{B}$ a line representing a light source, where **A** is a point on $\mathbf{L}(\lambda)$, and **B** is a vector defining the direction of the line. The signed perpendicular distance d(u,v) between the normal **N**(u,v) at a surface point **S**(u,v) and the linear light source is:

$$d(\mathrm{u},\mathrm{v}) = \frac{[\mathbf{B}\times\mathbf{N}(\mathrm{u},\mathrm{v})]\cdot[\mathbf{A}-\mathbf{S}(\mathrm{u},\mathrm{v})]}{|\mathbf{B}\times\mathbf{N}(\mathrm{u},\mathrm{v})|} \tag{1}$$

For a point on the highlight line d(u,v)=0 holds, which must be solved for the control points of **S**(u,v). To design high quality surfaces, this relation has to be computed with high accuracy. We developed a robust method for computing points on highlight lines, which is described in detail in [3].

The highlight lines are represented by curves constructed by interpolation in B-Spline form. For the calculation of $\mathbf{P}_i$ control points of C(t) curves system of equation is solved, where the unknowns are the control points [5].

$$Q_k = C(\bar{t}_k) = \sum_{i=0}^{n} N_i^r(\bar{t}_k)\, P_i \tag{2}$$

The parameter values tk of the highlight points Qk are set proportional to the chord distance between highlight points. To ensure C2 continuity of the curves, the degree r of the basis function N is set to 3.

3 Selection of the Defective Highlight Curve Segments

Selection identifies the location of the correction by fitting a sketch curve on the surface around the defective region. This is carried out by the interactive tools of the CAD system. For the identification of the affected highlight curve segments C_i, i=0...N, and the endpoints $\mathbf{A}_i$ and $\mathbf{B}_i$ intersection points are searched. The identification is carried out by an algorithm utilizing exhaustive search method [2]. The tangents $\mathbf{T}_{i1}$ and $\mathbf{T}_{i2}$ corresponding to the endpoints are also identified; they are utilized in a subsequent process of correction.

In Figure 2, the defective curve segments are shown in bold; the endpoints are marked by solid squares. The dashed curve represents the user drawn sketch curve.

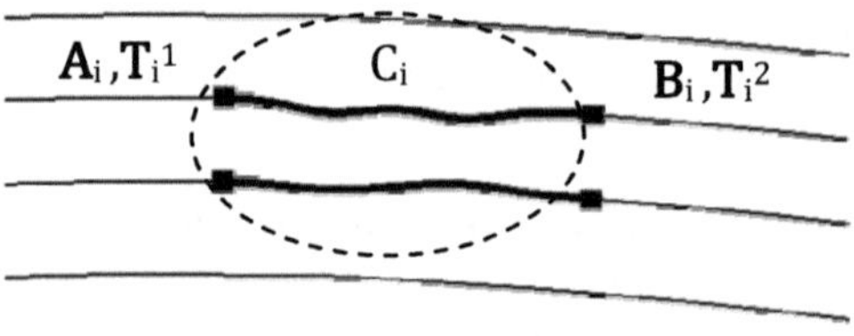

Fig. 2. Selection of the defective highlight curve segments

4 Evaluation of the Highlight Line Pattern

The structure of the selected highlight curve segments is evaluated on sequences sj, j=0...M of highlight points $\mathbf{E}_{0,0},\ldots\mathbf{E}_{i,j}\ldots\mathbf{E}_{N,M}$ spanning over the defective segment in crosswise direction. The sequences include correct highlight curve points $\mathbf{E}_{0,j}$, $\mathbf{E}_{1,j}$ and $\mathbf{E}_{N-1,j}$, $\mathbf{E}_{N,j}$ needed to ensure the continuity of corrected highlight segments with the adjoining unaffected region. We evaluate the structure error by d_j distance and $\propto_j$ angle functions defined on s_j, sequences The distance function represents the inequalities of the structure in crosswise direction; the angle function characterizes the structure error along the highlight curves.

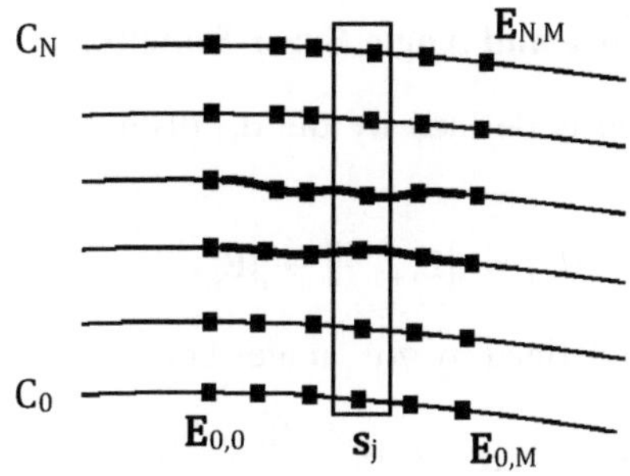

Fig. 3. Definition of the evaluation point sequences

4.1 Calculation of Evaluation Point Sequences

Point sequences start with points $\mathbf{E}_{0,0}...\mathbf{E}_{0,M}$ equally spaced on C_0. They are determined by the location of the furthest $\mathbf{A}_i$ and $\mathbf{B}_i$ endpoints and the M number of sequences. The subsequent points are calculated based on the shortest perpendicular distance between subsequent highlight curves.

Let $\mathbf{E}'_{i,j}$ a point on the C_i, and $\mathbf{E}'_{i+1,j}$ a point on the C_{i+1} highlight curve (Figure 4).

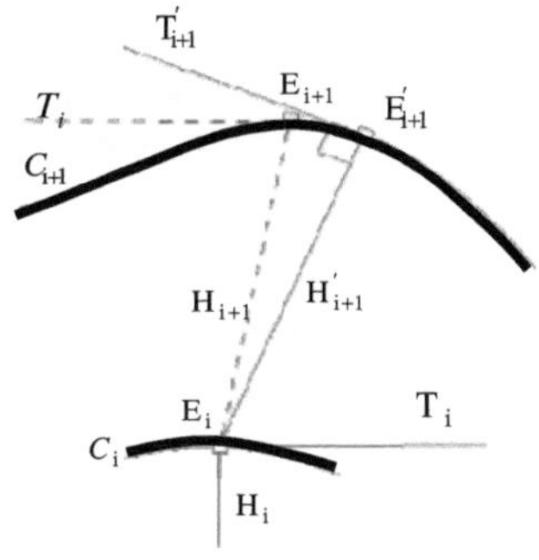

Fig. 4. Calculation of the evaluation points

Point $\mathbf{E}'_{i+1,j}$ is in the perpendicular direction if

$$\mathbf{H}'_{i+1,j} \cdot \mathbf{T}'_{i+1,j} = 0 \tag{3}$$

where $\mathbf{H}'_{i+1,j} = \mathbf{E}'_{i+1,j} - \mathbf{E}'_{i,j}$

$$\mathbf{T}'_{i+1,j} = C'_{i+1} \text{ at } \mathrm{H}'_{i+1,j}$$

The location of $\mathbf{E}_{i+1,j}$ evaluation point is in the surrounding of $\mathbf{E}'_{i+1,j}$ where:

$$\mathbf{T}_{i+1,j} \cdot \mathbf{T}_{i,j} = 1 \tag{4}$$

$$\mathbf{T}_{i,j} = C'_{i+1} \text{ at } \mathrm{H}'_{i,j}$$

4.2 Definition of Distance and Angle Error Functions

The distance error function is defined by the $d_{i,j}$ distances between the consecutive sequence elements:

$$d_{i,j} = \|\mathbf{E}_{i+1,j}\| - \|\mathbf{E}_{i,j}\| \tag{5}$$

The angle error function is defined by $\alpha_{i,j}$ angles between the consecutive H_i vectors:

$$\alpha_{i,j} = arccos\left(\frac{H_{i+1} \cdot H_i}{\|H_{i+1}\| \cdot \|H_i\|}\right) \tag{6}$$

In Figure 5, an error function constructed from points $E_{i,j}|$ i=0..N , N=5 is presented. The i=2…N-2 sequence of the error functions correspond to points on defective highlight curves. The rapid and irregular changes represent the defects in the highlight curve structure. The function values at i=0,1 and N-1,N correspond to points on highlight curves of the adjoining correct pattern.

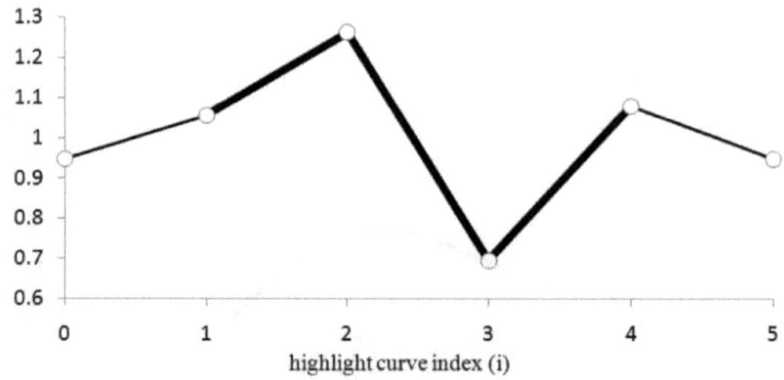

Fig. 5. Error function example

5 Calculation of the New Highlight Curve Segments

A proper highlight line pattern has smooth evaluation functions, without the disturbances seen in Figure 5 and maintains continuity with the adjoining correct pattern. For the correction of highlight curves new, smooth evaluation functions are calculated. Based on the new functions, points for the new highlight curves are obtained.

5.1 Calculation of the New Highlight Curve Points

The new function values are calculated by least square approximation method, applied on the original functions. The continuity with the highlight line structure of the adjoining region is ensured by constraints on function end tangents T_i^1 and T_i^2.

The tangents are calculated as: $\mathbf{T}_j^1 = \mathbf{E}_{0,j} - \mathbf{E}_{1,j}$ and $\mathbf{T}_j^2 = \mathbf{E}_{N-1,j} - \mathbf{E}_{N,j}$.

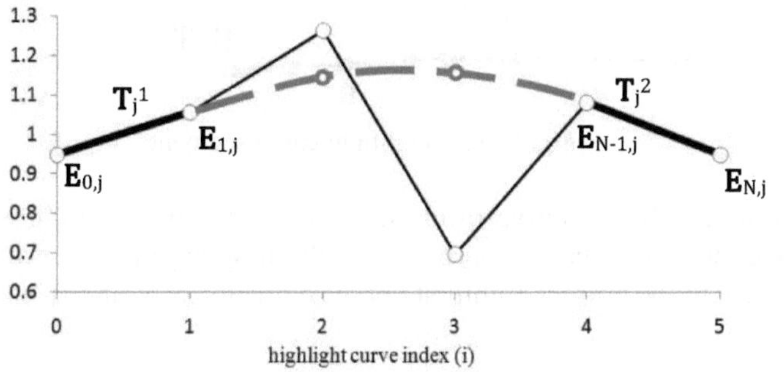

Fig. 6. The error function after smoothing

In Figure 6. the original (thin donuts) and the new function (thick donuts) values are shown. The $\mathbf{R}_{i,j}$ points of the new highlight curves are calculated by s_j sequences, starting from point $\mathbf{E}_{i,j|i=1}$ and adding the $\overline{\mathbf{V}}_{i,j} = d_{i,j} \cdot \alpha_{i,j}$ vectors defined by the angle and distance function values:

$$\mathbf{R}_{i,j} = \mathbf{E}_{1,j} + \sum_i \mathbf{V}_{i,j} \text{ where } i = 1..N-1, j = 1..M-1 \tag{7}$$

Figure 7 shows calculation of new $R_{i,j}$ points (indicated by solid squares).

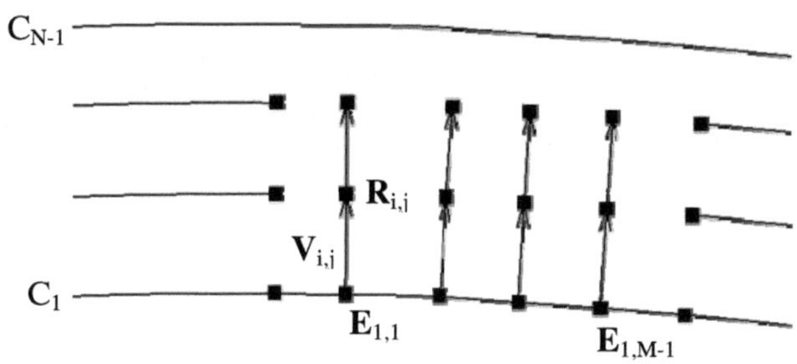

Fig. 7. Calculation of points for new highlight curve segments

5.2 Construction of the Corrected Highlight Curve Segments

The new C_i highlight curve segments are cubic B-Splines constructed from the new $\mathbf{R}_{(i,j)}$ points by constrained least squares curve fitting method [5]. The points to be approximated are $\mathbf{R}_{(i,0)},...\mathbf{R}_{(i,j)}...\mathbf{R}_{(i,M)}$ new highlight curve points, arranged by C_i, highlight curves. The constraints are $\mathbf{A}_i$ and $\mathbf{B}_i$ segment endpoints and the $\mathbf{T}_i^1$ and $\mathbf{T}_i^2$ endpoint tangents. The u_{Ai} and u_{Bi} parameter values of the new segments correspond to $\mathbf{A}_i$ and $\mathbf{B}_i$ segment endpoints. For the calculation of $\mathbf{P}_i$ control points, system of equations is solved. The u_k, parameters of the curve points $\mathbf{Q}_k$ are defined on $u_k = u_{Ai}...u_{Bi}$. The parameter values are set proportional to the chord distance between the highlight curve points.

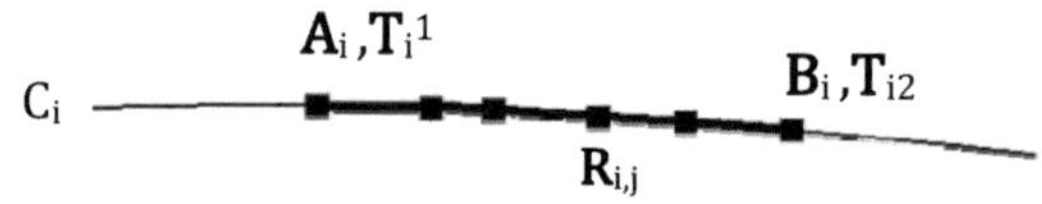

Fig. 8. New highlight curve segment

In Figure 8 the result of a curve fitting is shown. The corrected highlight curve segment (thick continuous curve) approximates the new highlight curve points and has C2 connectivity to the correct highlight curve segments in A_i and B_i endpoints.

6 Application and Examples

The method is implemented in Rhino 4 NURBS modeler. The calculation of new highlight curve points and the construction of corrected highlight curve segments is written in C++ code, the calculation and selection of highlight curves is realized in VBA. We tested our method on several industrial surfaces. In Fig. 9 and Fig. 10, two highlight curve structures before and after corrections are presented.

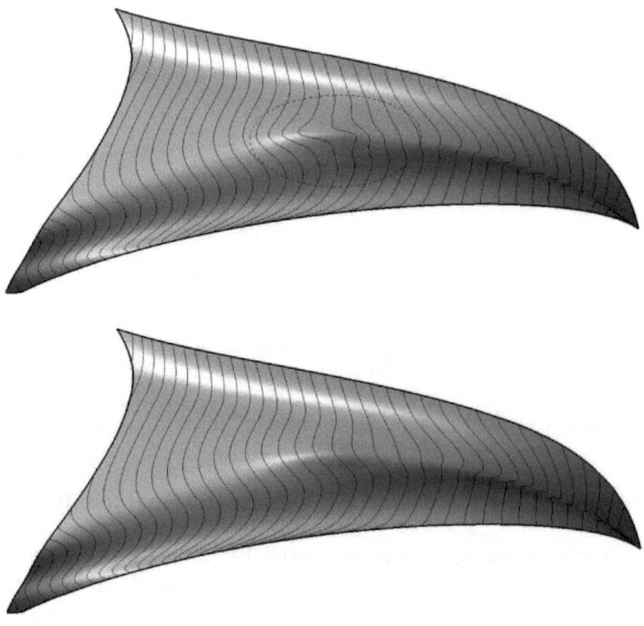

Fig. 9. Car body element before and after correction

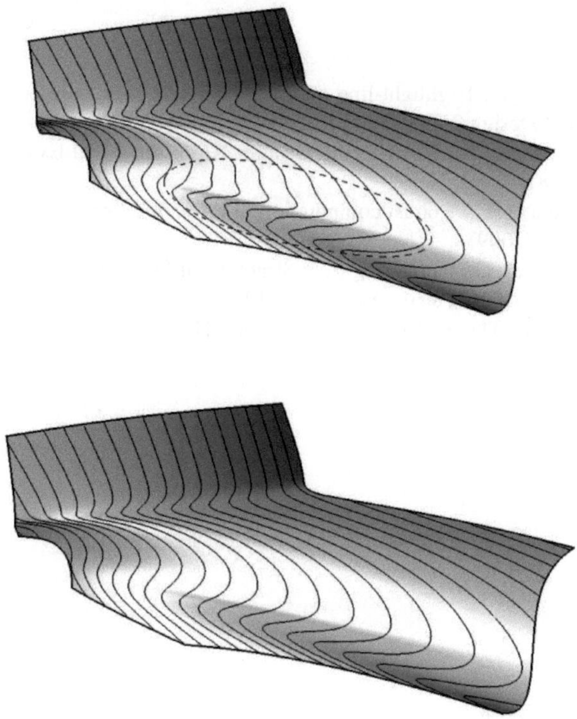

Fig. 10. Car body element before and after correction

7 Summary

A method for correcting highlight line structures is presented. The correction is result of two consecutive steps. In the first step, distance and angle evaluation functions are defined to characterize the error in the highlight curve structure. In next step, the functions are smoothed, and based on the smoothed functions new highlight curve points are calculated. The corrected highlight curves are constructed based on the new points.

The defective surface area is selected interactively, the evaluation and correction of highlight lines is automated. The parameters of the automatic correction can be adjusted by the designer.

The method is successfully implemented in the surface modeling software (Rhino 4) widely used in industrial shape design. The method is applicable to surfaces with uniform or changing highlight line pattern, and wide range of highlight line errors. The applicability of the method was proved on number of industrial surfaces.

References

1. Beier, K.P., Chen, Y.: Highlight-line algorithm for real time surface quality assessment. Computer Aided Design 26(4), 268–277 (1994)
2. Deb, K.: Optimization for Engineering Design: Algorithms and Examples. Prentice-Hall, New Delhi (1998)
3. Gyurecz, G., Renner, G.: Robust computation of reflection lines. Journal of Machine Manufacturing 1, 1–6 (2009)
4. Gyurecz, G., Renner, G.: Correcting Fine Structure of Surfaces by Genetic Algorithm. Acta Polytechnica Hungarica 8(6), 181–190 (2012)
5. Pigel, L., Tiller, W.: The NURBS Book. Springer, Heidelberg (1995)

Re-make of Sheet Metal Parts of End of Life Vehicles - Research on Product Life-Cycle Management

George L. Kovács[1,2] and Imre Paniti[1]

[1] Computer and Automation Research Institute, Hungarian Academy of Sciences
[2] PMMIK, Pécs University of Sciences, Pécs, Hungary
{kovacs.gyorgy,paniti.imre}@sztaki.mta.hu

Abstract. Today it can be accepted as fact that mankind is running out and will soon have shortages of energy, row materials and water. Sustainable development (or more often used sustainability) means to use as little materials, energy and water as possible. Material saving causes energy and water saving as well. Product Life Cycle Management (PLCM) of any kind of product starts when the idea of a product is born and lasts until complete dismissal through steps as detailed design, assembly design, part manufacturing, assembly, testing, diagnostics and operation, advertisement, service, maintenance, etc. Then finally our topics, namely dismantling/ disassembly and dismissal are coming. Dismissal often can be substituted by re-cycling or re-use, or re-make.

Keywords: re-use, end of life vehicle, sheet-metal, incremental sheet forming, sustainability, product life-cycle.

1 Some Definitions and Abbreviations

The motivation of the example used during this study was born from the EU Directive 2000/53/EC [1a] according to which by 2015, the portion of each vehicle that should be recycled or reused have to increase to 95%. To avoid being too abstract in the paper we use the management of sheet metal parts of worn-out or crashed cars as example. We show a possible way of value evaluation and measurement of values during PLCM. On the other hand we deal with the complexity and problems of decision making during processing of used sheet-metal parts' 1, if the main goal is Remake (re-use). The decisions are about dismantling and remake as e.g. re-cycling or re-use with or without repair, how to repair, etc.

Some definitions and abbreviations will be given first to help better understanding some expressions in the context of worn out or broken or simply End of Life (EOL) vehicles. Remake, re-use and re-cycling should not be mixed up with waste management, as our goal is to use everything instead of producing waste.

- **ELV** (End of Life Vehicle, **EOL** vehicle): cars after collision or worn out cars
- **Shredder** (S): a strong equipment breaking, tearing everything into small pieces, to shreds, almost like a mill.

G.L. Kovács and D. Kochan (Eds.): NEW PROLAMAT 2013, IFIP AICT 411, pp. 239–253, 2013.

- **Dismantling/disassembly** (DM, DA): to take out parts for further manipulations, opposite to assembly.
- Delayed or **conditional decision** (DD, CD), a decision that depends on something, and will be done later on
- **Dismissal/liquidation**: to get rid of the whole or of certain parts using a shredder and then the rest is "buried" (disposed)
- **Re-use**: to use again certain parts after dismantling, as they are or applying correction (**Re-make**). Usage as second-hand parts.
- **Re-cycle**: melting the metals (and plastic) separately for getting useful raw material
- **Sheet metal** parts – we use it mainly to speak about car body parts.
- **Dangerous material**: materials, which have to be collected and managed separately in the case of every car. For example chemicals in the battery, or in the air conditioner, oil, gasoline, etc.
- **ISF**: Incremental sheet forming: a technique to form requested sheets by a series of small incremental deformations based on corresponding CAD/CAM information.
- **CAD/CAM** (Computer Aided Design/Computer Aided Manufacturing) data: digital information, describing car body sheet metal parts: gained from original design, or from 3D measurements, etc. It can be used to produce CNC code
- **CAD/CAM information** of car body sheet metal parts for correction of shape errors. Data scanned from the part to be corrected vs. data of the needed sheet.
- **Thickness** measurement: the most important information to qualify the sheet
- **3D shape** measurement (scanning): to measure the actual shape (in most cases with laser technology), and to produce CAD/CAM data for comparison.
- **Scanning**: optical (laser) way of 3D shape management, resulting in a cloud of surface data, describing the shape, input for CAD/CAM and CNC,
- **Decisions** at several levels: they may be Automatic, Semi-automatic, or Human (**DA, DSA, DH**)
- **Decision assisting** tools: optical and other measurements, AI programs, human considerations
- **Product Life-Cycle-Management** (PLCM): management of any kind of product for its whole life-cycle. Almost everything can be taken into consideration as a product.

2 Introduction and State-of-the-Art

Management of EOV (EOL vehicles) is a rather complicated task, which needs to process several technical and legal steps (paperwork). All parts of all vehicles have to be under control during their whole Life-Cycle, including permissions to run, licence plates, permissions to stop, informing local authorities, etc. In this paper we deal only with technical aspects of a restricted set of parts, namely with sheet-metal parts.

2.1 Some Sources (for Example [1]) Simplify the Procedure to the Following 3 Steps:

- Depollution; Dismantling; Shredding

The main goal is achieved, but several points (decision points) remain open and several questions unanswered, and nobody really knows what to do and how to do and what are the consequences of certain activities. What happens with the different parts?? Is there anything worth repairing, etc. ?

There are others, who claim that the process is a little more complicated. The following EU suggestion does not go into details; it is a straightforward average procedure including paper work, which is crucial if someone deals with EOV.

2.2 According to [2] the Procedure of Dismantling Should Be the Following

- Delivery of EOL vehicle
- Papers and documents are fixed, permissions issued (or checked if issued by others)
- Put the car onto a dry bed: remove dangerous materials and liquids, store everything professionally
- Select useful parts, take them out and store under roof
- Sell/offer for selling the tested parts
- Press the body to help economic delivery
- Re-use raw materials

A little more precise description of the removal sequence of different important parts/materials with more details, however still without decision points is the following:

- Remove battery and tanks filled with liquid gas
- Remove explosive cartridges of airbag and safety bells
- Remove gasoline, oils, lubricants, cooling liquid, anti- freezing, break liquid, air-conditioner liquid
- Most careful removal of parts containing quick silver
- Remove catalysts
- Remove metal parts containing copper, aluminium, magnesium
- Remove tyres and bigger size plastic parts (fender/bumper, panel, containers for liquids)
- Remove windshields and all glass products

2.3 The IDIS Web Site [3] has the Following Opinion

The International Dismantling Information System (IDIS) was developed by the automotive industry to meet the legal obligations of the EU End of Life Vehicle (ELV) directive and has been improved to an information system with vehicle manufacturer compiled information for treatment operators to promote the environmental treatment of End-of-Life-Vehicles, safely and economically. The system development and improvement is supervised and controlled by the IDIS2 Consortium formed by automotive manufacturers from Europe, Japan, Malaysia, Korea and the USA, covering currently 1747 different models and variants from 69 car brands.

The access to and the use of the system is free of charge. The basic steps of dismantling suggested by IDIS2 are as follows:

- Batteries -- Pyrotechnics -- Fuels
- AC(Air Conditioner) - Draining - Catalysts
- Controlled Parts to be removed - Tires -- Other Pre-treatment
- Dismantling

2.4 At GAZ Autoschool [4] in the UK the Following Are Underlined as the Most Important Steps to Follow

1. Removing vehicle doors, bonnet, boot, hatch. Removing these items early in the dismantling process enables easier access to vehicle interior, reduces restriction in work bays and minimises the risk of accident damage to potentially valuable components.
2. Removing interior panels, trim, fittings and components. This is a relatively clean and safe operation which maximises the resale opportunities available for items whose value depends on appearance/condition and which may be damaged if left on the vehicle. Components to be removed include dashboard, instrument panel, heater element, control stalks, steering column.
3. Remove light clusters: An easy process but one which needs care to avoid damage. Once removed items need to be labeled and stored to enable potential re-sale.
4. Removal of wiring harness: The harness should be removed without damage, meaning that all electrical components are unclipped and the wires pulled back through into the interior of the car so that it can be removed complete and intact. Harness should be labeled and then stored appropriately.
5. Removal of Engine and Gearbox: This will involve the use of an engine hoist, trolley jacks and axle stands, and will often necessitate working under the vehicle for a short period to remove gear linkages etc. Often the dirtiest and most physical task. Engine and gearbox oil together with engine coolant will need to be drained and collected for storage.
6. Engine dismantling: Engines are kept for resale where possible.
7. Gearbox dismantling: Gearboxes are kept for resale where possible.
8. Brakes and shock absorbers: Brake components are checked and offered for resale where they are serviceable.

2.5 Finally We Refer to [5], Which Is a Survey and Case Study on Serial Flow Car Disassembly

The main considerations and results of the survey are the following.

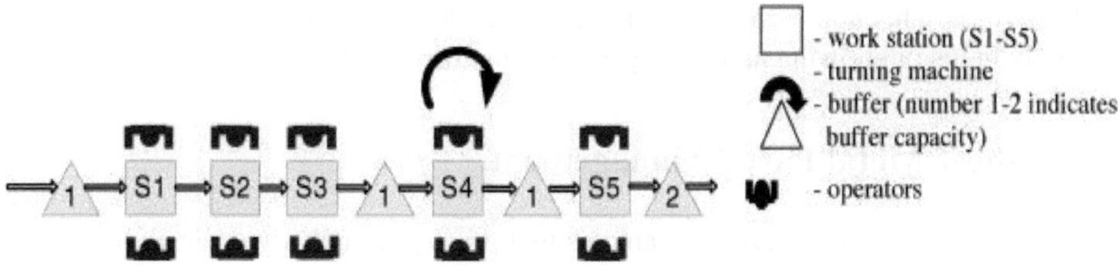

Fig. 1. Serial-Flow Car Disassembly

- In the European Union over 15 million passenger cars were sold in 2004 according to European Automobile Manufacturers Association (ACEA, n.d.), up 1.7% from 2003.
- Around 8 million end-of-life vehicles (ELVs) are disposed of annually in the EU (ACEA, 2004) and this number is expected to increase.
- An EU directive (2000/53/EU; European Parliament, 2000) [1a] requires that wastes from ELVs must be reduced to 15% of the total car weight by the year 2006 and to 5% by 2015.

The suggested technology can be represented by Fig. 1., where one can see, that the system used five stations and four buffers.

- At stations 1–3 glass, rubber, and interior are removed.
- At station 4 the "turning machine" rotated cars upside down to facilitate engine and gearbox unfastening
- At station 5 the engine and gearbox are removed .

The procedure is the following:

1. Take a lot of pictures before you begin the disassembly process, including pictures of the interior. This important issue is rarely mentioned by other authors
2. Get a box of zip lock plastic bags in each size available to store every nut, bolt, hinge, clip, shim, etc. Make color marks to all.
3. Make sure you have a pen and a notebook by your side at all times to document any helpful reminders, parts in need of replacement and to take inventory
4. Remove the fenders, hood and trunk lid with the assistance of at least one able body to avoid damage and personal injury
5. Remove the front windshield and the rear window by first removing the chrome molding from the outside of the car, being careful not to scratch the glass.
6. This would be a good point to gut the interior. Remove the seats, doors and interior panels, carpeting and headliner.
7. Clear the firewall and take all the accessories off the engine.
8. Go through your notebook and highlight all the parts that need to be replaced and make a separate "to do" list for ordering them.

The most characteristic in the above study is a practical approach: be very careful and document everything, use bags and color pens and notebook to keep track of all parts and activities. This non-stop bookkeeping may hinder effective and fast work, however surely helps in knowing and tracing everything, if needed.

3 Sheet-Metal Parts' Management

It can be seen from the previous examples that practically nobody deals with the special problems of sheet metal parts of EOL vehicles. On the other hand it is clear that sheet metal parts are only a certain percentage of an EOL vehicle. But it is clear, that almost every EOL vehicle has several sheet metal parts, which could be re-used with or without corrections, with or without re-paint. This makes us to believe that it is worthwhile to deal with sheet metal parts separately, moreover in the following part of our study this will be our only issue. To think and to speak about re-use (re-make, re-shape) as a practical issue, we need as minimum:

(a) proper dismantling technology to remove sheet-metal parts without causing damages to them
(b) a measurement technology to evaluate the dismantled part and a software to compare the measured values to requested values, to define whether the dismantled part is appropriate or needs correction and to decide its applicability to be used for another vehicle, and finally
(c) a technology to correct slightly damaged sheets, based on CAD/CAM information. This information may come from design (new or requested parts) or through measurements by a scanner (actual, dismantled part).

Our staff using ISF technology and the robotic laboratory of SZTAKI is able to perform the requested operations on our machines and on the software

3.1 A Sheet-Metal Decision Sequence

Our approach needs to follow a rather complicated decision sequence; it will be detailed only for sheet metal (SM) parts, supposing the available Incremental Sheet Forming (ISF) facilities.

It is the following, using the above defined abbreviations, emphasizing decision types and points:

1. Car arrives: on wheels or on a trailer, papers and documents are fixed, permission issued (or checked if issued by others)
2. It goes or it is taken to the dismantling bed (dry bed)
3. Remove liquids and dangerous materials (Unconditional)
4. Decision 1: Shredder (S) or dismantling (DM) or delayed decision after the beginning of dismantling (DD). Decision is done basically by human, eventually assisted by measurements, or even by 3D part modeling
   (a) if S: no more work to do: car goes to shredder and then burial (dismissal)
   (b) if DM: disassembly starts, parts are taken off one by one until the last one, based on a given protocol for all car types.

   (c) if DD: disassembly starts, parts are taken off one by one, and sorted and stored, until the next decision can be done
5. Decision 2: S or DM - done at any time. Decision may be partial shredder (PS) and partial dismantling (PD) after a while.
   (a) if PS&PD: certain parts are taken apart, the rest goes to shredder.
   (b) if DM or DD and PD are done, now we have a lot of parts organised somehow
6. Decision 3: Select sheet metal parts: automatically or manually or hybrid way, keep SMs, put away the rest
   (a) Examine all SM parts, first thickness (TH) measurement is done
   (b) if TH is too small, part goes to S. The rest goes to border measurement (SB), as borders (SB), of a sheet may easily be damaged during dismantling.
   (c) SB may be done by optics and AI and/or by human or both, after each other.
   (d) if SB is repairable or good, make shape measurement (SHM)
   (e) Compare measured sheet (MS) to standard shape (SH). SH can be taken by measuring a failure-free sample, or from any appropriate catalogue. For processing we need CAD data in both cases.
   (f) Compare SH with MS. and calculate differences somehow, it is the deviation from standard (DS)
7. Decision 4: if DS is small enough (defined by the customer, who will need the part, or it is an average value generally accepted), part goes to repair. The rest goes to shredder.
8. Decision 5: repair by hand, by ISF or combined, any sequence is possible
   (a) if ISF: part goes to ISF centre together with its CAD/CAM code, and processed
   (b) if Manual or Combined: part goes to worker, when needed, after or before ISF
   (c) if ISF is done – a final measurement is needed (SHM).
9. Decision 6: if accepted it goes to the shop or to a workshop for painting, and then to a shop.
   (a) if rejected it goes back to 6.6.
10. The part is accepted, sent to the shop or to business again

This decision sequence - naturally – can be taken into account as a small and timely short period of the PLMC, namely of the EOL cars' sheet-metal parts. Each move and activity have certain actual prices, which are commonly accepted, however we know that they are not really correct, they do not support sustainability, and on the other hand often increase negative effects.

3.2 Evaluation of Costs and Advantages of the Re-use and Re-make of Sheet-Metal Parts of EOVs

After such a long, or long-looking decision procedure we need a methodology for evaluation of all, what we do or do not perform to have re-usable sheet-metal parts from parts of EOL vehicles.

The simplest way would be simply compare costs and prices of all involved parts (good, to be corrected, etc.) and services (scanning, ISF, manual work, painting, etc.) and all conditions (shredder or repair, etc.).

Today that is the only way some people follow, if any. Generally a very fast view at the EOV is enough to send it to the shredder, as this is the simplest decision with the smallest risk. To be more precise, the risk is there, but rather hidden.

The cost/value estimations and comparisons can be performed relatively easily, however the results correspond only to the present economical – political situation and to the actual financial circumstances, and would not say anything about the future, what is embedded into the "sustainability", "footprint" and "side-effects" (see later on). We believe that there exist giving appropriate tools and means for "real", "future centric" evaluations, thus we need to find and use them.

Our choice are the KILT model and the TYPUS metrics, which will be explained and used for our study, for details see [6], [7], [8] and [8a].

The main goal of the above tools is to model and quantify the complete delivery (all products, side-products, trash and effects of them, i.e. all results) of a firm, and to model all interesting and relevant steps of the LC (or LCM). It is clear from the definitions (see later and the references) that any production steps can be evaluated and can be understood as cost values. If we speak about car production, the input is row material, machining equipment and design information, and people, who work, etc. Output (delivery) is the car. Sheet metal production is a little part of car making, generally prior to body assembly. For our study we take into consideration only sheet-metal parts.

We consider and make measurements, comparisons, re-make by using ISF and re-paint, and other actions. These can hardly be compared with the "simple" processes used in new car manufacturing. Every step of the decision sequence below can be investigated one by one, taking into account all effects and side-effects. For the sake of simplicity only the input (sheet metal part to be measured and perhaps corrected) and the output (sheet-metal part ready to be used again) may be enough.

Fig. 2. shows some qualitative relationships, which cannot be avoided if environmental issues, sustainability, re-use and our future are important points.

Fig. 2 gives a general picture of our main ideas, and it needs some explanation. See [8a] for more details. It is a rather simplified view of some main players in the production/service arena, however it still shows quite well certain main qualitative relationships. We believe that these can be used to understand what is going on in our (engineering-manufacturing-sustainable) world.

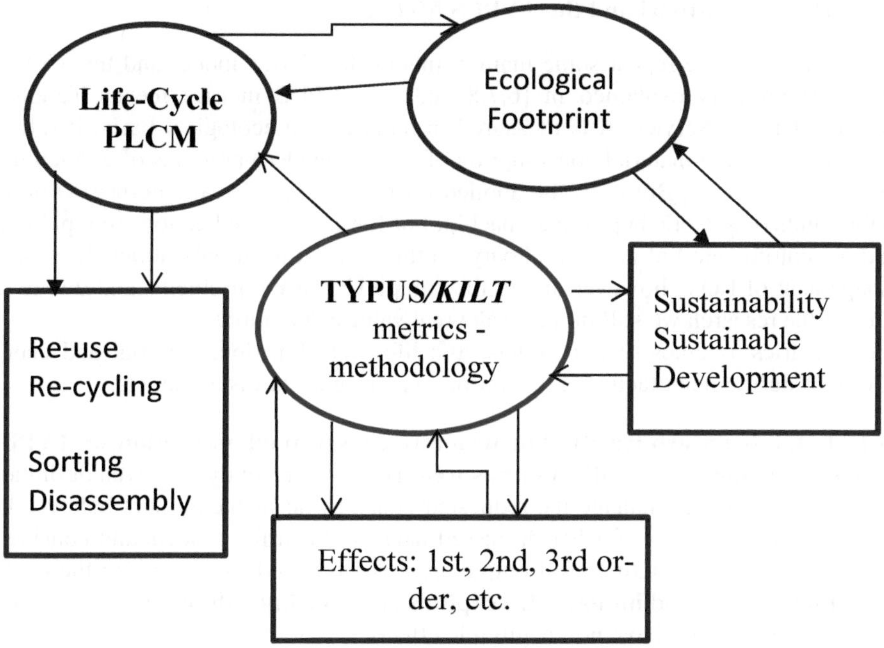

Fig. 2. Re-use, PLCM, ecology, sustainability and KILT

The TYPUS/KILT metrics, methodology and model give us a possibility to better understand and evaluate production results and their components in terms of the K, I, L, T values. They give us a method of calculations and comparisons based on realistic values. The side effects and 2nd and 3rd order effects, etc. mean the following: let us consider a simple and simplified example: to produce a hybrid car (today) means (among others) to produce and build in two engines, two engines need more metal than one (side effect), to produce more metal we need more electrical energy and more ores (2nd order side effect), to produce more electricity more fuel is necessary and to produce more ores needs more miners' work (3rd order side effect), etc., and it could be continued. It is a hard task to know how deep and how broad we should go with such calculations. And if we take a look at our example there are several other viewpoints that could be taken into account. Just one example: the increased water consumption during mining. We have to confess that in the recent study on sheet-metal parts of EOL cars we do not deal with the side-effects at all. The reason is simply that we are in the beginning of the research and only try to define what should we do in this aspect.

Today the whole world, all at least most countries understand the importance of natural resources, environment, and based on this understanding reuse and recycling are getting more and more important in everyday life, as well as the decrease of CO2 emission, etc. These all request to keep energy, water, natural resources, manpower, etc. consumption in a moderate, sustainable level. This leads to sustainable development, or even to sustainability.

3.3 The KILT Model and the TYPUS Metrics

Just to remember we repeat some main points of the KILT model and the TYPUS, which are properly explained in [6,7,8 and 8a].TYPUS metrics means Tangibles Yield per Unit of Service. It is measured in money – on ecological basis. It reflects the total energy and material consumption of (all) (extended) products of a given unit, e. g. of an enterprise. But it can be applied for bigger (e.g. virtual enterprises) or any smaller units (e.g. workshop or one machine) or for any selected actions (e.g painting, bending, cutting, etc.) of any complexity. In this study it is all only about sheet-metal management of EOV, however due to the complexity of the problems and due to the status of the research we still do not make real value calculations.

The metrics assumes several things, as : life-cycle function; material and energy provisions during manufacturing, operation, repair, reuse or dismissal, etc.

3.3.1 KILT is an arbitrarily, but properly chosen implementation of TYPUS, we could imagine other realizations as well, However, recently the given definitions seem to be the best to manage the requested goals, as far as the authors believe. The related TYPUS metrics is further discussed later on. In earlier models and considerations The delivered quantities (all outputs), Q, is assumed to depend on the contributed **financial (I)** and **human (L) capitals** , plus the **know-how (K)** (innovation) and the **tangibles (T)** have non negligible effects..

The relationships are still work as multiplication and looks the next:

$$Q = f(K, I, L, T)$$

Summarizing the different factors we get some content to all of them as capital, knowledge, activity, material, etc. at the same time:

K: Technical capital – knowledge, technology, know how, etc.- intangibles
I: Financial capital– investment, capital, etc.
L: Human capital– labor, traditional labor, human efforts, welfare charges, etc.
T: Natural capital – tangible resources: material, consumables, ecologic fees, utilities, commodities, etc.

All the contributed technical K, financial I, human L and natural T capitals are included, and there is a tetra-linear dependence, which assumes to operate nearby equilibrium assets. The KILT models reliably describe the delivered product quantities, Q. Lacking one contribution (any of the above factors has a value of 0), the balance is lame, and the reckoned productivity figures, untruthful or meaningless.

The tetra-linear dependence means the equivalence of assets alone, and their synergic cumulated action. The company return is optimal, when the (scaled) factors are balanced; the current scaling expresses in money the four capitals (the comparison of non-homogeneous quantities is meaningless; the output Q has proper value, with the four inputs homogeneity). The return vanishes or becomes loss, if one contribution disappears. The loss represents the imbalance between constituent (know-how, money, work out-sourcing, bought semi-finished parts, etc.) flows.

3.3.2 The TYPUS Metrics

TYPUS, tangibles yield per unit service: the measurement plot covers the materials supply chain, from procurement, to recovery, so that every enjoyed product-service has associated eco-figures, assembling the resources consumption and the induced falls-off requiring remediation. The results are expressed in money. The point is left open, but, it needs to be detailed, to provide quantitative (legal metrology driven) assessment of the "deposit-refund" balance.

The metrics is an effective standard, aiming at the natural capital intensive exploitation. The supply chain lifecycle visibility needs monitoring and recording the joint economic/ecologic issues, giving quantitative assessment of all input/output materials and energy flows.

We have to apply these considerations to the decision sequence of "remake – no remake" of sheet metal parts of EOV, taking into account all technological steps and human actions including their side effects, if we can understand, define, measure and quantify them. It will take some ore time and a lot of researchers' efforts. To be a little positive we are convinced that the above discussed metrics and model can be used between any two points of the PLC, i.e. all costs, outputs, results, effects and side effects can be measured, calculated and evaluated in the decision sequence of sheet metal parts of EOL cars in our case.

4 Some Technological Issues of ISF

There are several open issues concerning the ISF technology and 3D measurements. We have to make several experiments with the scanner system to have an exact view of the measured sheet, and with the software systems, which compare the different surfaces with each other and with the accepted shape's data. These may be results of scanning, but more often results from the design (CA/CAM) processes. Finally the software dictates – and the humans generally accept - what should be done using ISF. On the ISF side we still have problems with accurate shape and thickness measurements. These works are running recently with high efforts. Fig. 3 presents an ISF experiment with an industrial robot using a 50 cm x 50 cm frame for the sheet. For car parts we can use the same robot, but a larger, and different frame will be needed. It is on the design table already.

From the technological point of view, the ISF consists of the gradual plastic deformation of a metal sheet by the action of a spherical forming tool whose trajectory is numerically controlled. The interest in evolution of ISF is rather old; it started in 1967 with the patent of Leszak [10]. This idea and technology is still active today in the field of producing sheet metal and polystirol parts in small batch and one-of-a-kind production, rapid prototypes, in medical aid manufacturing and in architectural design. A specific forming tool is mounted on the machine spindle or on a robot, and it is moved according to a well-defined tool path to form the sheet into the desired shape. Several ISF strategies have been developed which mainly differ in equipment and forming procedure. In particular, the process can be divided into:

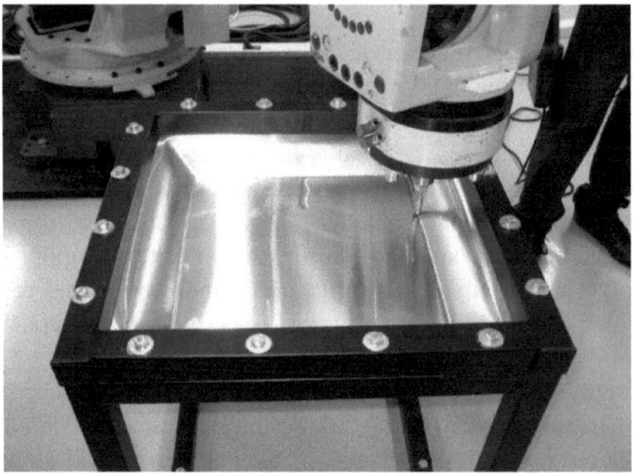

Fig. 3. ISF experiment with a FANUC robot

– **Single Point Incremental Forming (SPIF)**

Here the sheet metal is shaped by a single tool (with a faceplate supporting the initial level of the sheet).

– **Two Points Incremental Forming (TPIF)**

where the sheet metal shaping is ensured by: a) two counter tools or b) a local die support that is a sort of partial / full die.

In Full Die Incremental Forming the tool shapes the sheet alongside a die; this die could be produced from cheap materials such as wood, resin or low cost steel; the use of a die ensures a better and more precise shape of the final piece.

As repair tool only SPIF or TPIF with synchronised counter tools can be considered because the manufacturing of full or partial dies needs more time and money. On the other hand SPIF has some drawbacks compared to TPIF with two counter tools.

Experimental Investigations and Numerical Analysis were carried out by Shigekazu Tanaka et al. [11] to examine the residual stress in sheet metal parts obtained by incremental forming operations because distortion were observed after removing the outer portion of the incremental formed sheet metal part. Results showed that "tension residual stress is produced in the upper layer of the sheet and compression stress in the lower", furthermore the stress is increasing with the increase of the tool diameter [11].

Crina Radu [12] analysed the correlation between the accuracy of parts processed by SPIF using different values of process parameters and the distribution of the residual stresses induced in the sheets as results of cold incremental forming.

The hole drilling strain gauge method was applied to determine the residual stresses distribution through the sheet thickness. Experiments showed that the increase of tool diameter and incremental step depth increased residual stresses, which led to higher geometrical deviations [12].

J. Zettler et al. stated in their work that SPIF indicate "great residual stresses to sheet during the forming which lead to geometrical deviations after releasing the fixation of the sheet". They introduced a spring back compensation procedure in which an optical measurement system is used for measuring the part geometry after the forming [13].

Silva [14] et al. made some Experimental Investigations and Numerical Analysis to evaluate the applicability and accuracy of their analytical framework for SPIF of metal sheets. They stated that "plastic deformation occurs only in the small radial slice of the component being formed under the tool. The surrounding material experiences elastic deformation and, therefore, it is subject of considerably lower stresses."

In order to compensate springback more effectively (in-process) online residual stress measurements are suggested. Residual Stress Measurement Methods can be characterized according to the length detection over the stresses balance.

Feasible Non-Destructive Testing (NDT) Methods based on a summary of Withers et al. [15] are Ultrasonic and Magnetic Barkhausen noise (MBN) measurements. By comparing these methods we can say that ultrasonic solutions can be used for non-ferromagnetic materials too, but for the evaluation of Multiple Residual Stress Components the Barkhausen noise (BN) testing is preferable. The work of Steven Andrew White showed that "BN testing is capable of providing near-surface estimates of axial and hoop stresses in feeder piping, and could likely be adapted for in situ feeder pipe inspection or quality assurance of stress relief during manufacture" [16].

By adapting the MBN solution of Steven Andrew White to Incremental Forming of sheet metals we can realize an enhanced concept of J. Zettler et al. [13] where the optical measurement system is replaced/extended by a MBN measurement device integrated into a forming tool. This solution may allow finishing the manufacturing/repairing of a part with high geometrical accuracy however, without releasing the fixation.

5 Conclusions and Further Plans

Our real goal is to give some means and tools to calculate different values which correspond to different phases of the life-cycle of a product (PLC). We specially emphasize re-use and re-cycling as important LC phases, due to the approaching water-, energy- and raw material-shortages. Generally on product we mean anything which is used by simple users (a car, a cup, a bike, or a part of them, etc.), or which are used by dedicated users to produce or manage other products (a machine tool, a robot, a house, a test environment, etc.), or which are used to manage everything else (a firm, a factory, a ministry, etc.). We differentiate between simple products and extended products (as traditional and extended enterprise) and between tangible and intangible parts (aspects) and service is taken into account as a product, too.

In the recent study we restrict ourselves to a very narrow part of the PLCM of cars: to evaluate EOL cars' sheet-metal parts, then to decide whether to re-make (repair, re-use) them or let them go to the shredder to be dismissed.

During our research to assist re-use of sheet metal parts we had several problems to solve to make waste as little as possible, and to prefer re-use, or re-make. There were several machine- and human decisions, which need support. An important assistance is 3D modelling and visualisation to help human decision making if a simple view is not enough, as it cannot be exact enough. The set and methods of decision making drove us to the way of cognitive info-communication. This way should be extended and explained in more details in the future.

We showed that the above explained simple multiplications forms of KILT cannot yet be used for economically useful calculations, they contain only several ideas and qualitative relationships to go on a right way. We plan to find proper relationships to use our ideas and formulae for real world situations to assist not only designers and engineers in their work, but politicians and other decision makers as well. These studies and their resulting calculations, values and suggestions how to proceed will be in a following study. Specific applications to ISF technology may mean simplifications and easier understanding and using of the metrics and the model.

References

1. Gaillot, O., McCormack, I.: Depollution and Shredder Trial on End of Life Vehicles in Ireland, RPS Consulting Engineers, West Pier, Business Campus, Dun Laoghaire, Co., Dublin, Ireland (study)

1a. Directive 2000/53/EC of the European Parliament and of the Council of 18 September 2000 on end-of-life vehicles, http://www.municipalwasteeurope .eu/summary-current-eu-waste-legislation

2. http://www.autobonto.eu/bontas.html
3. http://www.IDIS2.com
4. http://www.gazautoschool.org.uk/home-2/vehicle-dismantling
5. Kazmierczak, K., Patrick, W., Winkel, J., Parkallé, L.: A Case Study of Serial-Flow Car Disassembly: Ergonomics, Productivity and Potential System Performance. Human Factors and Ergonomics in Manufacturing 17(4), 331–351 (2007), http://onlinelibrary.wiley.com/doi/10.1002/hfm.20078/pdf
6. Michelini, R.C., Kovács, G.L.: Integrated Design for Sustainability: Intelligence for Eco-Consistent Products and Services. In: EBS REVIEW, Innovation, Knowledge, Marketing and Ethics, pp. 81–94 (Winter 2002/2003) ISSN 1406-0264
7. Michelini, R.R.: Knowledge Enterpreneurship and Sustainable Growth, 325 pages. NOVA Science Publ. (2008)
8. Michelini, R.R.: Knowledge Society Engineering, A Sustainable Growth Pledge, 350 pages. NOVA Science Publ. (2010)

8a. Kovács, G.: Understanding enterprize delivery using the TYPUS metrics and the KILT model. In: Proc. of CSIT 2011, Garmisch, Germany, September 27-October 02, pp. 1–7 (2011)

9. Dernbach, J.C. (ed.): Stumbling towards sustainability. The Environmental Law Institute, Washington, D.C., pp. 1–968 (July 2002) ISBN: 1-58576-036-6
10. Leszak, E.: Apparatus and Process for Incremental Dieless Forming. US Patent 3342051A1 (September 19, 1967)

11. Tanaka, S., Nakamura, T., Hayakawa, K., Nakamura, H., Motomura, K.: Residual Stress in Sheet Metal Parts Made by Incremental Forming Process. In: Proceedings of the 9th International Conference on Numerical Methods in Industrial Forming Processes, Porto, Portugal, June 17-21, pp. 775–780 (2007), http://dx.doi.org/10.1063/1.2740904
12. Radu, C.: Analysis of the Correlation Accuracy-Distribution of Residual Stresses in the Case of Parts Processed by SPIF. In: Mathematical Models and Methods in Modern Science - 14th WSEAS International Conference on Mathematical Methods, Computational Techniques and Intelligent Systems (MAMECTIS 2012), Porto, Portugal, July 1-3, pp. 195–199 (2012)
13. Zettler, J., Rezai, H., Hirt, G.: Springback Compensation for Incremental Sheet Metal Forming Applications, 7. LS-DYNA Anwenderforum, Bamberg (2008), http://www.dynamore.de/de/download/papers/forum08/dokumente/C-I-03.pdf/view
14. Silva, M.B., Skjoedt, M., Bay, N., Martins, P.A.F.: Revisiting single-point incremental forming and formability/failure diagrams by means of finite elements and experimentation. Journal of Strain Analysis for Engineering Design 44(4), 221–234 (2009), doi:10.1243/03093247JSA522
15. Withers, P.J., Bhadeshia, H.K.D.H.: Residual stress part 2-nature and origins. Materials Science and Technology 17, 366–375 (2001)
16. http://www.phispace.net/spiffyprod/COGfeedworkshop2009.pdf

Enabling the Crowd Sourcing of Very Large Product Models

Martin Hardwick[1], David Loffredo[2], Joe Fritz[2], and Mikael Hedlind[3, 4]

[1] Rensselaer Polytechnic Institute, Troy, NY 12180, USA
hardwick@steptools.com
[2] STEP Tools, Inc., Troy, NY 12180, USA
[3] KTH Royal Institute of Technology., SE-100 44 Stockholm, Sweden
[4] Production Engineering R&D., Scania CV AB, 151 87 Södertälje, Sweden

Abstract. Part 21 is a public domain file format for the geometry of assemblies and is widely used by industry to describe design and manufacturing specifications for airplanes, automobiles, ships, buildings and other products. We describe a proposed new edition of Part 21 that includes URI's and JavaScript to enable the crowd sourcing of very large product models.

Keywords: Data exchange, Product Models, CAD, CAM, STEP, STEP-NC.

1 Introduction

Part 21 is a specification for how to format entities describing product data [1]. The format is minimal to maximize upward compatibility and simple to allow for quick implementation. It was invented before XML, though not before SGML, and it makes no special allowance for URL's [2].

Several technical data exchange standards use Part 21. They include STEP for mechanical products, STEP-NC for manufacturing products and IFC for building products. Over twenty years, substantial installed bases have been developed for all three with many Computer Aided Design (CAD) systems reading and writing STEP, a growing number of Computer Aided Manufacturing (CAM) systems reading and writing STEP-NC, and many Building Information Management (BIM) systems reading and writing IFC.

The data described by STEP, STEP-NC and IFC is continually growing [3]. STEP was first standardized as ISO 10303-203 for configuration controlled assemblies, and as ISO 10303-214 for automotive design. Both protocols describe the same kinds of information, and have taken turns at the cutting edge. Currently they are being replaced by ISO 10303-242 which will add manufacturing requirements, such as tolerances and surface finishes, to the product data [4].

STEP-NC is a related standard for manufacturing process and resource data. It has been tested by an industry consortium to verify that it has all the features necessary to replace traditional machining programs. They recently determined that it is ready for implementation and new interfaces are being developed by the Computer Aided Manufacturing (CAM) system vendors [5].

G.L. Kovács and D. Kochan (Eds.): NEW PROLAMAT 2013, IFIP AICT 411, pp. 254–272, 2013.

IFC describes a similar set of standards for building design and construction. IFC has made four major releases with the most recent focused on enabling the concurrent modeling of building systems. These include the systems for electric power, plumbing, and Heating, Ventilation and Air Conditioning (HVAC). The building structural elements such as floors, rooms and walls were already covered by previous editions. With the new release, different contractors will be able to share a common model during the construction and maintenance phases of a building [6].

All three models are being used by a large community to share product data but Part 21 has been showing its age for several years. In the last ten years there have been six attempts to replace it with XML [7]. To date, none has succeeded but there is a growing desire for a more powerful and flexible product data format.

This paper describes an extension to Part 21 to enable the crowd sourcing of very large product models. Extending the current format has the advantage of continuing to support the legacy which means there will be a large range of systems that can already read and write the new data. The new edition has two key new capabilities:

1. The ability to distribute data between model fragments linked together using URI's.
2. The ability to define intelligent interfaces that assist the user in linking, viewing and running the models.

The next section describes the functionalities and limitations of Part 21 Editions 1 and 2. The third section describes how the new format enables massive product databases. The fourth section describes how the new format enables crowdsourcing. The fifth section outlines the applications being used to test the specification. The last section contains some concluding remarks.

2 Editions 1 and 2 of Part 21

STEP, STEP-NC and IFC describe product data using information models. Each information model has a schema described in a language called EXPRESS that is also one of the STEP standards [3]. EXPRESS was defined by engineers for engineers. Its main goal was to give clear, concise definitions to product geometry and topology. An EXPRESS schema defines a set of entities. Each entity describes something that can be exchanged between two systems. The entity may describe something simple such as a Cartesian point or something sophisticated such as a boundary representation. In the latter case the new entity will be defined from many other entities and the allowed data structures. The allowed data structures are lists, sets, bags and arrays. Each attribute in an entity is described by one of these data structures, another entity or a selection of the above. The entities can inherit from each other in fairly advanced ways including AND/OR combinations. Finally EXPRESS has rules to define constraints: a simple example being a requirement for a circle radius to be positive; a more complex example being a requirement for the topology of a boundary representation to be manifold.

Part 21 describes how the values of EXPRESS entities are written into files. A traditional Part 21 file consists of a header section and a data section. Each file starts with the ISO part number (ISO-10303-21) and begins the header section with the HEADER keyword. The header contains at least three pieces of information a FILE_DESCRIPTION which defines the conformance level of the file, a FILE_NAME and a FILE_SCHEMA. The FILE_NAME includes fields that can be used to describe the name of the file, a time_stamp showing the time when it was written, the name and organization of the author of the file. The FILE_NAME can also include the name of the preprocessing system that was used to write the file, and the name of the CAD system that created the file.

```
ISO-10303-21;
HEADER;
/* Exchange file generated using ST-DEVELOPER v1.5 */

FILE_DESCRIPTION(
/* description */ (''),
/* implementation_level */ '2;
1');
FILE_NAME(
/* name */ 'bracket1',
/* time_stamp */ '1998-03-10T10:47:06-06:00',
/* author */ (''),
/* organization */ (''),
/* preprocessor_version */ 'ST-DEVELOPER v1.5',
/* originating_system */ 'EDS - UNIGRAPHICS 13.0',
/* authorisation */ '');

FILE_SCHEMA (('CONFIG_CONTROL_DESIGN')); /* AP203 */
ENDSEC;
DATA;
#10 = ORIENTED_EDGE('',*,*,#44820,.T.);
#20 = EDGE_LOOP('',(#10));
#30 = FACE_BOUND('',#20,.T.);
#40 = ORIENTED_EDGE('',*,*,#44880,.F.);
#50 = EDGE_LOOP('',(#40));
#60 = FACE_BOUND('',#50,.T.);
#70 = CARTESIAN_POINT('',(-
1.31249999999997,14.594,7.584));
#80 = DIRECTION('',(1.,0.,3.51436002694883E-15));
...
ENDSEC;
END-ISO-10303-21;
```

One or more data sections follow the header section. In the first edition only one was allowed and this remains the case for most files. The data section begins with the keyword DATA, followed by descriptions of the data instances in the file. Each instance begins with an identifier and terminates with a semicolon ";". The identifier is a hash symbol "#" followed by an unsigned integer. Every instance must have an identifier that is unique within this file but the same identifier can be given to another instance in another file. This includes another version of the same file.

The identifier is followed by the name of the entity that defines the instance. The names are always capitalized because EXPRESS is case insensitive. The name of the instance is then followed by the values of the attributes listed between parentheses and separated by commas. Let's look at instance #30. This instance is defined by an entity called FACE_BOUND. The entity has three attributes. The first attribute is an empty string, the second is a reference to an EDGE_LOOP and the third is a Boolean with the value True. The EXPRESS definition of FACE_BOUND is shown below. FACE-BOUND is an indirect subtype of representation item. The first attribute of FACE-BOUND (the string) is defined by this super-type. Note also that the "bound" attribute of face_bound is defined to be a loop entity so EDGE_LOOP must be a subtype of LOOP.

```
ENTITY face_bound
  SUBTYPE OF (topological_representation_item);
  bound       : loop;
  orientation : BOOLEAN;
END_ENTITY; -- face_bound

ENTITY topological_representation_item
SUPERTYPE OF (ONEOF (vertex, edge, face_bound, face, con-
nected_edge_set, connected_face_set, vertex_shell,
wire_shell, (loop ANDOR path)))
  SUBTYPE OF (representation_item);
END_ENTITY; -- topological_representation_item

ENTITY representation_item;
  name : label;
WHERE
  WR1: SIZEOF (using_representations(SELF)) > 0;
END_ENTITY; -- representation_item
```

There were two major design goals for Part 21 Edition 1.

1. Make it easy for CAD software engineers to create readers and writers for the new format.
2. Maximize the upward compatibility by minimizing the number of keywords and structural elements in the data exchange files.

The first goal was met by requiring the files to be encoded in simple ASCII, by requiring all of the data to be in one file, and by the requiring every identifier to be an unsigned integer that only has to be unique within the context of one file. The latter condition was presumed to make it easier for engineers to write parsers for the data.

The second design goal was met by minimizing the number of keywords and structural elements (nested parentheses). In most cases, the only keyword is the name of the entity and unless there are multiple choices there are no other keywords listed. The multiple choices are rare. They happen if an attribute can be defined by multiple instances of the same type. An example would be an attribute which could be a length or a time. If both possibilities are represented as floating point numbers then a keyword is necessary to indicate which has been chosen.

Making the Part 21 format simple was helpful in the early years as some users developed EXPRESS models and hand populated those files. However as previously mentioned there are thousands of definitions in STEP, STEP-NC and IFC and to make matters worse the definitions are normalized to avoid insertion and deletion anomalies. Consequently, it quickly became too difficult for engineers to parse the data by hand and a small industry grew up to manage it using class libraries. This industry then assisted the CAD, CAM and BIM vendors as they implemented their data translators.

The two design goals conflict when minimizing the number of keywords makes the files harder to read. This was dramatically illustrated when XML became popular. The tags in XML have allowed users to create many examples of relatively easy to understand, self-describing web data. However, for product models contain thousands of definitions the tags are less helpful. The following example recodes the first line of the data section of the previous example in XML.

```
<data-instance ID="i10">
   <type IDREF="oriented_edge">
    <attribute name="name"
        type="label"></attribute>
    <attribute name="edge_start"
        type="vertex" value="derived"/>
    <attribute name="edge_end"
        type="vertex" value="derived"/>
    <attribute name="edge_element"
        type="edge"><instance-ref
            IDREF="i44820"/></attribute>
    <attribute name="orientation"
        type="BOOLEAN">TRUE</attribute>
   </type>
</data-instance>
```

Six XML data formats have been defined for STEP in two editions of a standard known as Part 28 [6]. The example above shows the most verbose format called the Late Binding which tried to enable intelligent data mining applications. Other formats were more minimal though none as minimal as Part 21. In practice the XML tags add small value to large product models because anyone that wants to parse the data needs to process the EXPRESS. Plus they cause occasional update problems because in

XML adding new choices means adding new tags and this can mean old data (without the tags) is no longer valid.

The relative failure of Part 28 has been mirrored by difficulties with Part 21 Edition 2. This edition sought to make it easier for multiple standards to share data models. At the time STEP was moving to an architecture where it would be supporting tens or hundreds of data exchange protocols each tailored to a specific purpose and each re-using a common set of definitions. Edition 2 made it possible to validate a STEP file in all of its different contexts by dividing the data into multiple sections each described by a different schema. In practice, however, the applications have folded down to just three that are becoming highly successful: STEP for design data, STEP-NC for manufacturing data and IFC for building and construction date.

The failures of XML and Edition 2 need to be balanced against the success of Edition 1. This edition is now supported by nearly every CAD, CAM and BIM system. Millions of product models are being made by thousands of users. Consequently there is an appetite for more, and users would like to be able to create massive product models using crowd sourcing.

3 Massive Product Models

The following subsections describe how Edition 3 enables very large product models. The first two subsections describe how model fragments can be linked by URI's in anchor and reference sections. The third subsection describes how the transport of collections of models is enabled using ZIP archives. The fourth subsection describes how the population of a model is managed using a schema population.

3.1 Anchor Section

The syntax of the new anchor section is simple. It begins with the keyword ANCHOR and ends with the keyword ENDSEC. Each line of the anchor section gives an external name for one of the entities in the model. The external name is a reference that can be found using the fragment identifier of a URL. For example, the URL www.server.com/assembly.stp#front_axle_nauo references the front_axle in the following anchor section.

```
ANCHOR;
<front_axle_nauo> = #123;
<rear_axle_nauo> = #124;
<left_wheel_nauo> = #234;
<right_wheel_nauo> = #235;
ENDSEC;
```

Unlike the entity instance identifiers of Edition 1, anchor names are required to be unique and consistent across multiple versions of the exchange file. Therefore, although the description of the front_axle in chasis.stp may change, the front_axle anchor remains constant.

3.2 Reference Section

The reference section follows the anchor section and enables references into another file. Together the reference and anchor sections allow very large files to be split into fragments.

The reference section begins with the keyword REFERENCE and ends with the keyword ENDSEC. Each line of the reference section gives a URI for an entity instance defined in an external file. In this example, the external file contains references to the anchors given in the previous example. The file is defining a manufacturing constraint on an assembly [7].

```
REFERENCE;
/* assembly definitions for this constraint */
#outer_seal_nauo = <assembly.stp#outer_seal>;
#outer_bearing_nauo =
    <assembly.stp#outer_bearing>;
#right_wheel_nauo = <assembly.stp#right_wheel>;
#rear_axle_nauo = <assembly.stp#rear_axle>;

/* Product definitions */
#seal_pd = <assembly.stp#seal_pd>;
#bearing_pd = <assembly.stp#bearing_pd>;
#wheel_pd = <assembly.stp#wheel_pd>;
#axle_pd = <assembly.stp#axle_pd>;
#chasis_pd = <assembly.stp#chasis_pd>;
ENDSEC;
```

The example uses names for the entity identifiers. This is another new feature of Edition 3. Instead of requiring all entity identifiers to be numbers they can be given names to make it easier for casual users to code examples, and for systems to merge data sets from multiple sources.

```
DATA;
#124 = SPECIFIED_HIGHER_COMPONENT_USAGE ('id',
       'name', 'outer seal on outer bearing',
  #chasis_pd,#seal_pd,'outer', #125,
       #outer_seal_nauo);
#125 = SPECIFIED_HIGHER_COMPONENT_USAGE ('id',
       'name', 'right wheel on rear axle ',
       #chasis_pd,#bearing_pd,,'rear', #126,
       #outer_bearing_nauo);
#126 = SPECIFIED_HIGHER_COMPONENT_USAGE ('id',
       'name', 'right wheel on rear axle ',
       #chasis_pd,#wheel_pd,'outer',
       #rear_axle_nauo,#right_wheel_nauo);
ENDSEC;
```

Numbers are used for the entities #124, #125 and #126 because it is traditional but the rest of the content has adopted the convention of giving each instance a name to indicate its function and a qualifier to indicate its type. Thus "chasis_pd" indicates that this instance is the product_definition entity of the chasis.stp file.

3.3 ZIP Archives and Master Directories

The anchor and reference sections allow a single earlier-edition file to be split into multiple new files but this can result in management problems. The old style led to files that were large and difficult to edit outside of a CAD system, but all of the data was in one file which was easier to manage.

ZIP archives allow Part 21 Edition 3 to split the data and continue the easy data management. A ZIP archive is a collection of files that can be e-mailed as a single attachment. The contents of the archive can be any collection including another archive. A ZIP archive is compressed and may reduce the volume by as much as 70%. Many popular file formats such as ".docx" are ZIP files and can be accessed using ZIP tools (sometimes only after changing the file extension to .zip)

Edition 3 allows any number of STEP files to be included in an archive. Each file in the directory can be linked to the other files in the ZIP using relative addresses and to other files outside of the ZIP using absolute addressing. Relative addresses to files outside the ZIP are not allowed so that applications can deploy the zipped data at any location in a file system.

References into the ZIP file are allowed but only via a master directory stored in the root. This Master directory describes where all the anchors in the ZIP can be found using the local name of the file. Outside the archive the only visible name is that of the archive itself. If there is a reference to this name then a system is required to open the master directory and look for the requested anchor.

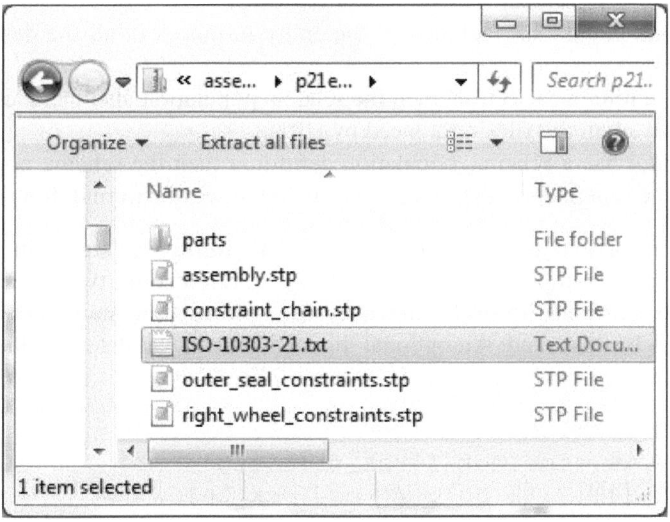

Fig. 1. Zip Archive

In Figure 1 the file ISO-10303-21.txt is the master directory. It contains the forwarding references to the other files and it is the file that can be referenced from outside of the archive using the name of the archive.

3.4 Schema Population with Time Stamps

The support for data distribution in Part 21 Edition 3 gives rise to a problem for applications that want to search a complete data set. If the data is distributed and normalized then there may be "orphan" files that contain outbound references but no inbound ones. For example, Figure 2 shows how a file may establish a relationship between a workplan and a workpiece by storing URI's to those items but not have a quality that needs to be referenced from anywhere else.

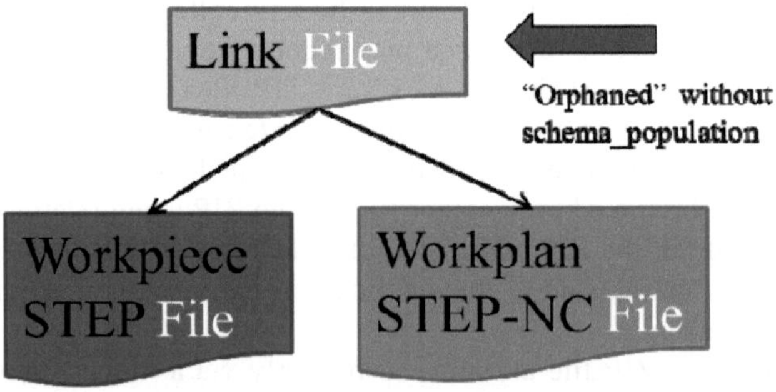

Fig. 2. Link files for massive databases

Part 21 Edition 3 solves this problem using a Schema Population which is defined as follows:

- The schema population includes all the entity instances in all the data sections of the file.
- If there is a reference section, then the schema population also includes the schema populations of all the files referenced by URIs.
- If the header has a schema_population definition then the schema population also includes the schema population of each file in the set of external_file_locations.

The last inclusion catches the "orphan files". The following code fragment gives an example. In this example the STEP file shown is referencing two other files. There are two other attributes in each reference. An optional time stamp shows when the reference was last checked. An optional digital signature validates the integrity of the referenced file.

```
SCHEMA_POPULATION(
('http://www.acme.net/first_file.stp',
 '2013-02-11T17:35:00', /* padlock below */      /* Time
stamp */
 '30dd879c-ee2f-11db-8314-0800200c9a66');
```

```
('http://www.giant.com/second_file.stp',
 '2013-02-11T17:30:00', /* time stamp left */
 '')); /* no padlock for this instance*/          /* not
locked */
```

The time stamp and signature enable better data management in contractual situations. Clearly if the data is distributed there will be opportunities for mistakes and mischief.

4 Intelligent Interfaces

The following subsections describe how Edition 3 enables crowdsourcing using intelligent interfaces. The first subsection describes how JavaScript has been added to the model. The second subsection describes how some of the data rules have been relaxed for easier programming. The third subsection describes how application specific programming is enabled using data tags. The last subsection summarizes the options for making the programming more modular.

4.1 JavaScript

The goal of adding JavaScript to Part 21 Edition 3 is to make managing the new interfaces easier by encapsulating the tasks that can be performed on those interfaces as methods. For example, the following code checks that a workpiece is valid before linking it to a workplan.

```
function link(workplan, workpiece, name)
{
    with workpiece.anchor.credentials.tag {
       if (stage != 'manufacturing')
    return null;
       if (status != 'released')
          return null;
    }
    if (name != 'as_is shape'
  && name != 'to-be shape')
return NULL;
    if (workpiece.anchor.shape.tag.type
  != 'product_definition_shape')     // alternate is to
check using an EXPRESS compiler
        return NULL:
    if (workpiece.anchor.shape.tag.geometry
  != 'brep')
        return NULL;
    name = type;
    shape = workpiece.anchor.shape;
    exec = workplan.anchor.executable;
    return model;
}
```

In Edition 3 a file can include a library of JavaScript functions to operate on an object model of the anchors and references but not necessarily the data sections. In many cases the volume of the data sections will overwhelm a JavaScript interpreter.

The following three step procedure is used for the conversion:

1. Read the exchange structure and create a P21.Model object with anchor and reference properties.
2. Read the JavaScript program definitions listed in the header section.
3. Execute the JavaScript programs with the "this" variable set to the P21.Model object.

For more details see Annex F of the draft specification at www.steptools .com/library/standard/. The procedure creates one object for each exchange file and gives it the behaviour defined in the JavaScript. The execution environment then uses those objects in its application. For example, in the code above workplan and workpiece are object models for two exchange structures and the application is checking for compatibility before linking them.

4.2 Data Relaxation

In order to be reusable a product model needs to flexible and extensible. The information models defined by the STEP, STEP-NC and IFC standards have been carefully designed over many releases to achieve these qualities.

Interface programming is different because an interface can be created as a contingent arrangement of anchors and references for a specific purpose. The information model has not changed so application programming for translation systems stays the same, but for interface programming the requirements are different. Therefore, two relaxations have been applied to the way data is defined for interfaces in the new edition.

1. The instance identifiers are allowed to be alphanumeric.
2. The values identified can be lists and literals.

Editions 1 and 2 of Part 21 restricted the format so that every identifier had to be an unsigned integer. This helped emphasize that the identifiers would not be consistent across files and at the time it was thought to make it easier for parsers to construct symbol tables. The symbol table reason is false. Every modern system uses a hash table for its symbols and these tables are agnostic with respect to the format of the identifier.

Requiring numbers for identifiers has always made hand editing harder than necessary. Therefore, Edition 3 supports alphanumeric names. The following example of unit definition illustrates the advantage. Each unit definition follows a pattern.

```
ANCHOR;
<NEWTON> = #newton;
<PASCAL> = #pascal;
ENDSEC;
```

```
REFERENCE;
#metre =
<http://www.iso10303.org/part41/si_base_units.stp#METRE>;
#kilogram =
<http://www.iso10303.org/part41/si_base_units.stp#KILOGRA
M>;
#second =
<http://www.iso10303.org/part41/si_base_units.stp#SECOND>
;
ENDSEC;

DATA;
/* Content extracted from part 41:2013 */
#5_newton=DERIVED_UNIT_ELEMENT(#meter,1.0);
#15_newton=DERIVED_UNIT_ELEMENT(#kilogram,1.0);
#25_newton=DERIVED_UNIT_ELEMENT(#second,-2.0);
#new-
ton=SI_FORCE_UNIT((#5_newton,#15_newton,#25_newton),*,$,.
NEWTON.);

#5_pascal=DERIVED_UNIT_ELEMENT(#meter,-2.0);
#25_pascal=DERIVED_UNIT_ELEMENT(#newton,1.0);
#pas-
cal=SI_PRESSURE_UNIT((#5_pascal,#25_pascal),*,$,.PASCAL.)
;
```

Unit definitions also show why a more flexible approach to defining literals is desirable. The following is a file that defines some standard constants.

```
ANCHOR;
<ARCHIMEDES_CONSTANT_PI >= #pi;
<EULER_NUMBER_e> = #e;
<GOLDEN_RATIO> = #golden;
ENDSEC;

REFERENCE;
#pi = 3.14159265359;
#e = 2.71828;
#golden = 1.61803398874;
ENDSEC;
END-ISO-10303-21;
```

The new identifiers can be used in the data section as well as the anchor and reference sections.

4.3 Tags for Fast Data Caching

In many cases the JavaScript functions operating on the interfaces need additional data to be fully intelligent. Therefore, the new edition allows additional values to be tagged into the reference and data sections.

Each tag has a category and a value. The category describes its purpose and the value is described by a literal. The following example shows the data checked by the JavaScript function of the previous example.

```
ANCHOR;
<credentials> {stage:'manufacturing'}
              {status:'released'}
    = $;
<shape>       {geometry:'brep'}
   = #217652;
ENDSEC;
```

The tag data may be initialized by a pre-processor or created by other means. In the above example two pieces of data are being linked and it is important to know that the workpiece is ready for use by manufacturing.

Another role for tags is as a place to cache links to visualization information. Again this data may be summarized from the STEP, STEP-NC or IFC information model and the tags allow it to be cached at a convenient place for rapid display.

```
ANCHOR;
<tool_tip> {visualization:<facets.xml>}
           {picture: <item.jpg>}
    = #197;
ENDSEC;
```

The last example shows tags being used to document the STEP ARM to AIM mapping. Those who have worked on STEP and STEP-NC know that they have two definitions: a requirements model describing the information requirements; and an interpreted model that maps the requirements into a set of extensible resources [3]. The tags can become a way to represent the mapping between these two models in the data.

```
REFERENCE;
#1234 {x_axis:(#3091,#3956)}{y_axis: (#2076)} = <#ma-
chine_bed>;
#4567 {z_axis:(#9876,#5273)}= <#tool_holder>;
ENDSEC;
```

A quick summary of the new data options in Edition 3 is that it allows URL's to be placed between angular brackets ("<>") and application specific data to be placed between curly brackets ("{}").

4.4 Modularity Options

Part 21 Edition 3 has three options for modularizing STEP, STEP-NC and IFC data.

- Continue using traditional files, but surround those files with interfaces referencing into the data
- Create a product data web linked by URL's.
- Create a ZIP archive of replaceable components.

The traditional approach to STEP implementation creates a massive symbol table using an EXPRESS compiler and then reads the exchange data into objects described by the table. This is expensive both with respect to processing time and software investment, but efficient if all of the data is being translated into a CAD system.

The new edition allows the Part 21 data to be arranged into interfaces for light weight purposes such as checking tolerances, placing subsystems and running processes. Therefore, alternate implementation paradigms are possible. Three being considered include:

1. A JavaScript programming environment can process just the data in an interface. In this type of implementation the Part 21 files are rapidly scanned to create the objects required for the anchor and reference sections.
2. A web browser environment can be activated by including an "index.html" file in the ZIP archive along with code defining the P21 object model of the interface. This type of implementation will be similar to the previous one but with steaming used to read the Part 21 data.
3. The third type of implementation is an extended Standard Data Access Interface (SDAI). The SDAI is an application programming interface for Edition 1 and 2 data that can be applied to Edition 3 because of upward compatibility.

One option for an SDAI is to merge all the Edition 3 data into one large file for traditional CAD translation processing. Another option is to execute the JavaScript and service web clients

5 Applications

5.1 PMI Information for Assemblies

The first in-progress application is the management of Product Manufacturing Information (PMI) for assemblies. Figure 3 shows a flatness tolerance on one of the bolts in an assembly. In the data, a usage chain is defined to show which of the six copies of the bolt has the tolerance.

The data for the example can be given many organizations. One is the traditional single file which will work well for small data sets. For large assemblies the new specification enables a three layer organization. The lowest layer is the components in the model. The second layer is the assemblies and sub-assemblies. The third layer is the PMI necessary to manufacture the assembly.

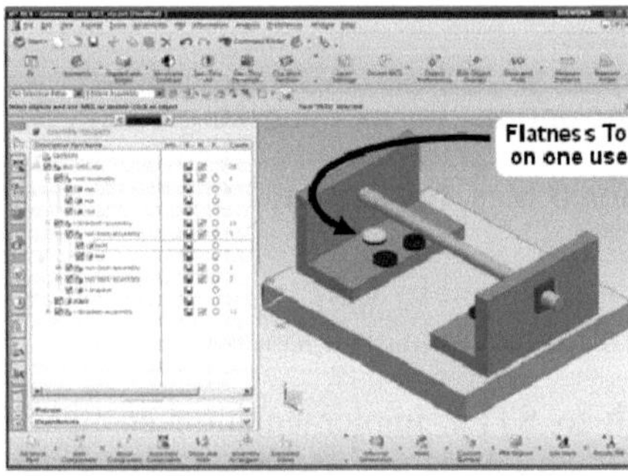

Fig. 3. Assembly tolerances

In order for this organization to work the product components must expose their product coordinate systems to the assembly modules and their faces to the PMI modules. Similarly the assembly modules must expose their product structure to the PMI modules. The following shows the resulting anchor and reference sections of the PMI module.

```
ANCHOR;
<as1-pe> = #141;
END_SEC;

REFERENCE;
#bolt = <bolt.stp#bolt>;
#nut = <nut.stp#nut>;
#rod = <rod.stp#rod>;
#plate = <plate.stp#plate>;
#l-bracket = <l-bracket.stp#l-bracket>;

#bolt_shape = <bolt.stp#bolt_shape>;
#nut_shape = <nut.stp#nut_shape>;
#rod_shape = <rod.stp#rod_shape>;
#plate_shape = <plate.stp#plate_shape>;
#l-bracket_shape = <l-bracket.stp#l-bracket_shape>;

#bolt_wcs = <bolt.stp#bolt_wcs>;
#nut_wcs = <nut.stp#nut_wcs>;
#rod_wcs = <rod.stp#rod_wcs>;
#plate_wcs = <plate.stp#plate_wcs>;
#l-bracket_wcs = <l-bracket.stp#l-bracket_wcs>;

#bolt_top_face = <bolt.stp#bolt_top_face>;
ENDSEC;
```

This code is then referenced in the data sections of the PMI modules.

5.2 Data Model Assembly

STEP and STEP-NC are related standards that share definitions. The two models can be exported together by integrated CAD/CAM systems, but if different systems make the objects then they must be linked outside of a CAD system.

In STEP-NC a workplan executable is linked to the shape of the workpiece being machined. The two entities can be exported as anchors in the two files and an intelligent interface can link them on-demand. The following code shows the interface of a linker file with open references to the two items that must be linked. The linker JavaScript program given earlier sets these references after checking the validity of the workpiece and workplan data sets. It also sets the name of the reference to indicate if workpiece represents the state of the part before the operation (as-is) or after the operation (to-be).

```
REFERENCE;
#exec = $;
#shape = $;
#name = $;
ENDSEC;

DATA;
#10=PRODUCT_DEFINITION_PROCESS(#name,'',#exec,'');
#20=PROCESS_PRODUCT_ASSOCIATION('','',#shape,#10);
ENDSEC;
```

A number of CAM vendors are implementing export interfaces for STEP-NC[5]. They export the process data which needs to be integrated with workpiece data to achieve interoperability. The workpieces define the cutting tools, fixtures and machines as well as the as-is and to-be removal volumes.

5.3 Next Generation Manufacturing

The last application is the control of manufacturing operations. Today manufacturing machines are controlled using Gcodes generated by a CAM system [8]. Each code describes one or more axis movements. A part is machined by executing millions of these codes in the right order with the right setup and the right tooling. Change is difficult which makes manufacturing inflexible and causes long delays while engineers validate incomplete models.

Part 21 Edition 3 can replace these codes with JavaScript running STEP-NC. The broad concept is to divide the STEP-NC program into modules each describing one of the resources in the program. For example, one module may define a toolpath and another module may define a workingstep. The modules can be put into a ZIP archive so the data volume will be less and the data management easier. The JavaScripts defined for each module makes them intelligent. For example for a workingstep, the script can control the tool selection and set tool compensation parameters.

Before a script is started other scripts may be called to make decisions. For instance an operation may need to be repeated because insufficient material was removed, or an operation may be unnecessary because a feature is already in tolerance. Such functionalities can be programmed in today's Gcode languages but only with difficulty.

The JavaScript environment is suited to manufacturing because it is event driven. Performance should not be an issue because in practice machine controls operate by running look-ahead programs to predict future movements. Changing the look-ahead to operate on JavaScript instead of Gcode is probably a better use of resources.

For an example of how such a system might operate see Figure 4 which is a screen capture of the following WebGL application:

http://www.steptools.com/demos/nc-frames.html?moldy/

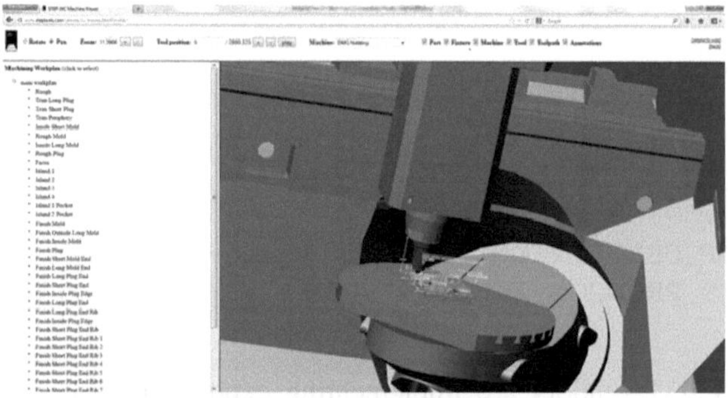

Fig. 4. WebGL Machining

6 Conclusions

Part 21 Edition 3 is an upward compatible enhancement to Part 21 that enables data linking and intelligence. Editions 1 and 2 have been highly successful. Nearly every CAD, CAM and BIM system has a Part 21 interface.

The new edition adds URI's and JavaScript to enable the crowdsourcing of massive product models. Other supporting features include a schema population to keep track of all the components, ZIP archives to enable better data management, data relaxation to enable easier interface programming, and data tags to allow application specific programming.

The new specification is not yet finished. Additional extensions are still being considered. They include allowing the files in a ZIP archive to share a common header section, merging the anchor and reference sections into an interface section and loosening the syntax to allow for lists of URI's.

The current specification can be accessed at:
http://www.steptools.com/library/standard/p21e3_dis_preview.html

Implementation tools are being developed to enable testing. They include libraries to read and write the data into a standalone JavaScript system called NodeScript, libraries to stream the data into web browsers, and libraries to run the scripts in an SDAI. The NodeScript implementation is available as open source at the following location.

http://www.steptools.com/library/standard/

The next steps (pun intended) are to:

1. Complete the prototype implementations so that they can be used to verify the specification.
2. Submit the specification to ISO for review as a Draft International Standard (DIS).
3. Respond to the international review with additional enhancements for the additional requirements.
4. Begin the development of common object models for design and manufacturing applications. For example, object models for the execution of machine processes, and object models for the definition of assembly tolerances.
5. Create applications to demonstrate the value of the specification.

The applications will include attention grabbing ones that use kinematics to show the operation of products and machines, value added ones that use the JavaScript to link data sets and create massive product models, and manufacturing ones to verify tolerances while processes are running.

References

1. ISO 10303-21: Industrial automation systems and integration — Product data representation and exchange — Part 21: Implementation methods: Clear text encoding of the exchange structure (2002)
2. Uniform Resource Identifiers (URI): Generic Syntax. Internet Engineering Task Force RFC 2396 August 1998, Available from World Wide Web (1998), `http://www.ietf.org/rfc/rfc2396.txt`
3. ISO 10303-1: Industrial automation systems and integration — Product data representation and exchange — Part 1: Overview and fundamental principles (1994)
4. ISO/CD 10303-242: Industrial automation systems and integration - Product data representation and exchange — Part 242: Application protocol: Managed Model-based 3D Engineering (2012)
5. ISO 10303-238: Industrial automation systems and integration — Product data representation and exchange — Part 238: Application Protocols: Application interpreted model for computerized numerical controllers (2007)
6. ISO 16739: Industry Foundation Classes for data sharing in the construction and facility management industries (2013)

7. ISO/TS 10303-28: Industrial automation systems and integration — Product data representation and exchange — Part 28: Implementation methods: XML representations of EXPRESS schemas and data, using XML schemas (2007)
8. Hardwick, M., Zhao, Y.F., Proctor, F.M., Nassehi, A., Xu, X., Odendahl, S.V.D., Xu, L., Hedlind, M., Lundgren, M., et al.: A roadmap for STEP-NC-enabled interoperable manufacturing. The International Journal of Advanced Manufacturing Technology 66(1-4) (March 2013)
9. ISO/IEC 16262, Information technology – Programming languages, their environments and system software interfaces – ECMAScript language specification (2011)

Knowledge Retrieval in Complex Information Landscapes Related to Products and Production

Marcus Schichtel

BMW Group
markus.schichtel@bmw.de

Abstract. Developing and manufacturing complex vehicles today creates an increasing amount of information and knowledge that is by far not stored in a single data source. The challenge of knowledge retrieval in complex information landscapes has been approached at BMW with the help of semantic search technologies and a user oriented design of the "Knowledge Finder", in order to truly connect unstructured and structured data sources semantically.

Keywords: Information Landscapes, Semantic Search, Knowledge, PLM, Manufacturing.

1 Introduction

Developing and manufacturing complex products such as vehicles today creates an increasing amount of information and knowledge that is by far not stored in a single data source like a PLM-system. Over time the number of isolated information islands has grown to a point where nobody oversees anymore how particular pieces of information are linked together that actually form a complex information landscape. In [1] a good example is given explaining what consequences could arise, if a specific dependency e.g. between a safety sensor and an environmental side condition remains hidden in a complex information landscape describing the functions of a car. That article points out in particular that the missing information that could have resolved the problem before product launch was simply located at an altogether different location.

To put this observation into another context it shall be pointed out (see fig. 1) that nobody can possibly know all available data sources in a large enterprise. Rather each of the people involved either has the needed piece of information in his memory or searches for it in those data sources that are personally known. Looking at the pie chart reveals that this way approximately only 10% of information and knowledge is tapped into. By far the largest portion is knowledge (approx. 90%), that an individual is not even aware of that it exists ("I don't even know that I don't know"). In fact making it possible for the individual to "sail this (to him) unknown sea of knowledge" is a major challenge today.

G.L. Kovács and D. Kochan (Eds.): NEW PROLAMAT 2013, IFIP AICT 411, pp. 273–285, 2013.

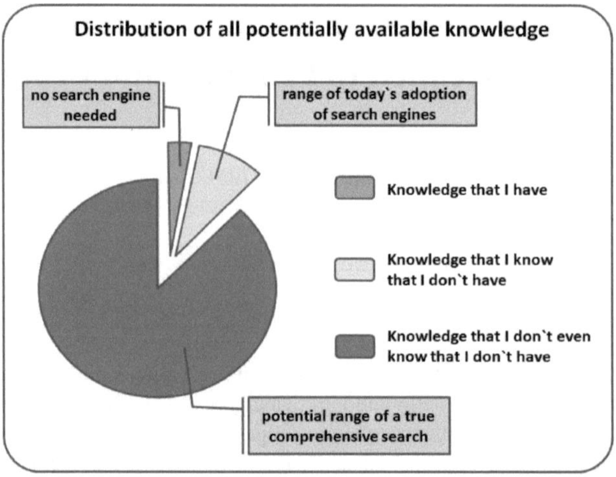

Fig. 1. The Threefold way of categorizing potentially available knowledge in a large enterprise

An interesting and strong corroboration of this observation is the following quotation from [2]:

"Gelesene Bücher sind längst nicht so wertvoll wie ungelesene. Eine Bibliothek sollte so viel von dem enthalten, was man nicht weiß, wie der Besitzer angesichts seiner finanziellen Mittel … hineinstellen kann"[1]

The challenge of dealing with isolated information islands (or information silos) has drawn top management attention as pointed out in [3]:

"… in order to answer a question information silos are a hindrance. … How many different locations do I have to visit in order to put many single pieces together for meaningful information? Information must be accessible to all decision makers throughout the enterprise."

Figure 1 can also be interpreted as the vast offer of knowledge in any enterprise that is waiting to be uncovered again by the user that has a great potential of matching the pressing demand for information and knowledge from prior experiences. The following short list shall illustrate this demand showing a great bandwidth of complexity:

- What does "CFK" stand for?
- Who is the right contact to discuss an issue with the fender?
- What module does part "4711" belong to?
- What would be an alternative material for this alloy?
- What issues did we have with the headlight in former car projects?
- What would be the environmental impact of using this coolant instead of the one in use today?

---

1 ***Translation by author:*** *"Books that have already been read are not as valuable as those unread. A library should always contain as much of those things not known to its owner as the owner can possibly shelve into his library according to his financial abilities"*

But digital information alone is not the whole story. Lots of information and knowledge is also buried in hardware prototypes! In former times the classical way of validating a product was putting hardware prototypes together. The process of putting together gave rise to insights into problems right away and naturally gave answers to questions not explicitly asked before, i.e. the physical prototype answered the question of "Does it work?" implicitly as a whole. The people carrying out the validation process gained a tremendous treasure of knowledge of experience with no particular need to store this knowledge digitally in data sources. However today, with the increasing pressure to reduce hardware prototypes more and more validation is carried out with virtual prototypes that are not put together like physical prototypes and therefore cannot reveal important issues right away.

Rather, validating a virtual prototype implies the need to know issues from the past so that the engineers validating a product can inspect the virtual prototype accordingly. Complex questions during validation such as "What implications on production processes would be incurred if we used a different blank thickness in car body design?" would resemble such a typical situation.

In order to make this new paradigm work we need to first digitize older knowledge and second find a way to make it accessible efficiently.

With the increasing pace of business (faster processes, increasing fluctuation among employees) less time is available to find missing vital information for the task at hand.

The classical idea of connecting IT-systems by interfaces that are hard and expensive to develop, to maintain and sometimes to even redesign may have worked in the past but cannot catch up with the increasing speed and complexity of today`s engineering processes. In fact, today it is not uncommon that business processes in general are expected to happen in real time. Hence new and more sophisticated IT-tools are needed guiding the user through the complex information landscape to find the right knowledge in acceptable time.

2 The BMW-Approach: Knowledge Finder

2.1 Business Context

The principle ideas of the Knowledge Finder originated from a close collaboration between business (department in charge of the car validation process) and IT while creating a new form of review method in order to detect issues early on in a car development project. During the first pilot of the new review method it became evident from the beginning that the method stimulated the open flow of knowledge among the participants across all disciplines. Though, as it has been noted evaluating the first reviews, the flow of knowledge couldn't be captured for future usage.

In other words knowledge was elusive and the immediate question was whether this knowledge could be easily captured by a knowledge database. Soon it became clear that creating a new knowledge database was not the right answer to the challenge at hand for a number of reasons:

- The creation of a new data source would have meant to implement a very challenging, if not impossible, change management task considering the vast amount of business processes that would have to be changed in order to store data into the new database rather than using the existing data sources
- Further it would have meant to create a new silo of information in addition to the existing ones that still continue to contain very valuable information
- It would have been totally unclear at what point in time the amount of "new information" would have reached a sufficient level so that the users would have reaped a benefit tapping into the new knowledge database compared to the existing data sources

Hence the fundamental design decision for the Knowledge Finder came to rest on three pillars (see fig.2):

- Reuse of existing data sources rather than creating an additional new knowledge database
- The idea of Knowledge Finder acting as an agent networking in the "community of data sources" similar to a human that networks among his coworkers
- Exploitation of "state of the art"- semantic search technologies in order to follow a user-centric approach and to create an intuitive and "fun to use" – tool to find knowledge rather than search for it

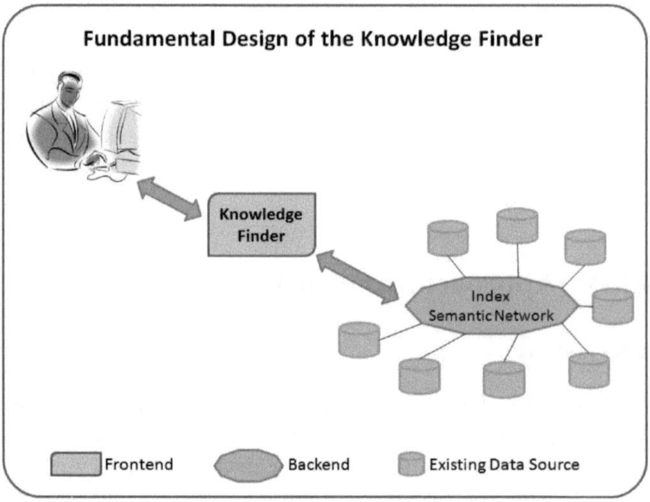

Fig. 2. - The fundamental design idea of the Knowledge Finder

Applying the fundamental design of the Knowledge Finder to the new review method mentioned before the following description of one of its main use cases could be described as follows

- The discussion in a review session stalls because a vital piece of information/knowledge is needed and none of the participants knows this piece of information
- The Knowledge Finder is triggered in order to answer the question "What (else) do we know about the issue being discussed?"
- The Knowledge Finder taps into all available data sources as a single access portal and presents its findings in a user friendly form thereby harnessing the flood of information
- The participants of the review either obtain the missing information at once or at least get a clue in which direction they should proceed with their discussion

2.2 Exploration of the Idea

With this fundamental design idea in mind it was decided to apply a stage gate innovation management process to explore the potential of the Knowledge Finder while keeping risks at a minimum at the same time.

A steering committee consisting of business and IT managers across development and production department alike was formed to decide at each stage whether to proceed in the process or not (see fig. 3).

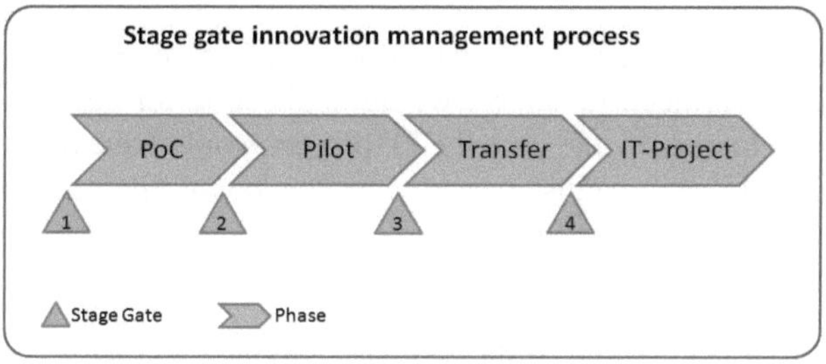

Fig. 3. Stage gate innovation management process

The stage gate innovation management process was carried out as follows

- First a proof of concept was carried out with the intention to find out whether the idea of the Knowledge Finder would really create additional business value and would be accepted by the users. Note here that at this point a truly business-centric approach has been followed. No data sources were connected to the Knowledge Finder online in order to avoid efforts considered as too risky at this early stage.
- After an initial research for possible solutions a list of six IT-partners was compiled by a very small core team and proposed as participants in a first "Proof of Concept" (PoC) to the steering committee, which approved the selection for the PoC.

- Once the findings of the PoC in terms of additional business value and user acceptance turned out to be very positive, the steering committee approved the proposal to proceed with a pilot phase where eight different data sources (structured data bases and unstructured data on a file server) should be truly connected to the Knowledge Finder.
- The core team then selected three IT-partners to be invited to the bidding process for the pilot solution. The winner of the bidding was then contracted to implement the Knowledge Finder as a pilot.
- Again the results of the pilot proved to be very positive and corroborated the earlier assessments of the Knowledge Finder. The steering committee consequently approved the proposal to proceed with the so called "transfer phase" which was intended to fulfill all the prerequisites for a formal proposal to initiate an IT-project to implement a productive solution.

2.3 IT-Architectural Considerations

While progressing along the stage gate process we also noted quite early on that the market for semantic search technologies is highly dynamic, i.e. the solution that seems appropriate today may not be the appropriate solution anymore in the future. Therefore we established two guiding principles:

- The solution we chose for the pilot would not necessarily be the natural choice for the prioritized IT-project, rather we would again look for the best suited solution available on the market by the time the IT-project would have been started
- The solution architecture must provide for the possibility to exchange the core semantic search technology while leaving the connectors to the various data sources unchanged (see fig. 4).

We designed a strict decoupling of the semantic engine together with its search index and the data sources with the corresponding crawler database views, such that at any future point in time the single investments in setting up the database views remain protected while the semantic engine is changed over to a new technology.

Furthermore we decided to decouple the **application** "Knowledge Finder" from the semantic engine such that other applications can reuse the semantic engine and the underlying search index for their purposes. This way we can both provide a semantic extension to a search engine inside a legacy system or provide other (new) applications (e.g. applications that have a focus on categorizing information) in need of semantic information processing with a common platform.

In fact, while being in the transfer phase another department (lab technology) requested to pilot the Knowledge Finder in order to assess possible improvements for their business process. And indeed, looking again at the market we found that a different solution provider could offer more added value to our Knowledge Finder (2.0) (see Table 1), which has been contracted for our second pilot.

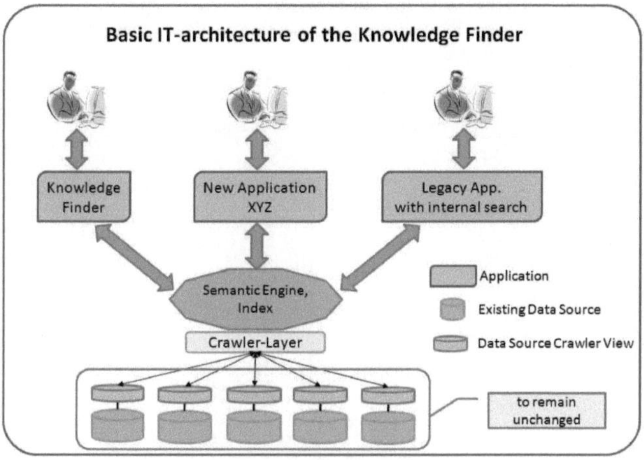

Fig. 4. Basic IT- Architecture of the Knowledge Finder

Table 1. The evolution of the Knowledge Finder

Knowledge Finder 1.0	Knowledge Finder 2.0 (more added value)
Connecting structured and unstructured data sources	True semantic search
Transparent interactive and iterative search with backtracking	Bilingual search (German / English)
Fast localization of most relevant hits	Search for semantically similar phrases
Facetted search	Keyword in context search
First approaches to semantic search	Extraction of information objects from free text fields
Search for BMW-abbreviations and BMW-jargon	Suggestion – Autocomplete

Figure 5 below gives an overview of what is possible with today's available semantic search technology. Given the search term "Plastik" Knowledge Finder 2.0 returns hits such as "Kunststoffanbauteile" and "plastic" demonstrating the capability of a semantic and bilingual search.

In addition Knowledge Finder 2.0 also extracts information objects such as "GS 97045-2" from free text, which denotes a so called "group standard" used at BMW that is mentioned within a lab technology report.

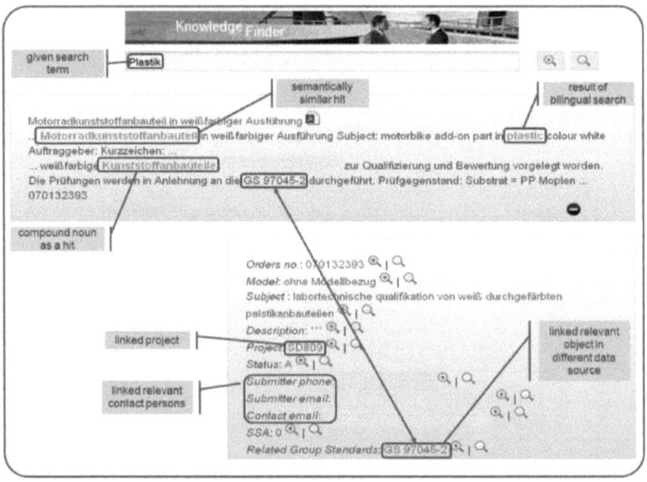

Fig. 5. Features of semantic search

3 Key Findings

The two successful pilots resulted in a number of very interesting findings and observations which are outlined in the following sections.

3.1 User Experience

The possibility to ask a variety of different systems in a single consistent way rather than filling out different clumsy query interfaces obeying the internal data base schema was very well received by the users.

The Knowledge Finder's interactive and intuitive interface enables a search process where the human brain and the search engine work together in finding the best hits thereby also avoiding the "Google-Syndrome" of obtaining thousands of hits that a user is incapable of dealing with. In fact it has been found from a cognitive information processing point of view that it is much easier for the human brain to start with a single search phrase first and then to successively add more search phrases thereby reducing the number of hits very fast.

It has been found empirically that in the vast majority of cases the user could reach a list of less than 20-30 meaningful hits with no more than three steps (either adding another search term or selecting a filter value)

Apparently the list of found knowledge matching the first search term stimulates the brain to pick the next search term by association. There is no need for the user to come up with more or less complicated boolean expressions to define the search, which in many cases cannot be stated anyway in the first place since the user is not necessarily 100% clear about what he is really searching for in the beginning.

3.2 New Opportunities

Finding experts on specific issues has become much easier than before as the Knowledge Finder helps employees and organizational units alike in extending and strengthening their networks.

Particularly in light of the increasing fluctuation of employees this opportunity cannot be valued highly enough, because for a knowledge worker it is essential that the cost of obtaining missing information is minimized.

The Knowledge Finder allows to ask questions or at least to explore possible solution spaces which has not been an option before because of the high hurdles to retrieve information hidden in all the various data sources.

Another new opportunity that the Knowledge Finder may offer is the possibility to address the challenge to put all information islands together mentioned by Randy Mott, CIO of General Motors in [3] (see also introduction). By pulling information from various data sources together into one single search index, the Knowledge Finder might open up a new road to circumvent all the challenges and problems of classical tedious system interfaces, such that the search index could be viewed as a new form of information bus.

3.3 Challenges

Of course, piloting the Knowledge Finder has revealed a number of challenges, some of which could very well be characterized as fundamental in terms of how do we manage information as a very valuable asset in a large enterprise in general. Within the scope of this paper three aspects of managing large quantities of information (quality, protection, and comprehension) shall be addressed and explained briefly why they are important.

3.3.1 Quality of Information

It is common knowledge that for a communication to be successful it is necessary that sender and receiver have the same understanding of coding and interpreting information. If for example the German word for "expanding rivet" is written incorrectly in the data source (i.e. "Spreitzniet" instead of "Spreizniet") then that piece of information is not found when entering the correct spelling even drawing upon semantic search technologies. In fact during the lab technology pilot we both found hits when searching for "Spreitzniet" (incorrect spelling) and searching for "Spreizniet" (correct spelling) but the sets of hits were disjoint!

In much the same way no semantic search technology today would be able to discern a variety of different ways to abbreviate (longer) words within free text (e.g. "homologation" = "hom." = "homol." = "homolog.").

Most search engines today include a spell checker with the auto suggestion function when entering a search term but they cannot cope with misspelled words in the data sources. Even worse, a spell checker applied to "Spreitzniet" would change the search term to "Spreizniet" thus missing (possibly) the most valuable information hits associated with the misspelled word "Spreitzniet".

Another type of lacking quality in information showed up in our first pilot. One of the test users turned very excited when he found a long missing Powerpoint-document only to turn disappointed when opening the document and discovering that its content didn`t match the search term at all. After a short investigation of this "false positive hit" it became clear that the meta information of the Powerpoint -document did match with the search term but by no means coincided with the content. What happened was, that (and that is typical) the Powerpoint-document originated from a template whose author and title had not been changed appropriately by the author of the found Powerpoint-document.

From a more general point of view the root cause of these observations can be explained by taking the approach that we are actually dealing with a flow of information.

With regard to a flow of material there exist a number of quality assurance techniques such as value stream analysis, (Ishikawa) fishbone diagrams or the application of Poka-Yoke principles in order to prevent wrong assembling of parts. For flows of information however, we do not possess corresponding quality assurance techniques nor have we transferred and adapted the well proven techniques for flows of material to flows of information.

One possible approach to do so could be to apply the so called DMAIC -cycle known from the six sigma toolset as indicated in [4] and [5]. However this is quite a generic practice which would lack specific tools in our context.

One such tool or best practice could be that the moment information is originally entered into any data source a spell checker automatically corrects input. This way we would also apply one of the lean production best practices known as "responsibility on site", meaning that any error that might occur should be corrected instantaneously right at the location where it occurred.

3.3.2 Protection of Information

Naturally the idea of an all encompassing semantic search that links across various data sources triggers concerns with regard to intellectual property protection. In fact, in almost all data sources rules are in place to discern users that are allowed to read particular information from users that are not. Today it is common practice to grant (read) privileges on a "need-to-know-basis". However, we have found throughout the pilots that this principle has a number of disadvantages, of which some are listed below

- Once a read privilege is granted on a personal basis in most cases it will never be removed, i.e. the read privilege might still be in place even though the personal need is not given anymore
- Granting privileges on a need to know basis can be tedious, in particular if the number of information objects to be protected or the number of users is large. (we came across a situation where about 800 different folders on a file server needed to be overseen resulting in a full time job to do all the administration)

- Typically read privileges on a file server can only be granted at the level of folders. Hence if we grant read privileges to allow a user to read a particular file in that folder we inadvertently give read privileges to other files residing in the same folder for which that particular user should not have read privileges.
- Last but not least the responsibility to grant any kind of privilege does not lie with the owner of the information object but rather with the (system) administrator, who even might not be employed by the enterprise

Therefore one of the key findings of the pilots was to propose a different principle to protect information which we call the "need-to-protect-principle". With this principle we strengthen the role of the information owner in the first place. The basic idea is that the information owner decides whether a certain piece of information can be read by anyone within the enterprise or shall only be visible to a restricted circle of users. In the latter case this information is not even presented to the Knowledge Finder, i.e. it is never been indexed and therefore cannot be found at all.

If, on the other hand, certain information can be read by anyone inside the enterprise then a fair balance of interests can be established between the urgent need to share information for better collaboration within the enterprise and the need to protect (strictly) confidential information.

As a consequence the unproductive efforts of granting read access on a need-to-know-basis are minimized if not eliminated.

Finally applying semantic technologies it is well conceivable to scan documents in order to help classify them into the right protection level.

3.3.3 Comprehension of Information

Piloting the Knowledge Finder we soon discovered that extending the scope of data sources beyond those known to the user (compare fig. 1) it is likely that users might get overloaded with the sheer number of information objects and their interdependencies.

In other words the Knowledge Finder gives rise to unveiling the complex information landscape that we inherently create while progressing within our product development process today.

First of all, making interdependencies in a complex product (recall [1] in the introduction) visible is e.g. highly advantageous to an engineer who needs to validate a complex function of a complete vehicle and cannot do this anymore because he is lacking a physical prototype.

On the other hand (see fig. 6) it is questionable whether such a complex information landscape is always correctly mapped to the user`s brain so that the user is having the correct understanding of all the information and thus being able to draw the right conclusions.

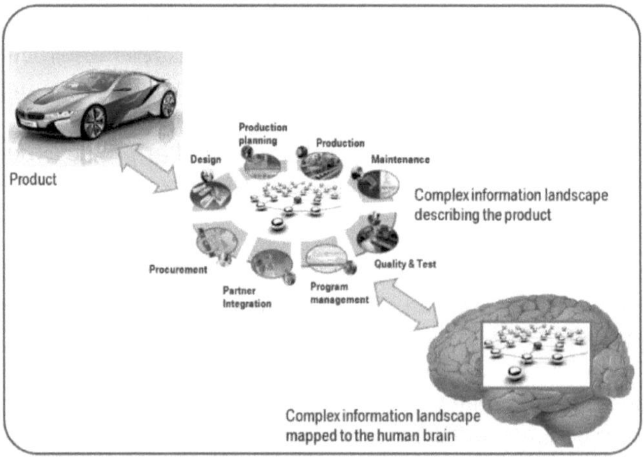

Fig. 6. Complex information landscapes vs. processing capabilities of the human brain

4 Conclusions

In this paper it was shown how the increasing need to find buried information more efficiently again has led at BMW to the development of the Knowledge Finder and what road has been chosen to do so. During two pilot phases in two different disciplines (domains) many very valuable findings have been collected that influence the design of the next evolution of the Knowledge Finder for productive use.

In order to address the challenges mentioned before (quality, protection and comprehension of information) further research is necessary resulting for example in advancements with respect to semantic technologies.

As far as the aspect of comprehending complex information landscapes is concerned it is believed that new cross-discipline input from cognitive psychology and neuro sciences is needed in order to answer the following two questions.

- What is the appropriate design of new man-machine interfaces such that a minimum level of stress is imposed on the user when exploring complex information landscapes?
- What is the appropriate design of new user trainings so that the brain can comprehend complex information landscapes better?

These questions cannot be addressed by computer science alone. Rather an interdisciplinary approach together with cognitive psychologists and neuroscientists is needed that could even lead to a new branch of computer science that might be called "Psycho-Informatics" in the future much like "Bio-Informatics" today.

Such an interdisciplinary approach should look at new forms of interaction with information landscapes that go beyond the currently prevailing paradigm of interacting with information by screen, keyboard and mouse clicks.

It shall also be noted here that the further development of the Knowledge Finder will be a continuous ongoing process with adding more data sources and functional improvements and the design of the user interface step by step. We aim actually at offering the Knowledge Finder and its steady improvements as a service rather than rolling out software releases from time to time.

The idea is "Finding knowledge as a service" where suggestions from our users are added in an agile manner.

References

1. Siemens, P.L.M.: Herstelleralbtraum: Ein Produkt mit Funktionsfehlern. In: Interface - Das Magazin für ProductLifecycle Management, 15, Jahrgang (January 2012)
2. Taleb, N.: Der Schwarze Schwan. Seite 17 DTV 34596, 2, Auflage (2010)
3. Bongard, A.: Interview with Randy Mott, CIO General Motors. In: Automotiveit, Ausgabe (October 2012)
4. Schmidt, T.: Systematisches DQ-Management: den Daten bloß nicht immer hinterherlaufen. In: BI Spektrum, Ausgabe (May 2011)
5. Schmidt, T.: Der DMAIC-Zyklus für ein systematisches Datenqualitätsmanagement. In: BI-Spektrum, Ausgabe (April 2012)

Advanced Engineering Visualization with Standardized 3D Formats

Christian Emmer, Arnulf Fröhlich, and Josip Stjepandic

PROSTEP AG, Darmstadt, Germany
josip.stjepandic@opendesc.com

Abstract. When products are developed in 3D for the engineering domain, the data is initially stored in the native format of the used CAD software. If this 3D CAD data is to be made available to people who do not have this software, neutral 3D formats are needed. For visualization of product data in the engineering field - regardless of native CAD formats – are various 3D formats available. Among these are disclosed or standardized formats like PDF from Adobe, JT and also X3D, Collada and STEP. The choice of a format has many implications, including which options are available for using the data and what follow-up costs will result. This paper illustrates an overview of the state of development, the committees' activities as well as current examples from practice and gives outlook for future developments too. The record is completed by examples from the industrial practice in the automotive industry.

Keywords: Engineering Visualization, Interoperability, Data Exchange, JT Standard, Supply Chain, Engineering Collaboration.

1 Introduction

Although the CAD translation technology has gathered a high level of maturity and robustness, there is still a strong need for further development driven by crucial business processes like digital mock-up (DMU) which is established as core validation process in the past years. Furthermore the CAD models have already exceeded the level of complexity, where they are easy to handle with. Meanwhile various 3D visualization formats have entered the market on the broad front by combining the high information content with easy handling and excellent performance. Thus these formats can support many process chains from styling to product simulation, validation and production as well as downstream life cycle phases. By the extensive penetration of visualization formats in the market the old discussion was reopened, whether the unique format can fulfil the requirements of the engineering collaboration (internal and external). Finally the users need the recommendation too, which format meets their needs in the best way [1], [2]. Typical use cases for functional evaluation of such formats are the viewing of engineering data, the design in context, the data exchange between partners in the supply chain, the packaging and digital mock-up (DMU), the documentation and archiving, and use in the portable PLM document, i.e. use of 3D and additional information in domains related to engineering (Figure 1).

G.L. Kovács and D. Kochan (Eds.): NEW PROLAMAT 2013, IFIP AICT 411, pp. 286–295, 2013.

After the rough evaluation of the technical specification of the visualization formats an overall comparison can be drawn [3] (Figure 2).

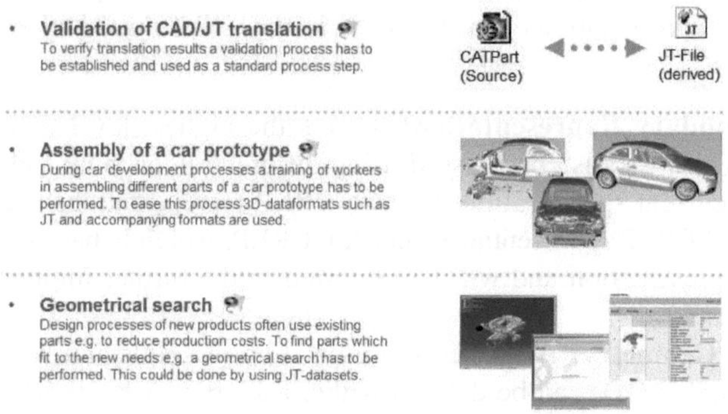

Fig. 1. Typical use cases of JT [1]

Use Case	STEP	3D XML	JT	3D PDF
Viewing	●	●●	●●●	●●●
Data Exchange	●●●	●	●●	●
DMU	●	●●	●●●	●
Documentation and Archiving	●●	●	●●	●●●
Portable PLM Document	●	●	●	●●●

Legend
●●● Highly suitable
●● Well suitable
● Suitable with reservations

Fig. 2. Comparison of four visualization formats [3]

However in the automotive and aerospace industry JT from Siemens PLM is currently under heavy investigation to apply it in downstream processes after it was established as leading format for the engineering visualization in many global companies [4]. For instance, there is the demand to exchange exact geometry, product data information and additional attributes. Moreover JT is also a potential data format for long-term data retention. Consequently, there is high user interest in quality and robustness of the applications, the interfaces in particular. Thus the further explanation in this paper will be focused on JT although other formats (STEP, 3DPDF, 3DXML) are in the different stages of standardization procedure.

Open standards enable reduction of total costs of ownership and ensure independence from specific vendors and competition. With regard to visualization data exchange there is a strong support to use JT and STEP AP 242 (ISO 10303-242) as complementing standards for lightweight visualization format for 3D industrial data as well as product structure, meta, kinematic data [5].

The publication of the JT specification as an ISO Publicly Available Specification (PAS 14306) laid the foundation for establishing JT as a binding process format. The ProSTEP iViP Association and the VDA (German Association of the Automotive Industry) subsequently launched three coordinated JT-specific projects: the JT Workflow Forum, the JT Implementor Forum and the JT Application Benchmark.

2 Technical Background and Standardization

JT is a binary format whose data model supports various representations of CAD geometry. The representations can be stored in a JT file individually or together [5] [6].

- **BREP (Boundary Representation):** Offers the highest level of representational precision. BREP data is compressed using different algorithms and stored without loss. In the current specification 9.5, two BREP representations are permitted: the traditional JT-BREP representation and XT-BREP, which is based on the Parasolid boundary representation and will be preferred in the further implementation of JT based software.
- **Tessellated Geometry:** A faceted representation of solids and surfaces. Different levels of detail (LOD) can be defined within a JT file. A low LOD means a lower level of precision but a smaller volume of data, while a very high LOD means an almost exact geometry but a large volume of data.
- **ULP (Ultra-Lightweight Precise):** The latest compression method is ULP. The ULP format enables a lightweight, semi-precise representation of the 3D geometry. The level of precision that ULP offers is significantly higher than for tessellated geometry while the file size is significantly smaller. The primary focus lies on providing high quality surface geometry that exhibits only minor deviations from the original BREP geometry.

JT version 8.1 has been initially published by the ISO as a publicly available specification (ISO PAS 14306). The ISO standardization process has been finished by endorsing the ISO standard (ISO 14306) in December 2012 for the recently published JT version 9.5, which has been expanded to include the specification for ULP and semantic product manufacturing information (PMI, product metadata) among other things.

Generating data in a neutral 3D format results in most use cases with a significant reduction in volume. In order to compare the file size of the 3D formats, fifteen test assemblies from different CAD systems were used to generate different 3D formats [7]. As a result the volume of data is determined more by data content then by the format itself.

The volume of exact BREP data after conversion to JT is approximately the same as after conversion to 3D PDF. The same applies to tessellated content, where the result for both formats is about the same as for 3D XML. In the case of simplified BREP data, the volume of data for both JT and 3D PDF are approximately the same. In the case of STEP data, use of an external compression algorithm achieves a marked reduction in size. As a consequence higher accuracy requires more storage space as a general rule.

3 Use Cases

An evaluation of the 22 use cases defined by automotive manufacturers and their suppliers indicated a particularly high priority for four use cases, which are presented below in more detail by way of example [7] (Figure 3).

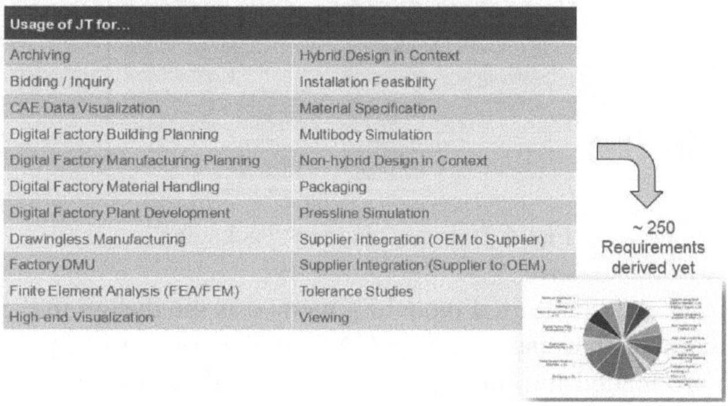

Fig. 3. Use cases for simplified formats [7]

Viewing: If the use of a CAD system is not desired, the visualization of engineering data using 3D viewers comes into play in a number of different situations: the presentation of product data, the representation of 3D models for information purposes (e.g. for a design review or marketing) and the realistic representation in virtual reality systems.

The use case can vary according to the concrete application context involved. While the simple viewing of the geometry is sufficient in many cases, in other cases metadata or high-performance viewing of huge assemblies plays a key role. The most important requirements are:

- Quality criteria for geometry accuracy must be fulfilled.
- Different level of details must be available.
- Detailed colour information must be available for entire part and single geometrical features.
- Metadata like attributes must be stored.

Packaging / DMU: In digital mock-up (DMU), the spatial properties of a product are examined and checked. This can involve checking the overall geometry with regard to dimensions and shape, interference checks, collision checks for assembly and disassembly, as well as design space checks (Figure 4).

For these purposes, the geometry, product structure and metadata are displayed and analyzed in a DMU application. The result of the DMU analysis is subsequently summarized and documented in a report. The most important requirements are:

- Use of models from different source systems (multi-CAD).
- High-quality examination of large assemblies.

- Transferability of kinematics from the original model to the target model for dynamic DMU analysis.

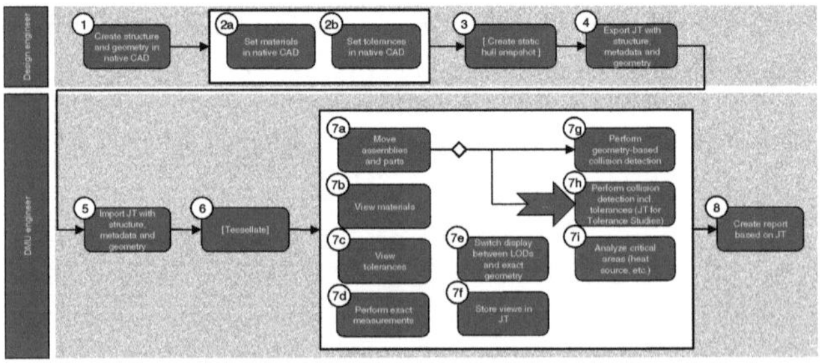

Fig. 4. Use case diagram "Packaging"

Design in Context: A typical used modeling method is the design in the context of the existing geometry. In that case existing CAD models (e.g. from previous design or from partners) are loaded into the CAD system. The design engineer then references new or existing JT-geometry to the loaded parts or assemblies. References from native CAD to JT are also in focus. These references can be between auxiliary geometry (point, axis, plane), geometrical elements such as edges, vertices or faces and exact geometrical references like curves. In the subsequent downstream processes it is required to use this assembled model as merged single unit. Hence the creation of technical drawings must be possible. The most important requirements are:

- Tessellated as well as exact B-Rep data must be stored.
- Links between JT and native data must be possible.
- Technical drawings must be derivable from hybrid models.

Partner Integration: The use case is divided into two sub sections. The aim of the "OEM to Supplier" use case is to send validated JT geometry with all needed metadata to the supplier, whereas the "Supplier to OEM" use case defines the exchange of JT files to the OEM.

Common to both sections is that the use cases provide a basis for further use cases. This means that by the Supplier or the OEM more specific use cases (viewing, design in context, etc.) will be applied subsequently. The most important requirements are:

- Quality check of JT and validation against native CAD model.
- Meta data must be contained in JT model.
- Removal of intellectual property (if necessary).

4 Evaluation and Testing

An important aspect in driving the application and development of JT is a coordinated approach between the different bodies on the one hand and a coordinated approach of the activities themselves on the other hand.

4.1 Involved Bodies

Basis for the standardization activities of JT were the enormous efforts of many international bodies. After the initial impact has be accomplished by SASIG [1], the four bodies ProSTEP iViP Association (PSI), the German Association of the Automotive Industry (VDA), the Automotive CIO´s and the Global Automotive Advisory Group for PLM (GAAG) have played a leading role (Figure 5). Therewith ProSTEP iViP (www.prostep.org) has driven the activities thoroughly.

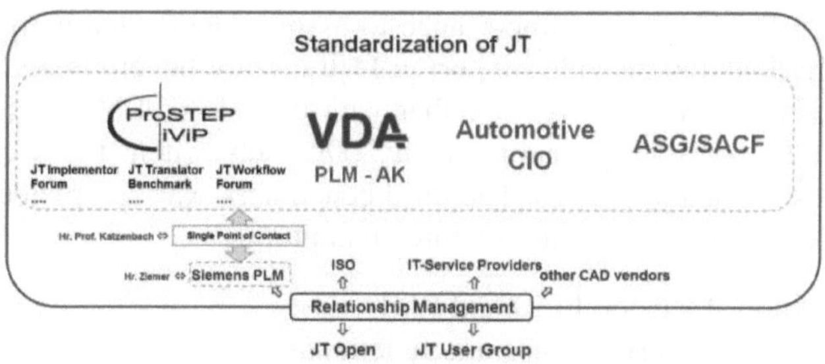

Fig. 5. International cooperation in the standardization of JT

Last but not least the MoU (Memorandum of Understanding) between Siemens PLM und ProSTEP iViP provides the basis between both parties in the area of JT, in order to foster a common understanding of goals and measures to establish JT as binding process format.

4.2 National and International Coordination

In 2009 the ProSTEP iViP Association, in a joint effort with the VDA (German Automotive Association), has started three different but interacting world-wide unique JT activities. Within the so-called JT Workflow Forum Use Cases for the application of JT are specified, describing the common view from industry. The JT Implementor Forum provides support in JT translator development maintains a platform for interoperability testing for the vendors. Within the JT Application Benchmark the requirements are comprehensively tested within neutral environment.

The project plans of these three project groups are well-harmonized, for driving the application of JT as a common effort.

JT Workflow Forum: Aim of the project is to define and prioritize key use cases and to identify relevant downstream processes for the JT based data exchange. For all use cases requirements are collected and test criteria defined e.g. under the aspects of exchanging of visualization data, the validation of 3D geometry, GD&T (Geometric Dimensions and Tolerances) and last but not least translator quality.

These tasks are supported by the Content Harmonization subgroup, which is responsible for providing a clear definition of which requirements should be satisfied by the JT format and which by STEP AP242 as backbone format.

JT Implementor Forum: Within the framework of the JT Implementor Forum Project Group, a platform for interoperability testing for the vendors is maintained. It is a forum where participating vendors can test the translators that they are developing "behind closed doors". The JT Implementor Forum is a neutral forum for software vendors, where they can perform tests in an atmosphere of mutual trust and exchange information on experience already gained.

In addition, from the vendor perspective, the activities within the JT Implementor Forum can be seen as a kind of preparation for the JT Application Benchmark.

JT Application Benchmark: Based on the recent standardization activities of JT and also STEP AP 242, it is of importance to assure data exchange quality within a neutral Benchmark and to provide support in JT translator development.

The ProSTEP iViP Association, together with the VDA PLM working group, has initiated a first JT translator benchmark in 2009. Subsequent the second translator benchmark was conducted in 2010. The third benchmark was completed by the end of 2012. Thereby the focus will be laid on three main topics:

1. CAD to JT export, focus on LOD and PMI.
2. JT viewing, focus on performance and functionality
3. Assembly conversion with STEP AP 242 XML and JT

Vendors participating in the benchmark are invited to present their latest functionalities in form of Showcases to the JT Workflow Forum members. In focus are cutting-edge solutions that demonstrate the possibilities of the JT format and applications. There are still small functional errors and therefore need for improvement (Figure 6).

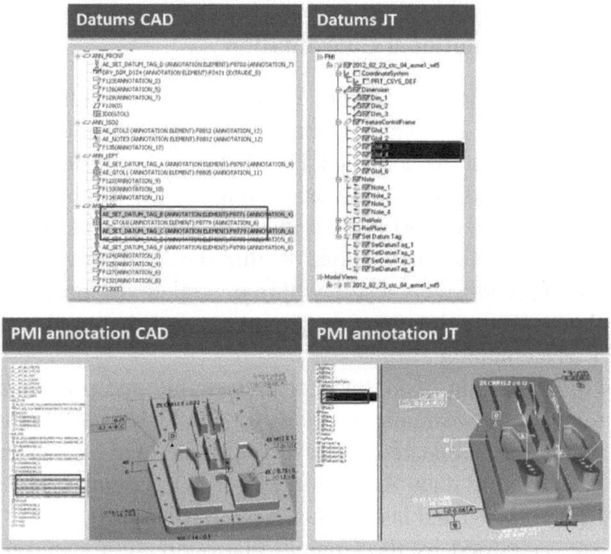

Fig. 6. Results of recent JT benchmark

5 Pilot Projects

As a further step towards the productive usage of JT in automotive downstream processes, first pilot projects are established. Volkswagen decided to use JT for internal downstream processes to eliminate the need for drawings [7] [8] (Figure 7). This application is limited to the use case "single parts" and shall be subsequently extended to the full assemblies to avoid the usage of huge JT monolithic files.

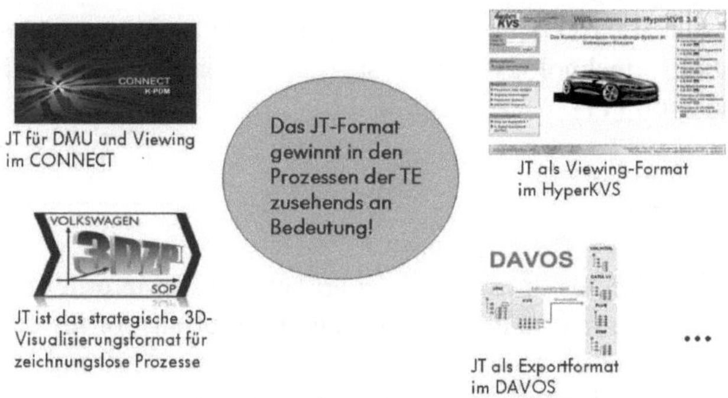

Fig. 7. Usage of JT within the Volkswagen Group

One of the most challenging project is the Daimler/Continental JT pilot project by exchanging an ESP Electronic Control Unit.

In the following the process landscape (Use Case: "Supplier to OEM") of the Daimler/Continental JT-pilot project is shown (Figure 8).

In order to fulfill the OEM needs (e. g. in case of data quality) it has to be ensured that the JT and the accompanying structure formats are created properly. In a first step the configuration settings of the JT-translator have to be adjusted to the OEM needs. Therefore the mentioned ProSTEP iViP JT Content Harmonization workgroup (JTCH) has created a first best practice document [9]. In the second step, the OEM specific data preparation has to be done. Therefore the Daimler JT Supplier Package (JTSP) is available for all Daimler suppliers [7].

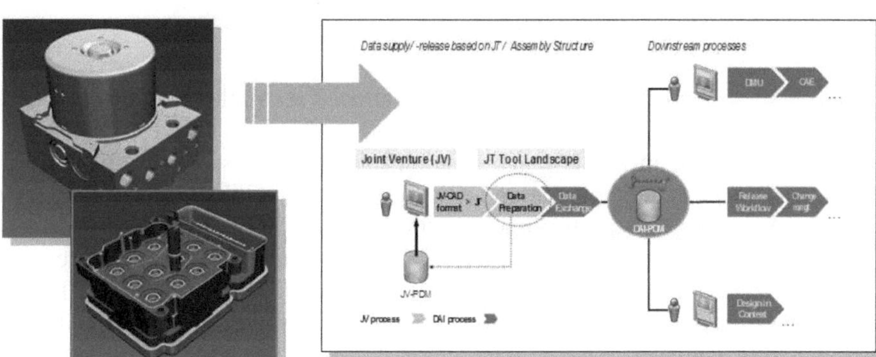

Fig. 8. Usage of JT in partner integration [7]

The Daimler JTSP offers the opportunity to edit the JT internal attributes and to prepare XML based structure formats to meet the OEM requirements. The third step in the process landscape is the data exchange to the OEM. In case of the Daimler AG the OFTP based SWAN system is used.

The final step of this process is the data import into the Daimler PDM system Smaragd. At this point in time the externally created JT datasets are released and available to all Daimler downstream processes such as Viewing, DMU or Design in Context [10].

Meanwhile the full CAD data exchange by using JT is going to be adopted as base CAD translation technology by supplier portals like OpenDESC.com (www.opendesc.com) [11].

6 Conclusions

This contribution addresses the state of the art activities in in order to establish JT as an universal process format. Industry needs simple solutions for the efficient use of IT technology inclusive consistency in processes. There are strong requirements for lightweight 3D formats for the visualization and downstream processes, complementary formats in order to exchange meta data, structure data and kinematics data as well as open and standardized formats to reduce total cost of ownership and to minimize dependency of single vendors [8], [9], [10].

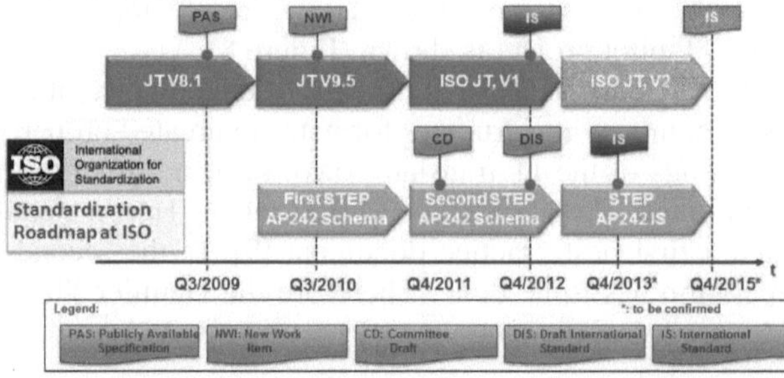

Fig. 9. The schedule of JT and STEP AP 242

After these very promising pilot projects mentioned above the activities on the subject of lightweight visualization based on JT have already reached a good level of maturity encouraging the wide range of companies and users, whereas it is a forthcoming task to push the integration with the accompanying format STEP AP242 (Figure 9). However first applications creating or using STEP AP242 XML as a out- or input format are already available today, the wide usage will be announced in the next years. The future data exchange process will adopt JT as well as STEP AP242 ensuring the exchange of whole complex products in all stages of product and process development (Figure 10).

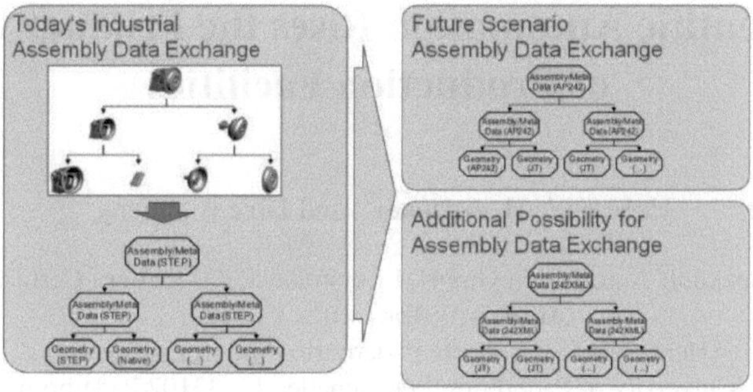

Fig. 10. Assembly data exchange with JT and STEP AP 242

References

1. N. N.: Digital Engineering Visualization, SASIG, SASIG Guideline (2005), http://www.aiag.org/source/Orders/index.cfm?search=D-21
2. N. N.: Collaborative Product Visualisation - General issues and use case description, Frankfurt, VDA, VDA Empfehlung 4966 (2007), http://www.prostep.org/fileadmin/freie_downloads/Empfehlungen-Standards/VDA/VDA_4966_Collaborative-Product-Visualization_1.0.pdf
3. Fröhlich, A.: White Paper 3D Formats in the Field of Engineering – a Comparison, PROSTEP AG, Darmstadt, http://www.pdfgenerator3d.com/nc/en/product/white-paper.html
4. Biahmou, A., Fröhlich, A., Stjepandic, J.: Universelle Formate für Visualisierung und Konvertierung in der Engineering-Kollaboration. In: International Conference GO 3D, Rostock (2009)
5. Handschuh, S.: Wertextrahierende Nutzung von offenen leichtgewichtigen Datenformaten in automobilen Kollaborations- und Entwicklungsprozessketten, Technische Universität Kaiserslautern, PhD thesis (2011)
6. Beckers, R., Fröhlich, A., Stjepandic, J.: Anwendung und Potenziale universeller Visualisierungsformate. In: 9. Paderborner Workshop "Augmented & Virtual Reality in der Produktentstehung", Paderborn (2010), http://www.transmechatronic.de/uploads/tx_vitramemberadmin/literature/Anwendung_und_Potenziale_universeller_Visualisierungsformate.pdf
7. Handschuh, S., Dotzauer, R.: ProSTEP Symposium 2012: JT on the Path of Enlightenment – Report of the JT working groups, Hamburg (May 10, 2012)
8. Röhl, H., Fröhlich, A.: Einsatz des JT-Formates in der Digitalen Fabrik. VDI Kongress "DigitaleFabrik", Regensburg (2012)
9. ProSTEP iViP, Whitepaper Applying JT – Guidance for Using JT in Practice Version 2 (2010), http://www.prostep.org/de/mediathek/veroeffentlichungen/white-paper-studien.html
10. Handschuh, S., Dotzauer, R., Fröhlich, A.: Standardized formats for visualization - application and development of JT. In: 19th ISPE International Conference on Concurrent Engineering, Trier (2012)
11. Bondar, S., Potjewijd, L., Stjepandic, J.: Globalized OEM and Tier-1 Processes at SKF. In: 19th ISPE International Conference on Concurrent Engineering, Trier (2012)

Scientific Automation Rises the Productivity of Production Facilities

U. Frank[1], H. Anacker[2], and Dirk Bielawny[3]

[1] Beckhoff Automation GmbH, Eiserstraße 5, 33415 Verl, Germany
u.frank@beckhoff.com
[2] Heinz Nixdorf Institute Paderborn, Product Engineering, University of Paderborn, Fuerstenallee 11, 33102 Paderborn
harald.anacker@hni.upb.de
[3] Heinz Nixdorf Institute Paderborn, Control Engineering and Mechatronics, University of Paderborn, Fuerstenallee 11, 33102 Paderborn
dirk.bielawny@hni.upb.de

Abstract. Manufacturer of production facilities and operating companies are subjected to increase competitive constraints. To be successful production facilities should work more productively and more environmentally friendly. Furthermore the quality of the produced goods should rise. To reach this manufacturer of production facilities ask for more efficient and intelligent automation technology. In the project "ScAut – Scientific Automation" the Beckhoff Automation GmbH – a producer and vendor of open automation systems based on PC Control technology – meets these requirements with Scientific Automation. Scientific Automation is the integration of automation software with findings from engineering science, which go beyond the limits of conventional control. The basis for this is the continually increasing performance of PCs. The power of the PC Control philosophy offers sufficient capacity to integrate numerous advanced functions beyond standard control. Scientific Automation complements the conventional areas of control technology such as PLC, Motion Control and control technology, for instance, with precise and fast measurement technology and the associated engineering algorithms.

Keywords: manufacturer, intelligent automation, scientific automation, control technologies, measurement technologies.

1 Introduction

Due to increasing competition and the approach of scarce resources, the manufacturers and operators of production facilities are attempting to develop more reliable, flexible, and more energy efficient manufacturing and assembly plants. The manufacturers and operators in automation technology see great potential in achieving these goals. They call for automation concepts and technologies that enable a self-contained anticipation of wear, in order to tie in maintenance procedures during pauses in the production line, to reduce emissions, to optimize energy consumption system wide and to avoid errors in production.

G.L. Kovács and D. Kochan (Eds.): NEW PROLAMAT 2013, IFIP AICT 411, pp. 296–309, 2013.

The industry and scientific community agree that the progressive development of information technology and communication technology are the crucial point to meet these requirements. In the near future there will be fascinating perspectives for modern ma-chines and plant production that will go beyond current standards, for instance Intelligent technical systems with inherent partial intelligence. Keywords such as "Things that Think", "Cyber-Physical Systems" or "Industry 4.0" express this perspective. [1], [2], [3]. The development of such systems is the overall aim of the leading edge cluster it`s OWL.

Information technology and also non-technical disciplines such as cognitive sciences, neurobiology and higher level mathematics bring forth a variety of methods, techniques, and procedures for developing technical systems that were previously only associated with biological systems. By integrating these properties, technical systems in the future will be able to react to altering environmental conditions, adjust their behaviour independently according to the situation, and to communicate, negotiate, cooperate with other technical systems in order to optimize its individual behaviour.

The intelligence of these systems is achieved through the constant integration of new functions. The realisation of these functions requires the following developments:

- new solutions for the electronic, software, and control technology as well as sensor technology especially for critical real-time applications;
- the integration of mathematical and scientific engineering methods, procedures, and algorithms as well as technology that is directly necessary for control;
- improved methods and procedures for data entry, data processing, data analysis, and data presentation (intelligent measuring technology and image processing systems;
- system-wide monitoring procedures.

As a provider of automation solutions, Beckhoff Automation is confronted with challenges to ease the applications of our solutions despite the increasing complexity of the challenges. The user provision that deals with new intelligent solutions and their integration is therefore critical (Figure1).

We confront these challenges in the ScAut Project, whose content and initial results are discussed in the framework of this paper.

In the innovation project "ScAut – Scientific Automation"[1] the Beckhoff Automation GmbH – a producer and vendor of open automation systems based on PC Control technology – meets the defined requirements with Scientific Automation. Scientific Automation complements the conventional areas of control technology such as PLC's, Motion Control and control technology, for instance, with precise and fast measurement technology and the associated engineering algorithms. The ScAut Pro-

[1] "This research and development project is funded by the German Federal Ministry of Education and Research (BMBF) within the Leading-Edge Cluster "Intelligent Technical Systems OstWestfalenLippe" (it's OWL) and managed by the Project Management Agency Karlsruhe (PTKA). The author is responsible for the contents of this publication."

ject started in mid-2012 and is one of three core projects of the top cluster it's OWL (Intelligent Technical System OstWestfalenLippe). The top cluster it's OWL is the first major project in the framework of the future Industry 4.0.

A detailed explanation of the project and its objectives are presented in Chapter 2. In Chapter 3 we will present the PC-based control. Chapter 4 is the core of this paper. We will explain the scientific approaches as well as the exemplary use of pre-defined solution elements. To conclude, we will sum up the major points and give a short outlook on our future work.

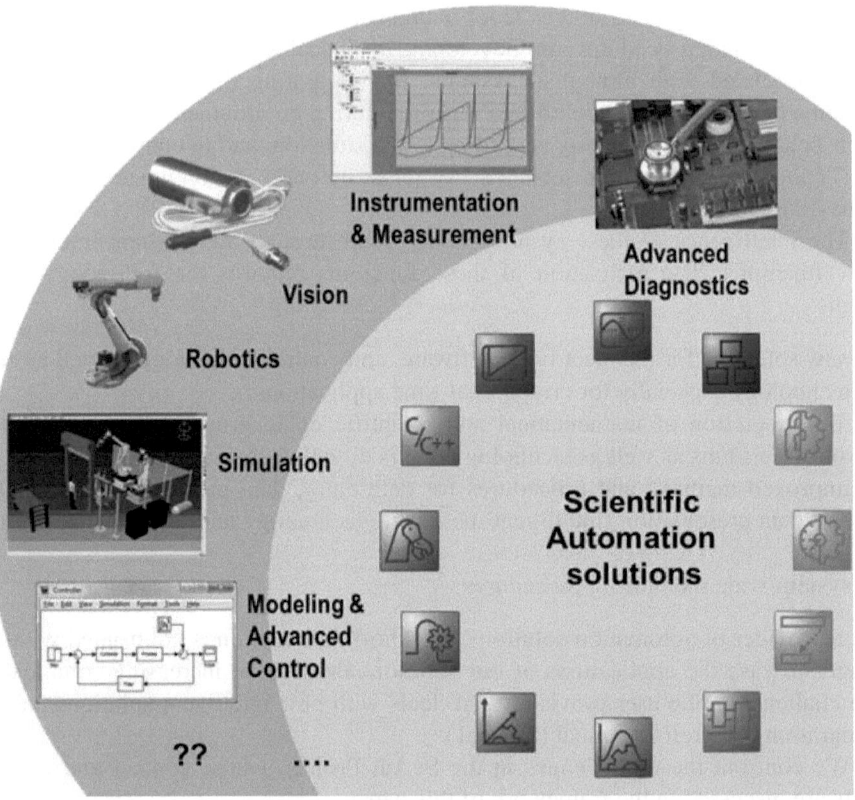

Fig. 1. Scientific Automation Platform

2 Project Scientific Automation

The aim of the project is to create a scientific automation platform for the development of real-time operating systems that have an inherent technical partial intelligence. The platform provides scientific automation solution elements in the form of hard-ware and software components, architectures of scientific automation systems, development tools, real-time run-time environment and a methodology. The platform is being developed by the interaction of three cooperative projects in cooperation with the cluster cross sectional project. The cross sectional cluster projects provide a successive comprehensive technology platform and broad base for personnel with knowledge and experience. They advise on application methods, processes and technologies and assist in the implementation of the platform. The overall objective is divided into the following sub-objectives:

Target 1: Reference Architecture of Scientific Automation Systems
Scientific Automation Systems are intelligent and adaptable. To realise these properties, the system must be built modular, be open and contain standardised interfaces. Scientific Automation Systems therefore contains modules that can be configured into a complete system. The systems have a hierarchical structure, which is composed of sub-modules. Each module implements one or more automation functions and has a specific behaviour and has defined standardised interfaces. The modules can be considered as solution elements. Scientific Automation Systems consist of pre-existing solution elements (e.g. flow control, numerical control, power measurement terminal) and Scientific Automation Solution Elements. The current and potential relationships of the solution elements of a system define the system structure, the possible system structure (configurations) and the system behaviour.

The structure, the behaviour and the necessary interfaces of the solution elements are derived from the structure and the behaviour of the system. In addition, the integration mechanism is made available in order to share existing solution elements.

Target 2: Construction Kit with Scientific Automation Solution Elements
The objective is to define the engineering science, mathematical methods, procedures as well as technologies in technological roadmaps for the development and operation of the pilot applications, to specify the software components and hardware components necessary for their application and to provide them as Scientific Automation Solution Elements in the construction kit.

Technology-Roadmap: for the implementation of relevant methods in pilot applications, methods and procedures as well as technology, Technology-Roadmaps are collected and evaluated for functionality, performance, feasibility and beneficial factor for automation technology, potential re-usability and for market acceptance.

Scientific Automation solution elements: In the first step, Scientific Automation solution elements are prepared from the measurement fields and the condition monitoring. In the second step of the technology roadmap, the methods, processes and technologies are implemented with the highest ratings from the solution elements.

Construction kits: The Scientific Automation solution elements are to be made available to the developers of the construction kits. The kits can be realized through the database which the developer has access to. The search for the optimal solution element should be supported through the use of a semantic search.

Target 3: Methods and Software Tools for the Design and Operation of Scientific Automation Systems
Design tools, a runtime environment and methodology should be provided for the development and operation of Scientific Automation systems.

Design tools: for the development of Scientific Automation systems, relevant design tools for modelling, simulation, analysis, optimization and implementation should be merged into a design environment. The solution elements are provided by the developers as construction kits. During the development, the developers use the construction kits depending on their design environment and in doing so develop efficient technical systems with inherent partial intelligence.

Run-time Environment: A real-time run-time environment is initialized for the operation of Scientific Automation systems. It is crucial that during operation of the system the real-time execution of the control program has access to the solution elements of the Scientific Automation construction kit as well as other solution elements of other construction kits and the appropriate software components of the solution elements.

Methodology: With the help of process models, methods and recommendations, third parties should also be able to develop Scientific Automation systems and thus enhance the competitiveness of their systems and their company.

The following pilot applications will demonstrate the power of the platform:

Intuitive, self-optimizing, energy efficient, profile processing machines (SCHIRMER Machines GmbH, Verl); Integrating intelligent measuring technology/software components in assembly and manufacturing plants (Hüttenhölscher Maschinenbau GmbH & Co. KG, Verl); Woodworking machinery for the furniture industry (IMA Klessmann Holzbearbeitungssysteme, Lübbecke, Maschinenbau); Intelligent kitchens manufacturing with lot size 1 (Nobilia-Werke J. Stickling GmbH & Co. KG, Verl).

3 PC-Based Control

In automation technology, the idea of converging information technology and automation technology with PC-based control was born in 1986 (see fig. 2).

As a pioneer of the industry standard IT hardware Beckhoff offers more than 25 years of PC-based control technology, which not only provides a huge increase in performance but has also exploited a medium for convergence of two worlds in the field of technology ever since. Already then it was recognized that the industrial PC facilitates high performance controls in a variety of form factors with an ideal IT orientation thanks to its openness. Presently and in the future, the increase in computing power will allow more and more tasks to be integrated in PC-based controls of a machine (figure 2). In addition to the classical sequence controller, steering control and traction control are already calculated directly by the PC. However in many cases, the available computing power is not used at all. There-fore the available computing power cannot exclusively be used for control but also for regulation, for measuring, and monitoring.

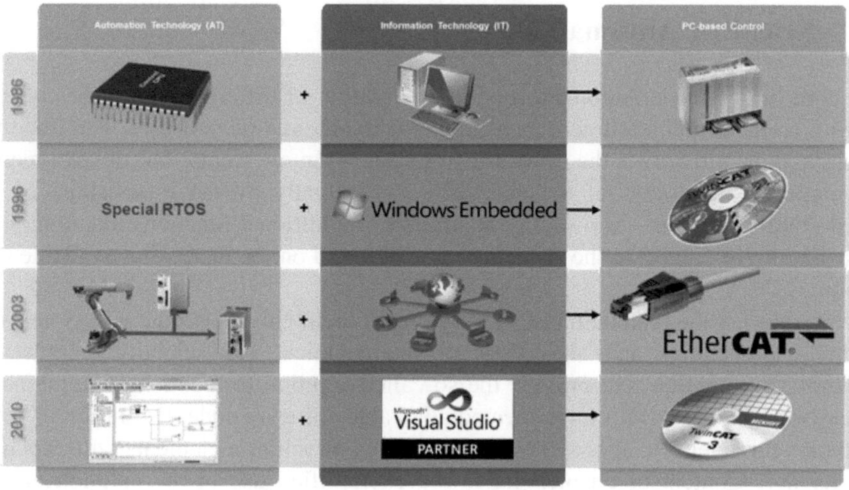

Fig. 2. PC-based Control uses the potential of the technology convergence of IT and AT for industrial applications

PC-based control systems consist of an Industrial PC, I/O modules (bus terminal), drive technology, and automation software. On the software side, the TwinCAT (The Windows Control and Automation Technology) automation suite is the heart of a Beckhoff PLC. The TwinCAT software systems turns virtually any PC-based system into a real-time control PC with many SPS, NC, CNC and other robotic run-time systems. TwinCAT serves as a run-time environment and also as a development platform.

On a PC, TwinCAT 3 can realise both the development as well as the simulation of technical systems. In enabling TwinCAT 3, a consistent model-based design from concept to implementation into an environment can be performed, which reduces costs and development times [2], [5], [6].

The PC-based platform, with its high performance processors and hardware platform, makes it principally possible to easily develop large projects, in which machines and control systems are simultaneously and reciprocally developed [7]. In practice, the steps in development look different up to now because the design of motion functions and control functions in the automation technology usually takes place independently from system design. The reason for this is that many industrial control systems and their software and hardware are largely incompatible with the development tools from other domains. A further complication is that currently the established development tools that do support aspects in the development of intelligent systems, such as modelling, simulation and optimization of flow controllers drive controllers, as well as the sup-port for methods, procedures and technologies for implementation such as self-optimization, multi-objective optimization, learning procedure or condition monitoring are not adequately considered in the area of design for automation systems. The project ScAut is to make a significant contribution to addressing this deficit.

4 Scientific Automation

The core of Scientific Automation is the integration of different approaches of engineering sciences and intelligent measurements in automation software. The basic task of automation software is the controlling of system`s movements. The development of automation technologies of the last years – especially by the establishing of PC-based control systems – shows the integration of additional functions like contouring control or drive control in the sequence control based on the increasing available CPU resources.

The basis for the realization of such autonomous intelligent production systems are methods, procedures and algorithms of mathematical and scientific engineering. Beckhoff Automation will provide the so called solution elements by a combination of hardware modules with implemented adaptable software code. Definition of terms, examples for scientific approaches and examples of solution elements will explain in chapter 4.1. Content of chapter 4.2 is the exemplified use of the solution element „balancing".

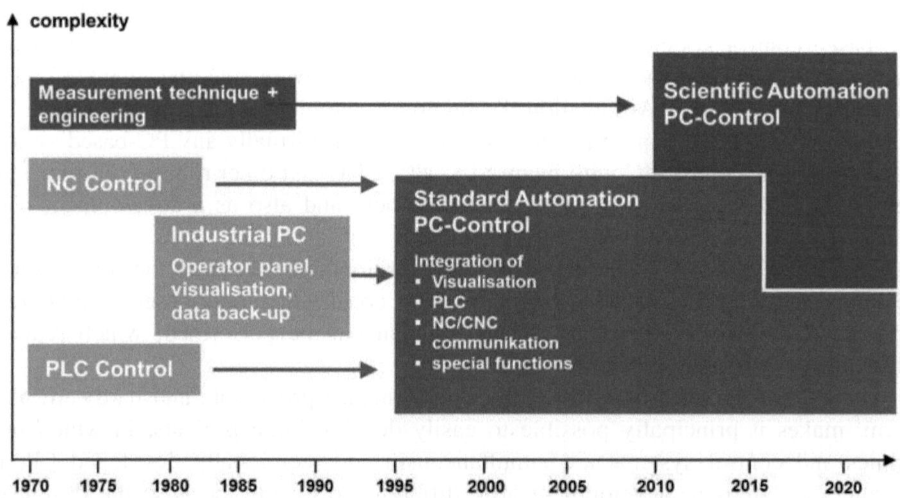

Fig. 3. From Standard Automation to Scientific Automation

4.1 Scientific Approaches

Scientific Automation provides new and enhanced functionalities for many application areas in the context of automation. Advanced methods, algorithms and technologies from various disciplines such as engineering, mathematics and computer science are explored and analyzed with regard to their possible use in Scientific Automation solution elements.

A method is "a particular procedure for accomplishing or approaching something, especially a systematic or established one" [9]. Here, we mainly – but not exclusively

– consider mathematical or technical methods and procedures to solve certain problems or achieve certain goals. Besides, we regard mathematical and technical approaches (in contrast to particular procedures) as methods as well.

According to CORMEN ET.AL, we generally regard an algorithm as a well-defined procedure, i.e. a sequence of computational steps that uses a set of input values to produce a set of output values [10].

Technology is commonly defined as "machinery and equipment developed from scientific knowledge" [9]. In the context of ScAut, by technology we mean either specific machinery or scientific knowledge and methods closely related to certain machinery or equipment.

4.1.1 Identification of Scientific Approaches

The basis for identifying applicable methods for use in Scientific Automation solution elements is a comprehensive overview of the latest developments and research in disciplines that are related directly or indirectly to automation. "Scientific" approaches from various disciplines are to be exploited for their use in Automation.

ScAut will offer completely new features to automation systems. Thus a survey with the purpose of identifying methods, algorithms and technologies that are suitable for ScAut solution elements must not be limited to the classical areas of automation (e.g., control theory, signal processing and electrical drives). In order to realize intelligent and more efficient factories, new approaches for scientific automation solution elements are considered to be found in areas such as artificial intelligence, medical technology or economics. However, the focus is mainly on the areas of signal processing, cybernetics, computer vision, mathematics (e.g. optimization, stochastic), and data processing.

Starting with a specific application or a subject area, comprehensive inquiries are made based on expert interviews, literature and research reports. Thus, methods, algorithms and technologies are identified to create an overview of the state of the art regarding specific areas or applications. To find suitable approaches for ScAut solution elements, the identified methods and technologies are then analyzed and checked for their applicability.

Classification, Evaluation and Selection

So far about 500 methods, algorithms and technologies have been identified during the surveys for new approaches to ScAut solution elements. The identified items have been described, assigned to categories as well as potential application areas and merged to groups if possible. In a next step the methods are analyzed in order to obtain more detailed knowledge of their suitability for use in ScAut solution elements. There are some requirements towards ScAut solution elements allowing for decision whether an approach is suitable. Technical requirements such as real-time capability and computational complexity are of great importance. Usability and economic aspects have to be considered as well, among other things. Methods that meet these requirements can be further analyzed, described in detail and stored in a knowledge base. Thus, ScAut developers as well as users get quick access to information needed for their applications.

4.1.2 Promising Approaches and Examples

Both the evaluation of the approaches already identified and the research for further methods are currently in progress. Nevertheless, there still are some promising approaches which show a great potential for the use in ScAut. In the following, we briefly present some examples. An more detailed example is given in chapter 4.2.

Model-based Control

Model-based approaches in both feedforward and feedback control offer interesting possibilities for applications in automation. Commonly established methods such as state regulators might serve as a useful basis for ScAut solution elements, which can be used in combination with more advanced approaches. Especially non-linear approaches, e.g., gain scheduling (see [11]) show great potential.

Another important concept is model predictive control. In model predictive control, optimal feedforward control signals are calculated on the basis of a model used to predict the future behavior of the system. An interesting example is given in [12]. The authors present an approach for model predictive control on the basis of differential dynamic programming. Possible applications in automation would be the efficient control of industrial robots.

A major drawback of the use of model-based approaches is the need for sufficiently accurate system models. ScAut can overcome this problem in different ways. Of course, users can create their own models using common tools, and import them to their Scientifc Automation applications.

But solution elements also could directly provide suitable (partial) models and support the user in selection and parameterization of the models. Other promising features are automatic generation and parameterization of plant models. An example for automatic model identification is the approach of Box-Jenkins [11]. Based on time series analysis, plant models as well as disturbance models can be generated automatically by this method through measuring of input and output signals. Although applying predefined input signals while measuring the outputs is preferred, the identification can take place even during normal operation of a plant.

Examples in Context "Condition Monitoring"

To ensure consistent production quality, a provision is used to monitor the conditions of machinery and the changes in the systems behaviour in order to introduce accompanying measures. Using sensors and supervisory equipment, the state of the machine and its components are collected and the data is analysed in information processing systems. The decisive factor for the reliability of the data analysed is the quality of the data collected. This depends entirely on the temporal accuracy of the data collection, the amount of data, the accuracy of the data and the amount of interference analysed.

The project ScAut found it necessary to filter out interferences and those with time stamps for collection, saving and analysis of data for solution elements such as bus terminals or data with high resolution and frequency. Secondly, insubstantial solution elements in the form of software are used. The Fast-Fourier-Transformation, the Hilbert envelope, or the digital Filter are for example employed. The algorithms are encapsulated in function blocks and are made available as solution elements in libraries for general use.

A typical process for the realization of a digital filter is the Wattmeter method [13]. Both the amplitude and the phase of a sinusoidal signal can be deter-mined at a particular frequency at a mixed-voltage. The method is implemented as a function block named "Watt meter" and initialized as a solution element. This includes the state machine shown in Figure 5.

The evaluated values are assigned in the form of a pointer to an array of data; in addition, the number of available data is transferred into this array. The function block requires the values of the time stamp of a reference pulse detector and the next conversion cycle and the time of cycle in order to assign the time of the reference pulse and the function block.

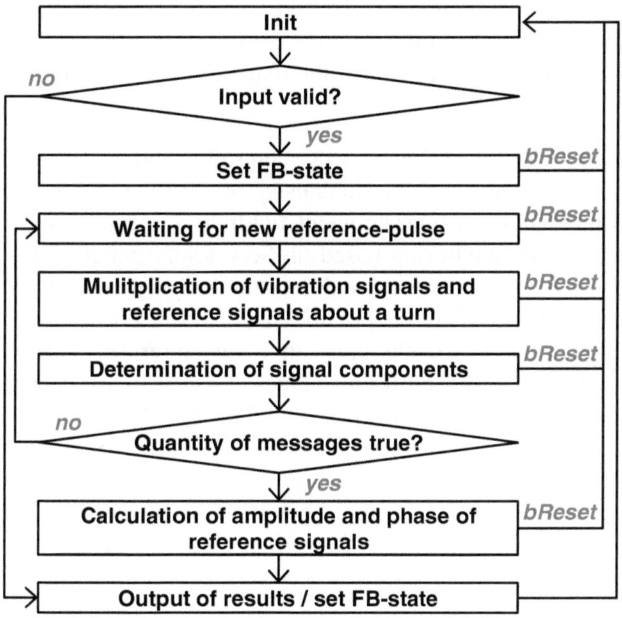

Fig. 4. Program sequence: Function block „Wattmeter" [14]

4.2 Application Example "Balancing"

The use of ScAut solution elements are emphasized in balancing the use of spindles. The productivity, quality, durability, and cost-effectiveness of modern milling machines and work centres are rising steadily. Therefore motorized spindles with high revolutions are used. To achieve a high quality of machining operations, the spindles have to round and are not allowed to vibrate during the process.

Upon completion, all of the spindles will be tested, measured and balanced upon their dimensional accuracy, concentricity, balance, and durability. This takes place on a specially designed testing rig. The spindles are then put into operation and subjected to various tests.

To control the testing rig, former stand-alone devices are used as well as dedicated systems that work in parallel with its own controls. In addition to its own configuration software and programming software, the system also has its own bus system in order to capture data from the connected sensors. These however lack flexibility due to the limited amount of analogue and digital input channels. Even the correlation of data from the "normal" control can only be realized with cross communication and heightened engineering effort.

The alternative to these "black box" systems is the integration of a condition monitoring system in the machine controls. Therefore the following two conditions have to be met:

1. The controller has to be a high performance controller in order to calculate the partially complex analysis algorithms and the control tasks.
2. The fieldbus system used has to allow the required data throughput and have a high determinism.

These requirements can be met with a PC controller and ScAut solution elements. EtherCAT meets the demands on the fieldbus system. EtherCAT has established itself as a fieldbus control and as a measurement fieldbus. Complex applications, like the integration of condition monitoring based on ScAut solution elements, are realizable with these ethernet-based high speed and deterministic fieldbus protocols. The functional principle of Ether-CAT makes possible payload rates of well over 90% and Fast Ethernet full-duplex and bus cycle times of a few microseconds. With EtherCAT oversampling technology, the buffering values directly in the EtherCAT slave are increased way beyond the actual bus cycles: Digital input bus terminals can deliver up to one million samples per second. Analogue signals may be sampled up to 100 kHz. Distributed clocks in EtherCAT slaves provide a network-wide time-synchronized data acquisition. The jitter is significantly below one microsecond, in most cases less than 100 nanoseconds. Moreover, the flexibility in the network topology and modularity of the bus terminals has to be mentioned. Using this process, each application is realized cost effectively.

The automation suite TwinCAT 3 (The Windows Control and Automation Technology) from the Beckhoff Automation realizes the real-time controls on an industrial or embedded PC, and also takes advantage of today's multi-core CPUs (Figure 2). TwinCAT can outsource different arithmetic operations on different cores of control. For example, one core can be used for machine control and another core can be dedicated to a condition monitoring application. Even the calculation of a single algorithm in a separate core is possible.

With the help of scientific automation, the balancing process that includes the necessary measurement equipment can be integrated directly into the controls of the spindle testing rig. The basis for this is the use of PC-based control technology that is composed of an industrial PC, the EtherCAT fieldbus, bus terminal and the Automation suite TwinCAT 3. The system data can be recorded very precisely with a time stamp and analysed with the aid of oversampling and distributed clocks technology.

The balancing of spindles is made in several steps. First the vibration sensor (encoder) and the spindle are aligned. These are then calibrated. In addition, several tests are performed with and without test weights. Then the balancing is performed. Any errors that occur are recorded and analyzed.

The calibration has to be done before testing new spindles. Otherwise, the settings might be taken from existing measuring routes. Figure 6 shows the structure of the main program for balancing (see. [14]).

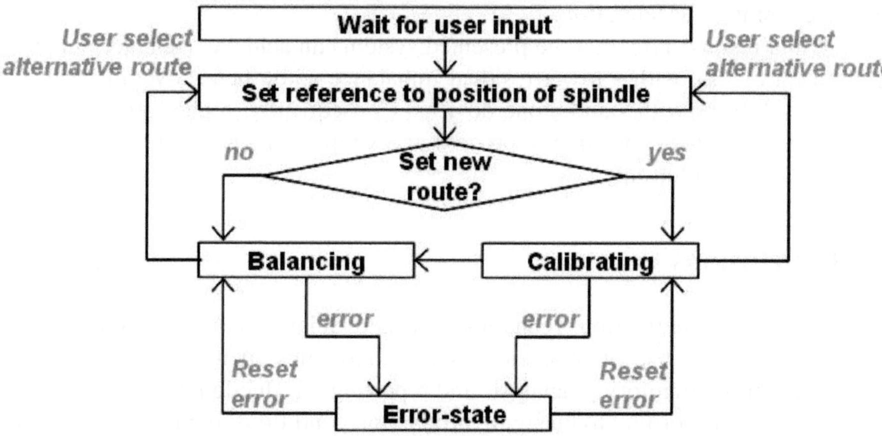

Fig. 5. Program sequence for the balancingoperation [14]

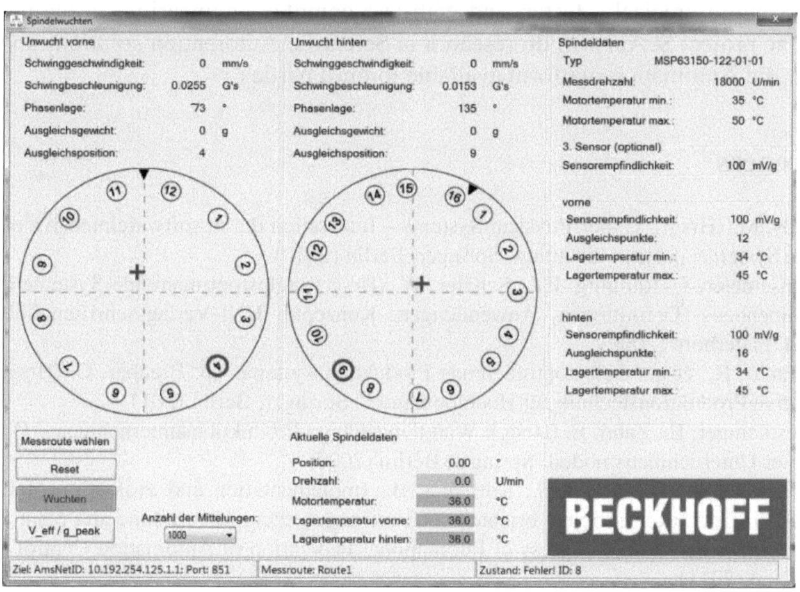

Fig. 6. User interface: Description of the results of the balancing operation [14]

In order to detect the imbalance, the vibrations of the spindle and the frequency of the vibrations have to be filtered out. This way the amplitude is the reference for the imbalance and the phase indicates the spatial position of the imbalance. This filtration is performed with the ScAut solution element "watt meter". The ScAut solution element can be integrated into the main program for balancing, which in turn can be interpreted as a ScAut solution element.

After performing the balancing process, the results are presented according to figure 6. The phase of the imbalance, the necessary balance weight and the balancing position can be specified along with the information on the measured vibration velocity and the vibration acceleration.

The initial results show that the presented system can achieve a smaller number of measurement cycles, with a lower residual imbalance and a better overall result than the former solutions with stand-alone devices. Consequently, higher quality spindles at lower costs and fewer balancing process and simpler engineering can be reached through integrated solution.

5 Conclusions

The expansion of Automation Technology around Scientific Automation has many advantages. With the use of procedures and technologies such as self-optimization, learning, condition monitoring, analysis and reporting, production can be improved to react intelligently according to changing production conditions and to adapt its behaviour appropriately to its situation. This can lead to dramatic energy savings, reduced machinery wear, just like in the example of the balance; the balancing process can be executed more precisely, faster, and with less complex engineering. In the next four years, the project ScAut will do research in Scientific Automation solutions and make a Scientific Automation platform available to third parties.

References

1. Broy, M. (Hrsg.): Cyber-Physical Systems – Innovation durch softwareintensive eingebettete Systeme. acatech diskutiert. Springer, Berlin (2010)
2. Gausemeier, J., Rammig, F.J., Schäfer, W. (Hrsg.): Selbstoptimierende Systeme des Maschinenbaus. Definitionen, Anwendungen, Konzepte. HNI-Verlagsschriftenreihe, Band 234, Paderborn (2009)
3. Schmitt, R., et al.: Selbstoptimierende Produktionssysteme. In: Brecher, C. (Hrsg.): Integrative Produktionstechnik für Hochlohnländer. Springer, Berlin (2011)
4. Westkämper, E., Zahn, E. (Hrsg.): Wandlungsfähige Produktionsunternehmen – Das Stuttgarter Unternehmensmodell. Springer, Berlin (2009)
5. Vogel-Heuser, B., Braun, S., Kormann, B.: Implementation and evaluation of UML as modeling notation in object oriented software engineering for machine and plant automation. In: 18th World Congress of International Federation of Automation Control (IFAC), Milano (2011)
6. Papenfort, J.: Effektives Software-Engineering. In: Elektro Automation. Konradin Mediengruppe, Leinfelden Echterdingen, 11S (2012)
7. Papenfort, J.: Steuerung und Motion Control auf dem PC, geht das? In: Krammer, E. (Hrsg.): Automatisierungstechnik Aktuell-Ausgabe 2005-Trends, Produkte und Entscheidungshilfen, Oldenbourg-Industrie, München (2005)

8. Haufe, J., Schneider, P., Donath, U., Reitz, S.: Simulationsgestütztes Prototyping und Optimierung von Maschinensteuerungen. In: Advances in Simulation and Logistics Applications. Fraunhofer IRB Verlag, Stuttgart (2008)
9. Stevenson, A. (ed.): Oxford Dictionary of English, 3rd edn. Oxford University Press (2010)
10. Cormen, T.H., Leiserson, C.E., Rivest, R.L., Stein, C.: Introduction to algorithms, 2nd edn. The MIT Press (2001)
11. Rugh, W., Shamma, J.: Research on gain scheduling. Automatica 36(10) (2000)
12. Julier, S., Uhlmann, J., Durrant-Whyte, H.: A New Approach for Filtering Nonlinear Systems, Seattle (1995)
13. Lingener, A.: Auswuchten – Theorie und Praxis, 1. Auflage. Springer, Heidelberg (2007)
14. Dresselhaus, P., Frank, U., Papenfort, J.: Scientific Automation erhöht Zuverlässigkeit – Integration von Condition Monitoring in die Automatisierungstechnik. In: WT Werkstatttechnik Online Jahrgang 103. Springer VDI-Verlag, Düsseldorf (2013)

Prospective Evaluation of Assembly Work Content and Costs in Series Production

Ralf Kretschmer[1], Stefan Rulhoff[2], and Josip Stjepandic[2]

[1] Miele & Cie. KG, Carl-Miele-Straße 29, 33332 Gütersloh, Germany
ralf.kretschmer@miele.de
[2] PROSTEP AG, Dolivostraße 11, 64293 Darmstadt, Germany
stefan.rulhoff@prostep.com, josip.stjepandic@opendesc.com

Abstract. Strategic decisions in early production planning phases have a high impact on various production aspects. Decision making is often based on vague expert knowledge due to lack of a reliable knowledge base. Implications of this problem are especially observable in the field of assembly planning, which integrates results from various planning disciplines. The proposed paper introduces a new concept and the corresponding data model for application of Data Mining (DM) methods in the field of production assembly planning and product design. The concept presents assistance potentials for development of new products variants along the product emergence process (PEP).

Keywords: Product Realization, Manufacturing, Digital Factory, Assembly, Process Planning, Data Mining.

1 Introduction

Today globally operating companies face additional challenges due to the increasing variability of products and complexity of processes. Therefore there are growing demands on the flexibility in the production system on the economic dispatch of new products in an existing production line. In the modern product emergence process production planning gets increasingly important and has to be executed in parallel to the product development [1]. In this early phase of the product creation a first start for planning processes is a cost calculation for the industrialization of the product in existing production lines regarding basic conditions [2]. The economic feasibility of series production must be assured with vague information on the product and given general conditions, e.g. shift model. This is a great challenge especially to the planning of the cost-intensive assembly of the product.

In order to meet this challenge PROSTEP AG supports Miele & Cie. KG, one of the leading manufacturers of domestic appliances, developing innovative methods in the research project ProMondi. Aim of this project is the accurate estimation of the expected assembly work content and the resulting costs in an early stage of the product development as well as the additional support of the design process with assembly knowledge for the specific design. The approach contains the usage of existing

G.L. Kovács and D. Kochan (Eds.): NEW PROLAMAT 2013, IFIP AICT 411, pp. 310–319, 2013.

planning data in order to extrapolate assembly processes. Especially linked product and process data allow the innovative usage of Data Mining methods. New processes appropriate to assemble the given new product shall be designed based on this existing linked product and process data. Automatic analysis with a specific Data Mining model can be used to create a first draft of the assembly process and estimate the expected costs. Additional use cases can be addressed. Following production planning processes can be supported by automatic proposals of adequate assembly processes, which then can be customized. Moreover the design engineer can be supported at the selection of appropriate joining elements. With this approach an assembly knowledge based support of the designer in series production can be achieved using innovative data mining methods.

2 Use Cases Miele

In order to address the challenges of data mining the integration of various planning tasks within the PEP, new concepts are necessary. Though, as a part of integrated product and process development there are different definitions for various phases and aspects of planning activities along the PEP. Regardless of the specific definition of these phases and aspects, however, based on the analysis it is certain that great amount of their containing information and knowledge are either utilized insufficiently and ineffectively or remain unused [3]. In this regard, the presented concept focuses on product design and production assembly planning. Subsequently, for the product designer and production planer, there are varieties of applications, which can assist the design or the planning process through information gathered by data mining.

Enrich CAD Data with Assembling Information: Derived from similar previously constructed products assembly information such as time data about the actual design situation can be identified and provided in order to support the designer. These additional information can be used to enrich the CAD data to support the current design and be updated in later assembly planning processes.

Suggesting Assembly Connections: An assisting option for the designer is a suggestion list of similar previously constructed assembly connection variations. These lists give a quick overview of possible and already implemented connection types in the assembly.

Assembly Process Estimation: The focus is on the creation of an assembly process for a new product. Based on existing product and process data compilation of a first approximated assembly process for a new product could be developed. From this, the production planner can specify further details and thus determine a first estimation regarding assembly time. Based on the assembly time and associated calculation scheme, the planner can perform the first cost estimation in a very early planning phase.

The information in production planning and engineering processes can mutually enrich each other. Additionally intelligent interconnecting information from both areas creates added value. The newly obtained information supports the workflow throughout the PEP. Therefore, as part of this concept some requirements need to be met. Thus the pre-conditions attached to both systems as well as their respective processes have to be fulfilled [4].

Data Mining methods can be used for data clustering and classification, however criteria for comparison of data sets have to be identified [5]. To determine these criteria, within the scope of ProMondi project, a survey of users as well as an analysis of various tools of the DM was performed. The objective of this analysis was to identify attributes relevant for assembly processes that could be assigned to products and parts in CAD [6], PDM and production planning systems. In CAD systems attributes assigned to parts contain mainly geometric information including volume and weight. The PDM systems contain organizational information, such as creator, version and revision as well as the mentioned parts information form CAD [7]. In addition to the conventional systems for design and stacking product parts and assemblies, systems for process planning and time measurement were also taken into account. They sustain a comprehensive portfolio of information and therefore can be used to distinguish different product parts and assemblies. The results of this analysing are capsulated as an object oriented data model, further described in chapter 3.

The necessary enrichment of product and process data on the fly for the presented concept requires additional efforts in the design. This additional expenditure also relates to the assembly connections and includes the acquisition of new information form the designer's know how. The designer usually defines assembly connections either implicitly through formed locked joints by the shaping of the parts or explicitly by connecting elements such as in screwed fasteners.

The designer of the assembly connection considers all these information in the design but cannot store them in the CAD model because the CAD tools for the most part are not able to define the necessary attributes.

To overcome this problem as part of the concept presented in this paper, the designer will be provided with an additional tool in the CAD system. It can be used to create assembly connections and gives additional information and explicit design possibilities. These additional assembly information are named below as "product assembly information". Thus, data will be collected in the source system, the CAD system in particular. Since the defined objects are not part of PDM systems an extension is necessary in order to implement connections as objects and to store them after the transfer in the PDM system persistently. In the further processing, the product information will be linked to the planning processes. Unless the storage of product data are in the same system as for production planning, the information flow from the PDM system to the planning system as well as the Data Mining tool, for further analysis, has to be ensured.

In current planning systems often direct linking of processes to products is possible [8]. Thus an allocation of to be assembled product and the associated assembly processes can be realized. In the assembly, however, parts are joined with other parts or product. These assembly connections with their additional information have no digital equivalent object yet. However by means of an object such as the "product assembly information" it is possible to store useful additional connection information, which relates directly to the respective assembly connection. As part of this concept, the combination of the products and processes does not take place directly but through this object "product assembly information". The linking of product and process does not necessarily need to occur at the part level.

3 Concept

The concept presented in this paper describes an assisting workflow to support the designer (Figure 1). As part of a new or modified design the designer creates new product data. In creating the assembly connections a software assistant supports the designer and enriches the CAD model with product assembly information for each connection. This product assembly information includes additional connection information including e.g. the torque screwed fasteners or the type and the form of a welded joint and information about other connection types. In the on going design process the designer can trigger an evaluation regarding the assembly connections in the model.

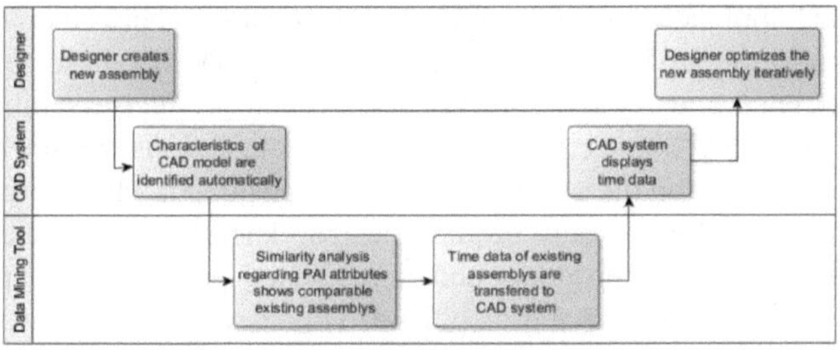

Fig. 1. Design optimization with additional time data

For this purpose, the characteristics of the CAD model are first prepared and analysed with Data Mining. The analysis focuses on the product assembly information and their properties. The parts associated with the product assembly information and their geometric properties, are also included in the analysis as additional information set. Furthermore, an extended database is also provided and consists of product and process data of existing products, which are linked via the product assembly information. The characteristics of the product assembly information of the new product are compared with the properties of the product assembly information of the existing product in the extended database. Then the most similar product assembly information is determined from the existing products. This analysis can be restricted by a class of the connection types (screwed, weld, rivet) or deliberately left open to widen the solution space and to provide the designer with information about other assembly connections.

A limitation on the particular type of connection yields as a result of the closest realized assembly connection of the same kind. Depending on the properties of the parts, other mounting connections can also be found and proposed to the designer in a proposal list.

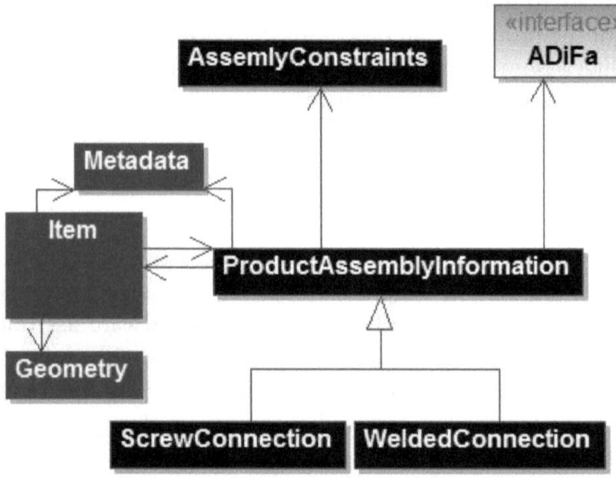

Fig. 2. Data model overview

The presented application for the support of the design process uses the product assembly information identified in the analysis of the PDM database to determine the respective associated and related sub processes. Each product assembly information represents an assembly connection. By multiple connections within the assemblies multiple sub-processes for the assembly can be determined. These processes contain the time data relevant for the new product design. Therefore the corresponding time information of the existing products and if requested an alternative proposal list is transferred in the CAD System and displayed. This assembly time information of the existing product represents a first approximated assembly time for the new product. So the designer is provided with this additional information regarding the assembly time and with an enterprise specific factor the corresponding cost of the current design solution. In the final step the designer is able to optimize the product iteratively on the base of anticipated assembly time and costs for each design.

Based on determined assembly characteristics a range of attributes is derived to classify the assembly of the parts. Figure 2 shows an overview of the generated data model for the data mining analysis.

The ProductAssemblyInformation (PAI) is the central element in this data scheme and represents the assembly of the product parts. References for time analysis, assembly requirements, designed parts or products as well as a wide range of meta data including the assembly department and other information are lodged. This element is supplemented with attributes of different connection types (see Figure 2).

Further connection types can be added to the data model. To provide the required information for the time analysis a standardized data model is applied. In this regard, ADiFa project's "Application-specific data models", so called ADiFa Application Protocols, were used, which offer the integration of processes and data for different DM systems [8].

The second fundamental object in the data model is the Item. It contains references to existing sub-assembly units, geometrical characteristics as well as ProductAssemblyInformation. Each Item refers to the ProductAssemblyInformation, which also refers to further used Items. This construct is chosen to enable Data Mining methods to determine exact similarities between new parts and/or products and other existing parts. Furthermore, it makes comparison parts and products the new and existing ones, in any order and combination interchangeably possible.

In the first approach the attributes for screw connections are clustered and evaluated regarding the relevance for assembly operation. Figure 3 shows the identified attributes classified in the categories fasteners, installation / assembly situation, tools / equipment, installation regulations and additional assembly elements. These attributes are represented in the data model in different object classes here illustrated by colour. The evaluation regarding the influence on the assembly time provides a first indication for the relevance in the data mining analysis. Which attributes really are significant for the similarity of assembly connection have to be determined in a data mining analysis with a large quantity of product data.

Not all of these attributes can be identified in the mechanical design in the CAD system. Some can be determined in production planning workshops in order to optimizes the current design. Experienced designers can preallocate some parameters with estimated values which can be reviewed later. Other parameter and the corresponding value data can be extracted out of other systems e.g. the attributes of standard parts.

After aggregating and appending the data subsets from different sources and systems, it is necessary to remove redundant data sets [9] for the data mining process. The next step is converting and porting data in the presented data model. Depending on data source the conversion is either fully automated or partially automated with further manual adjustment. Often value and scale of different attributes are often heterogeneous. In these cases a normalization of ratings prevents the undesired high or low impact of certain attributes on the results and evaluation process. In this regard a [0, 1] linear normalization has been used. Additionally, a further attribute prioritizing via weighting can be necessary to define the importance of each attribute for the evaluation. An automated learning of the weights via machine learning methods depends on the existing data sets and their quality possible. Otherwise they are determined based on expert knowledge or a combination of both methods. To prevent further expansion of scope and the complexity of existing problem expert knowledge was applied to determine the attribute weights. Furthermore, it is possible to have more than a single weight vector. This approach is useful, if there are various object types or parts, which have different prioritization for their attributes. [10] To identify the objects with most similar product assembly information for a new object the classification algorithm k-nearest neighbour (kNN) [11] with Euclidean distance as evaluation function is used. From the identified objects a list is generated and the most related one can be manually chosen, which passes its assembly process data to new object. To assure the reliability of the presented method and prevent over fitting a cross validation [12] is used.

Category	Product Assembly Information (PAI)	influence on the assembly time
fasteners	screw head diameter	low
	thread type	medium
	number of thread transitions (used)	low
	screw diameter	low
	screw length (thread)	medium
	screw type	medium
	material	low
	magnetic screw	high
installation/assembly situation	chamfer on screw	high
	output of the screw	low
	additional elements	high
	lack of space	high
	visual disability	high
	risk of injury	low
	additional fix the add part	medium
	working both hands feasible	medium
	screwing in	medium
tools, equipment	threaded sleeves used	high
	glove used	low
	equipment used	low
	tool	high
installation regulations	additionally tighten	low
	tightening torque	low
	check torque	low
	assembly sequence	medium
	multi-stage screwdriving	medium
additional elements	flat washer	medium
	nut	medium
	LocTite	low
	grease	low

Fig. 3. PAI attributes example screwing connection

The implementation of the presented approach is challenging due to high requirement for interconnection and the overall quality of the existing data in different source system. In particular the pure number of realized and existing assembly connections and thus necessary instances of an product assembly information as well as the quality of the data regarding their attributes are important.

Momentarily the fulfilment of these high requirements have to be verified. In particular the quality of the linkage of product data with the corresponding assembly processes poses a challenge. Methods to improve the for this concept important quality will be evaluated. Is this task solved the selection of the properties and attributes for the data mining analysis also has to be determined based on production data to ensure the reliability of generated results. In this scope a special focus is on the characteristics of the parts and of the connection itself. In conformity with the presented objectives and concept a utilization of the methodology is described as follows.

Suggesting Assembly Connections and Enrichment of CAD Data: The designer creates a new module with already known and new assembly connections in the CAD system. He designs assembly connections and complements these connection properties in the context of the new module. Via the automated Data Mining process, he is provided with various information about the assembly connections. Moreover, for each assembly connection a list of alternative or ever realized connections can be created. Depending on the product properties the five most similar product assembly information are made available to the designer as a prepared proposal list, which is generated through cluster analysis of existing product data. These information can be used directly and enhance the CAD model in order to use it later or in an extended context of the product. If the analysis is dispensed with the filtering of associated connections with the product assembly information, the designer can also be provided with other not associated connections as alternatives.

Estimation of Assembly Process and Information: The production planner drafts an initial assembly process for a new assembly at an early stage of product development. Analogous to the use case of the designer, for known assembly connections that are implemented in the new product as well as in the old product data, the right product assembly information and thus the assembly processes are found. For new unknown connections the most similar product assembly information and related assembly processes from the database are determined and duplicated. Each of the founded product assembly information represents a single connection and the linked process represents precisely the assembly work content for this connection. The sum of the individual connections for the new product is its first assembly process. Thereby an initial draft of an assembly process of the new module can be generated. The founded individual connections, the individual process, as well as the overall process can be used in different ways to assist the designer and the production planner. The planner and designer also get a first estimation for the expected assembly time and cost in the automated process. In addition, the production planner can increase the quality of the process by manual intervention. On the one hand he adapts the product assembly information, which are created by the designer, before the Data Mining analysis. On the other hand he can complete the product assembly information in the attributes with practical knowledge. Thus he has an impact on the input of the Data Mining analysis and increases the quality of the result thereby. Furthermore, the designer has a first draft for the assembly process at one's disposal and a first estimated assembly time in the current CAD system. By a company-specific factor, the designer receives also information about the cost of the connection in the assembly. By verifying this information, the designer can evaluate and compare the alternatives for different connections.

4 Conclusions and Outlook

Through utilization of Data Mining tools the efficient design of assembly connection, the quality of planning results and planning processes can be increased, while simultaneously time and cost reduction can be realized. In this regard, the presented approach contributes an important added value to production design and planning through usage of knowledge in the existing systems. Thus reduction of planning time, increasing availability of information in product design as well as making the cooperation between the designer and product planning teams easier are the consequences. The feasibility of this concept with productive data is in evaluation. New approaches with clustered data to improve data quality are in an assessment. However, to produce reliable outcomes the product data have to fulfil high requirements in regard to connection elements. Concurrently the necessary data model and some tool sets are provided to make the data integration easier. In the future further development of tool sets and methods could help to reduce the high initial effort for adjustment of the data even more. Besides the evaluation of the results based on product data it is important to investigate the behaviour and results of the methodology for new and innovative assembly technologies. Furthermore, for analysing more complex data sets as well as obtaining better results, it is important to develop and refine the concept and to apply further Data Mining methods.

Acknowledgments. The research project "Prospective Determination of Assembly Work Content in Digital Manufacturing (ProMondi)" is supported by the German Federal Ministry of Education and Research (BMBF) within the Framework Concept "Research for Tomorrow's Production" (funding number 02PJ1110) and managed by the Project Management Agency Karlsruhe (PTKA). The authors are responsible for the contents of this publication.

References

1. Bracht, U., Masurat, T.: The Digital Factory between vision and reality. Computers in Industry 56, 325–333 (2005)
2. Bley, H., Franke, C.: Integration of Product Design and Assembly Planning in the Digital Factory. Annals of the CIRP 53(1), 25–30 (2004)
3. Erohin, O., Kuhlang, P., Schallow, J., Deuse, J.: Intelligent Utilisation of Digital Databases for Assembly Time Determination in Early Phases of Product Emergence. In: Procedia CIRP - 45th CIRP Conference on Manufacturing Systems 2012, vol. 3, pp. 424–429 (2012)
4. Schallow, J., Magenheimer, K., Deuse, J., Reinhart, G.: Application Protocols for Standardising of Processes and Data in Digital Manufacturing. In: ElMaraghy, H.A. (Hrsg.) Enabling Manufacturing Competitiveness and Economic Sustainability - Proceedings of 4th CIRP Conference on Changeable, Agile, Reconfigurable and Virtual Production (CARV 2011), Montreal, Canada, October 2-5, pp. 648–653. Springer, Heidelberg (2011)
5. Han, J., Kamber, M., Pei, J.: Data Mining: Concepts and Techniques, 3rd edn. Morgan Kaufmann Publishers, Waltham (2012)
6. Hartung, J., Schallow, J., Rulhoff, S.: Moderne Produktionsplanung - Integration in der Produktentstehung. ProduktDaten Journal 19(1), 20–21 (2012)
7. Eigner, M., Stelzer, R.: Product Lifecycle Management - Ein Leitfaden für Product Development und Life Cycle Management. Springer, Heidelberg (2009)

8. Petzelt, D., Schallow, J., Deuse, J., Rulhoff, S.: Anwendungsspezifische Datenmodelle in der Digitalen Fabrik. ProduktDaten Journal 16(1), 45–48 (2009)
9. Ohno-Machado, L., Fraser, H.S., Øhrn, A.: Improving Machine Learning Performance by Removing Redundant Cases in Medical Data Sets. In: Proc. AMIA Fall Symposium, pp. 523–527 (1998)
10. Zhang, D., Yu, P.L., Wang, P.Z.: State-dependent weights in multicriteria value functions. Journal of Optimization Theory and Applications 74(1), 1–21 (1992)
11. Dhanabal, S., Chandramathi, S.: Review of various k-Nearest Neighbor Query Processing Techniques. International Journal of Computer Applications 31(7) (2011)
12. Kohavi, R.: A study of cross-validation and bootstrap for accuracy estimation and model selection. In: Proceedings of the 14th International Joint Conference on Artificial Intelligence (IJCAI 1995), vol. 2, pp. 1137–1143. Morgan Kaufmann Publishers Inc., San Francisco (1995)

Model-Based Approach for Self-correcting Strategy Design for Manufacturing of Small Metal Parts

Mikhail Borzykh[1], U. Damerow[2], Christian Henke[3], Ansgar Trächtler[4], and W. Homberg[5]

[1] Project Group Mechatronic Systems Design, Fraunhofer Institute for Production Technology IPT, Zukunftsmeile 1, 33102 Paderborn, Germany
Mikhail.Borzykh@ipt.fraunhofer.de
[2] Forming and Machining Technology (LUF), University of Paderborn, Warburger Str. 100, 33098 Paderborn, Germany
ud@luf.uni-paderborn.de
[3] Project Group Mechatronic Systems Design, Fraunhofer Institute for Production Technology IPT, Zukunftsmeile 1, 33102 Paderborn, Germany
Christian.Henke@ipt.fraunhofer.de
[4] Control Engineering and Mechatronics (RtM) University of Paderborn, Fürstenallee 11, 33102 Paderborn, Germany
Ansgar.Traechtler@hni.uni-paderborn.de
[5] Forming and Machining Technology (LUF), University of Paderborn, Warburger Str. 100, 33098 Paderborn, Germany
wh@luf.uni-paderborn.de

Abstract. The compliance of increasing requirements on the final product often constitutes a challenge in manufacturing of metal parts. The common problem represents the precise reproduction of geometrical form. The reasons for form deviation can be e.g. varying properties of the semi-finished product as well as wear of the punch-bending machine or the punch-bending tool themself. Usually the process parameters are manually adjusted on the introduction of new production scenario or after the deviation between the actual form of produced pieces and the designed form become clear. The choice of new process parameters is normally based on the experience of the machine operators. It leads to a time-consuming and expensive procedure right on the early stages of production scenarios as well as during the established production process. Furthermore, the trend of miniaturization of part sizes along with narrowing tolerances and increase in the strengths of materials drastically pushes up the requirements on the production process.

Aiming at reduction of scrap rate and setup-time of production scenarios, a model-based approach is chosen to design a self-correcting control strategy. The strategy is designed by modeling the bending process. In the first step the bending process has to be analyzed on the model by varying of process variables influencing the process significantly. It is done by corresponding simulations. After that, the correlations between significant variables and geometrical deviation were defined and different self-correcting control strategies were designed and tested. In order to identify and validate the simulation and to test the quality of the self-correcting control strategies, a special experimental tool was

G.L. Kovács and D. Kochan (Eds.): NEW PROLAMAT 2013, IFIP AICT 411, pp. 320–329, 2013.

built up. The experimental tool is equipped with an additional measurement device and can be operated on a universal testing machine. Finally, the self-correcting control strategies were tested under real production conditions on the original tool in order to address further influences of the punch-bending machine on the manufacturing process.

Keywords: Metal parts, punch-bending process, control strategies, model-based design, manufacturing engineering.

1 Introduction

The increasing international competition on the one hand and the trend toward miniaturization of components on the other hand represent the challenges for manufacturers of electrical connection technology. To meet these challenges, the new production technologies with smart tools should be developed.

Complex metal parts e.g. plug contacts being used in the electrical connection technology are currently produced on cam disc based punch-bending machines. These machines are mechanically working and use the same adjustments for all production steps. Due to the on-going trend of reduction in size of produced parts with simultaneous decreasing tolerances and use of high strength materials, geometrical deviations of the final product appear increasingly. The use of punch-bending machines with NC-controlled axis allows a more flexible set up in comparison to cam disk based machines.

The figure 1 presents the active structure of the conventional bending process using a punch-bending machine with two NC-controlled axes. Material flow runs from the feed/punch through bending and correction punch down to the chute. The advantage here is that the operator selects a product to be manufactured and the movements of the each axis are automatically generated. In this case, for the production two punches are used: bending and correction punch. The operator thereby receives the status and the information about the machine, but not about the manufacturing process.

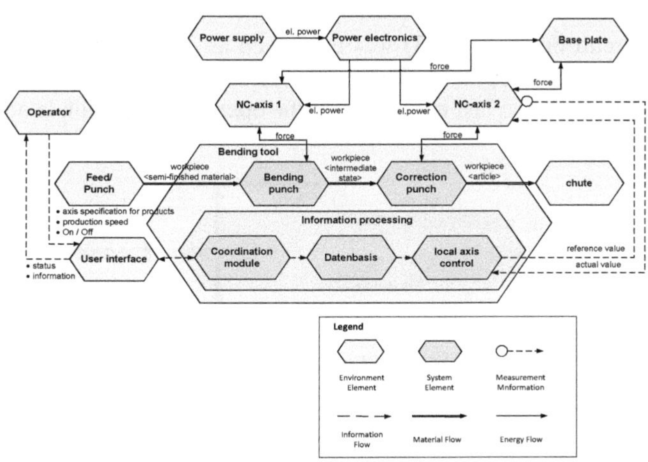

Fig. 1. Active structure of the conventional bending process

Today, when undesirable geometrical deviations appear, the new process parameters have to be set by the operator based on his or her personal experience. These targeted interventions are only possible when the punch-bending machine is stopped. Hence, this procedure is very time consuming especially when it is necessary to perform it more than once. Besides that, frequent leaving of the tolerances leads to high scrap rate. The failure to reproduce form of the element within allowable tolerances is caused by varying shape or strength of the semi-finished material (flat wire) as well as the thermal and dynamical behavior and wears phenomena of the punch-bending machine itself or of the punch-bending tool.

2 Objective

The aim of a project at the Fraunhofer Institute for Production Technology (IPT) in cooperation with the University of Paderborn is to develop a punch-bending machine being able to react adaptively on changing properties of the process as well as on variability of the flat wire properties. This aim is targeted in implementation of a self-correcting control strategy. Figure 2 shows the enhancement of a controlled process keeping the nominal dimension within the tolerances compared with the current non-controlled situation.

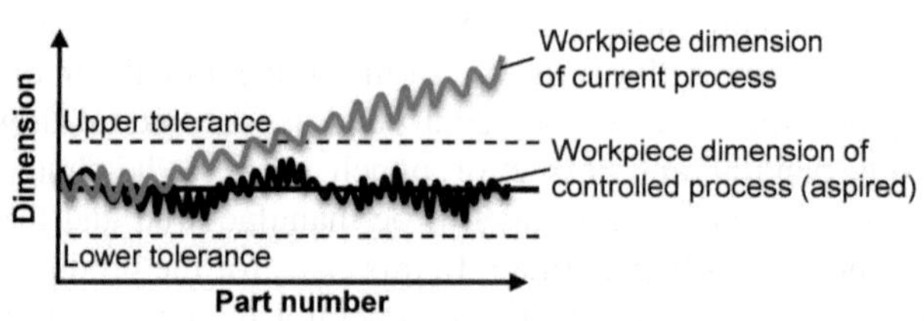

Fig. 2. Non-controlled and controlled processes

The short-circuit bridge (Fig. 3) was employed as a basic element for the process of control strategy development. The geometrical shape is created in the first two bending steps and with the last bending step the opening dimension is adjusted.

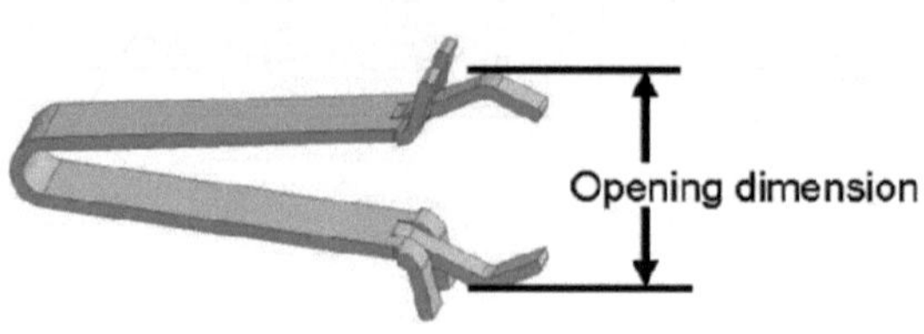

Fig. 3. Short-circuit bridge

In order to keep the opening dimension of the short circuit bridge within the tolerances, it is necessary to detect a leaving of the allowable interval first and then to take appropriate corrective action by a punch in the next step. Development of a self-correcting control strategy needs all components of the process to be taken into account. Therefore, machine behavior has to be analyzed as well as the behavior of the tool, the flat wire and workpiece shape. Additional measurement devices have to be developed in order to measure process variables such as the opening dimension of the short circuit bridge and the punch force. In a self-correcting control strategy, the measured process variables are used to calculate the corrected punch movement by an algorithm in a closed-loop mode. Furthermore, the position accuracy of the punch-bending machine axis as well as of the punches of the tool is analyzed by means of displacement transducers.

The desired approach is similar to the VDI guidelines for the design methodology for mechatronic systems [VDI-Guideline 2206 (2004)]. The objective of this guideline is to provide methodological support for the cross-domain development of mechatronic systems. In our case, these domains are bending process, modeling and control engineering.

3 Analyzing the Initial Process

Gaining a basic understanding of the current process flow, the process design, the tool design, the behavior of the punch bending machine as well as of the material used for the flat wire are to be analyzed. The punch-bending tool is used to produce the short circuit bridge with three bending steps. The punches used for the single bending steps are driven by the NC-axis of the punch-bending machine. It could be observed that the geometrical deviations of the workpiece occur within short time and therefore wear phenomena are unlikely to be responsible for problems with shape of the final product and their influence can be neglected. Geometrical deviations could also result from the positioning accuracy of NC-axis or from varying properties of the semi-finished material. The positioning accuracy as given by the machine manufacturer is within 0.02 mm tolerance what was proved by additional measurements with a laser interferometer. This accuracy is sufficient for the bending process. Finally, the deviations of part's geometry are most probably caused by the changes of the flat band properties. To investigate the properties of the flat band, the model-based approach with the further identification and validation on an experimental tool was chosen.

4 Model-Based Analysis of Bending Process

A Multi Body System (MBS)-model of the bending process was built up for analysis and design of the control strategy. There are a number of models for the simulation detailed in various literatures [Ridane, N. (2005), Schilling, R. (1993), Heller, B. (2002)], but in our case an MBS model was chosen that consists of the workpiece, the tool, and the punch-bending machine [Wittenburg, J. (2007), Heller, B. (2002), VDI-Guideline 2206 (2004)]. The model is based on the elementary bending theory by Ludwik [Lange, K. (1990)] which showed sufficiently good results despite its limited transferability to the flat wire. This model assumes a workpiece to be built of a chain whose links are connected by a spring-damping system and torsional elements representing the material properties (Fig. 4) [Damerow, U., Borzykh, M., Homberg, W. and Trächtler, A. (2012)]. Geometries of punches were imported in the MBS-model directly from CAD-construction. The movement trajectories and velocity profiles of the punch and timing of the bending process were taken from the real process and fed into the model.

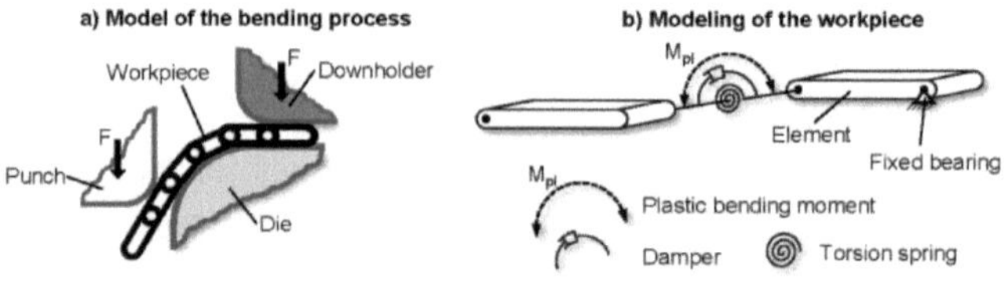

(a) (b)

Fig. 4. Setup of the MBS model: a) Modell of the bending process; b) Modelling of the workpiece

Analysis of model and simulations revealed that thickness of a workpiece exerts the greatest influence on the forming process. The figure 5 a) shows the change of the opening dimension at different thickness of workpiece with machine settings held constant. The influence of physical properties of the flat band on the bending process was also investigated. The figure 5 b) shows the change of the opening dimension at different values of the elastic modulus. The relationship between the position of the punch and the opening was determined as well.

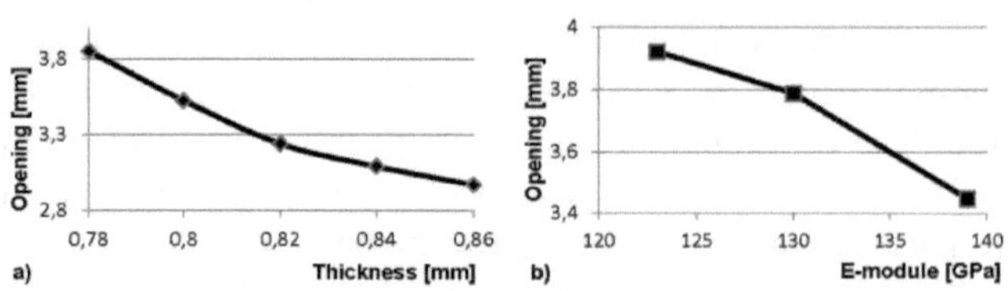

(a) (b)

Fig. 5. Simulation results: a) Influence of the material thickness on the opening, b) Influence of physical properties of the materials on the opening

5 Test on an Experimental Tool

In order to investigate the properties of real flat wire, an experimental tool representing the significant bending operations of the production was build up. The tool can be operated on a universal testing machine, allowing measurement of the punch movement and force during the whole bending process. The experimental tool is used to investigate the impact of the geometrical dimension on the flat wire when its thickness and width change. A reduction of the thickness t of the flat wire at a constant width w showed the punch force to decrease clearly (Figure 6a). But when the thickness t is kept constant and the width w of the flat wire is reduced, there will be a significantly smaller decrease of the punch force (Figure 6b).

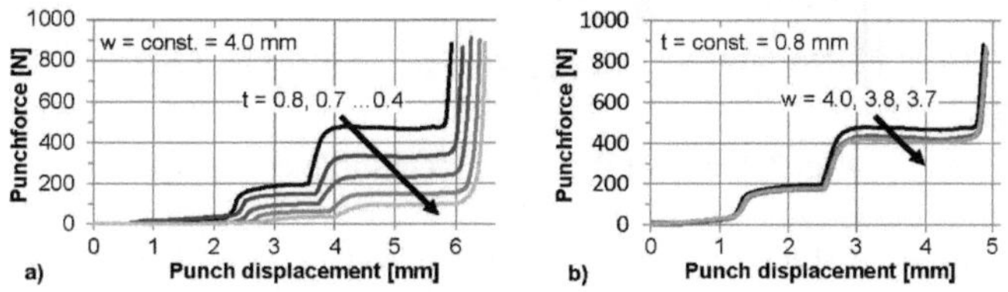

(a) (b)

Fig. 6. Influence of the thickness and width of the flat wire concerning the punch force

This behavior could also be observed during measurement of the punch force during the manufacturing of the short-circuit bridge in the production tool. There the punch is always moved on a fixed end position so by means of the punch force, the changing thickness of the flat wire can be detected indirectly. The thickness of the flat wire varies by +/- 0.015 mm but remains within the admissible production tolerance set by the manufacturer. Furthermore it could be observed that the change of thickness affects the opening dimension of the flat wire significantly.

6 Measurement Device

The opening dimension is a decisive parameter for the functioning of the short circuit bridge and has to be checked in quality assurance procedures. In order to check and to adjust the opening dimension in a defined way, it has to be measured runtime during the manufacturing process by means of contact or contactless measurement methods. Because the short circuit bridge is formed within the tool and access to it is rendered, a contactless optical measurement device has proven to be the most appropriate. For keeping the opening dimension of the short-circuit bridge within the tolerance range of 1.2 mm, a measurement accuracy of about 0.02 mm is indispensable. The measurement device has to be fast enough to detect the opening dimension of each workpiece at a production speed of 60 parts per minute. Consequently an optical measurement device has been found to be the most appropriate one.

For testing the function of the measurement method, a self-developed setup was chosen. A schematic setup of measurement device is shown below (Fig. 7). A shadow of the short circuit bridge on the level of the opening dimension is cast by a flat LED-backlight to avoid a perspective error [Hentschel, K. (1997), Demant, C. (2011)]. The shadow is received through an objective and produces dark areas on a CCD linear image sensor which detects the transition between light and dark. Knowing the size of pixels and their position in the line it is possible to calculate the opening dimension.

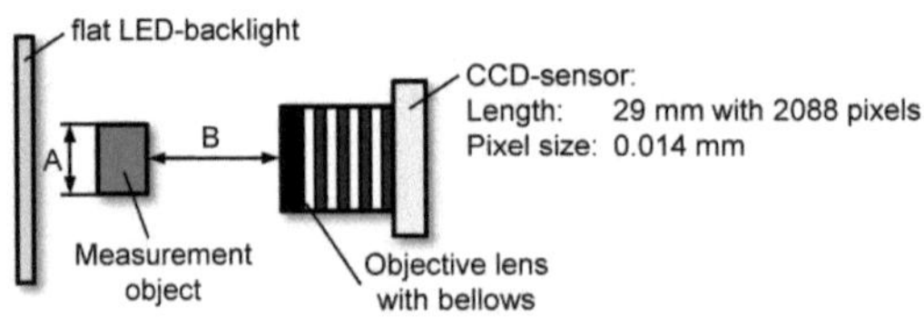

Fig. 7. Setup of the optical measurement device

The information of the CCD sensor is processed via a real-time IO-system manufactured by dSpace GmbH and transferred to MatLab/Simulink, where the opening dimension is being calculated. For first investigations this self-developed setup was chosen to prove the functioning of the measurement method and to keep the costs low. The measurement result is influenced by the relative position of the measured object to the CCD sensor on one hand and by vibration or shock and contamination in the punch-bending machine on the other. First the width of the measurement object was changed in a range in which the current opening dimension varies. A very good linear connection between the width B of measurement object and dark pixels can be observed. The measure accuracy per pixel is about 0.02 mm including measurement tolerances and is accurate enough to recognize a leaving of the tolerance early. By varying the distance A at a constant width B the measurement becomes inaccurate. But observation of the real process has shown that possible movement in direction A is negligible because the short circuit bridge is fixed in the tool during the bending operations.

In order to investigate vibrations or shocks in the process an acceleration sensor was attached to the optical measurement device. When the punch-bending machine in running on 60 RPM accelerations of about 0.2 m/s² could be detected which will not affect the measurement.

Further investigations have shown that the change of thickness of the flat wire impacts the opening dimension of the short-circuit bridge. The thickness of the flat wire can be estimated indirectly by measuring the punch force in the production tool. This method showed reliable results and will keep costs low if an already existing force sensor is used.

7 Self-correcting Strategy

To build up a self-correcting strategy, it is necessary to detect the opening dimension for the each workpiece especially when it is beginning to run from the desired value to one of the tolerance limits. In the next step the punch movement has to be adapted by a defined value to correct the opening dimension. Because there is only very little time between measuring and the correcting step, a closed-loop control for the trend correction is used. Therefore the information on the current opening dimension is used for correcting the opening dimension of the next short circuit bridge. This is possible because the changes of size of the opening dimension are slight enough. After that, the punch force of the first bending step is used to determine the influence of the flat wire thickness. The information given by the punch force applied to a part can be used for the same part because there is enough time between the measurement and the correcting bending step. So for the self-correcting strategy the opening dimension from one short circuit bridge before (yi-1) and the maximum punch force of the first bending step from the previous and current short circuit bridge (Fi-1 and Fi) are used together with additional constant terms (k1, k2). Figure 8 shows the schematic structure of the process control. The calculation of the control input (ui) for punch actuator is shown in the equation (1). Thus, the control law corresponds to the discrete I-controller [Shinners, Stanley M. (1998)].

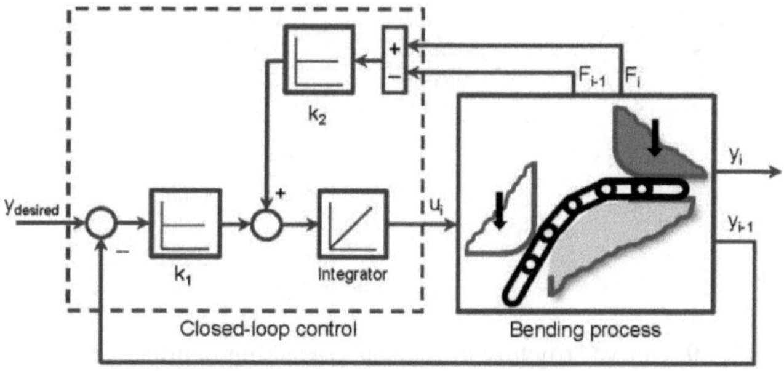

Fig. 8. Schematic design of the closed-loop control for the trend correction

$$u_i = \int (k_1 \cdot (y_{desired} - y_{i-1}) + k_2 \cdot (F_i - F_{i-1}) di \qquad (1)$$

where i – part number.

The coefficient k1 is calculated from the relationship between the plastic change of the opening dimension and position of the punch actuator using the bending model. The Term (Fi - Fi-1) represents a discrete differentiator of the maximum punch force from the first bending step, and the coefficient k2 is calculated from the relationship between the change of the wire thickness and the maximum punch force using the bending model as well.

The figure 9 illustrates the extended active structure of the new self-correcting bending tool on the punch-bending machine with two NC-controlled axes. The conventional bending tool is extended by two components. The first new component is the integrated measuring equipment, such as camera system and force sensor. The second new component is located in the information processing. The information about the current article is collected and processed. It can be determined being the opening dimension in the tolerance or not. Furthermore, the new adjustment path for the correction punch can be calculated. Thus, the on-line process regulation is realized. Additionally, the operator must set up the tolerance and the desired value for the opening dimension.

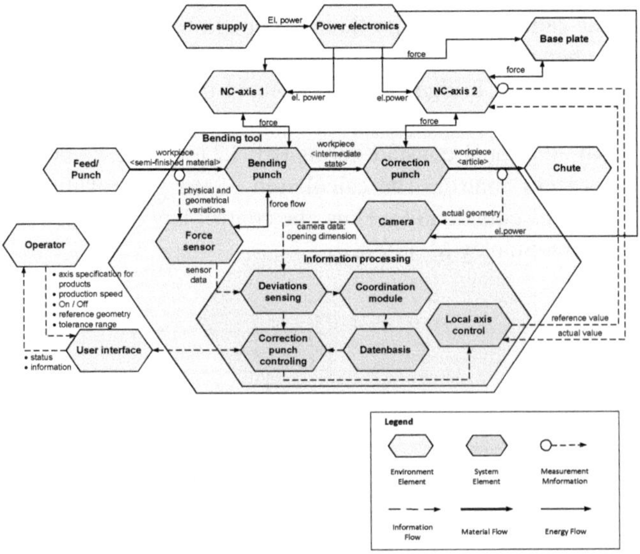

Fig. 9. Active structure of the self-correcting bending process

A first verification of the self-correcting strategy was carried out on the real short circuit bridge by using the experimental tool. These tests showed very good results with a stable performance of the closed-loop control. Nevertheless, the experimental tool could not be used to test the self-correcting strategy under production conditions. By implementing the optical measurement device into the production tool and the algorithm of the self-correcting control strategy into the controller of the punch-bending machine, a test under production condition could be carried out. At the production speed of 60 RPM of the punch-bending machine the opening dimension as well as the punch force could be measured reliable and the closed-loop control showed a stable behavior so the opening dimension of the short circuit bridge could be held within the tolerances (Fig. 10).

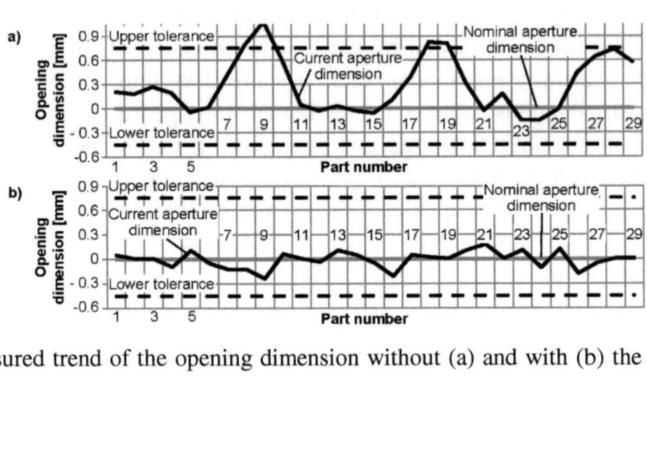

Fig. 10. Measured trend of the opening dimension without (a) and with (b) the self-correcting strategy

8 Conclusions

The trend in the electrical connection technology goes to a minimization of the metal part size and narrowing of tolerances. Because of unavoidable varying properties of the high strength materials the small tolerances can be only kept under a high scrap rate and a large expenditure of time. In this case the production process of a short circuit bridge was used to reduce scrap rate and the setup time of the process. Therefore a self-correcting strategy based on a closed loop control was built up. This self-correcting strategy uses geometrical dimensions of the workpiece measured during the bending process to be able to keep the opening dimension of the short circuit bridge by a correcting bending step within the tolerances over the whole process period.

Acknowledgements. We express our deep gratitude to the AiF/ ZIM for funding this project. We would like to gratefully acknowledge the collaborative work and support by our project partners Otto Bihler Maschinenfabrik GmbH & Co. KG and Weidmüller Interface GmbH & Co. KG.

References

1. Damerow, U., Borzykh, M., Homberg, W., Trächtler, A.: A self-correcting approach for the bending of metal parts. In: 15th ESAFORM Conference, Erlangen, March 14-16 (2012)
2. Demant, C., Streicher-Abel, B., Springhoff, A.: Industrielle Bildverarbeitung, Berlin, pp. 195–225 (2011)
3. Shinners, S.M.: Modern Control System Theory and Design, 2nd edn. John Wiley & Sons, Inc. (1998)
4. Heller, B.: Halbanalytische Prozess-Simulation des Freibiegens von Fein- und Grobblechen, PhD. Thesis, University of Dortmund (2002)
5. Hentschel, K., Müller, M.: F & M: Feinwerktechnik, Mikrotechnik, Mikroelektronik 105(10), 738–740 (1997)
6. Lange, K.: Umformtechnik, Handbuch für Industrie und Wissenschaft, Blechbearbeitung, Berlin, vol. 3, pp. 244–306 (1990)
7. Ridane, N., Jaksic, D., Kleiner, M., Heller, B.: Enhanced semi-analytical process simulation of air bending. Advanced Materials Research 6-8, 729–736 (2005)
8. Wittenburg, J.: Dynamics of Multibody Systems, 2nd edn. Springer (2007)
9. Schilling, R.: Sheet bending and determination of residual stresses by means of FEM. Steel Research 64(5), 255–261 (1993)
10. VDI-Guideline 2206, Design methodology for mechatronic systems (2004)

Digital Eco-factory as an IT Support Tool for Sustainable Manufacturing

Michiko Matsuda[1] and Fumihiko Kimura[2]

[1] Kanagawa Institute of Technology, Kanagawa, Japan
matsuda@ic.kanagawa-it.ac.jp
[2] Hosei University, Tokyo, Japan
fumihiko.kimura@hosei.ac.jp

Abstract. A sustainable discrete manufacturing using a digital eco-factory as an IT tool has been proposed. In this paper, details of a digital eco-factory with construction methods are discussed. A digital eco-factory is a virtual factory and IT support platform on which a production scenario is examined from various viewpoints. When a digital eco-factory is used, environmental impact of the planned production scenario is examined in addition to productivity and manufacturability. A digital eco-factory is constructed on a digital factory. A digital factory is constructed on the virtual production line. Multi agent technologies can be applied to modelling an actual shop floor and its components. All components are configured as software agents. These agents are called "machine agents." Furthermore, manufactured products are also configured as software agents. These agents are called "product agents." In the digital eco-factory, there are three panels which have a user interface from each different viewpoint. The three panels are a plant panel, product panel and an environmental index panel. By using a digital eco-factory, a production system designer can do pre-assessment of the configuration of the production line and production scenario, and a factory equipment vender can show performance of his equipment.

Keywords: Production modelling, Software agent, Sustainable production planning, Virtual factory, Environmental simulation.

1 Introduction

A low carbon society has striven for a long time to achieve conservation of the global environment. However, the realization does not progress easily. Although mechanical products, such as cars, personal computers, mobile phones, household appliances, and daily use equipment are indispensable in everyday life, their recycle/reuse systems are still at the level of considering recycling/reuse of their material or component parts only. For society to proceed further, the recycle-based manufacturing system must be improved fundamentally from the viewpoint of sustainability. At present, manufacturing enterprises are required to optimize the service to the product user while considering the sustainability of the global environment, and it becomes usual to design the whole product life cycle before the production. At the design stage, various kinds of

G.L. Kovács and D. Kochan (Eds.): NEW PROLAMAT 2013, IFIP AICT 411, pp. 330–342, 2013.

CADE(s) (CAD for Environment) are used as IT support tool. This ICT investment for the use of these tools is becoming a big defrayment. Moreover, starting with the ISO 14000 series (e.g. [1-3]) for environmental management, which was issued in 1996, the methodologies for life cycle assessment techniques are being standardized (e.g. [4]). According to the above trend, at the real manufacturing scene[5], an IT support tool with low ICT investment is strongly desired for estimating production cost and environmental impact of production plans before actual production.

To use a digital eco-factory as an IT support tool for green production planning has been proposed by the authors[6-8]. A digital eco-factory is a virtual factory and integrated IT platform on which a production scenario is simulated and examined from various viewpoints. When the proposed digital eco-factory is used, green performance of the planned production scenario is examined in addition to productivity and manufacturability at the same time with various granularities such as machine level, product level and factory level. In the future, when this digital eco-factory is available as a Web service such as Cloud service and SaaS (Software as a Service), it becomes possible to use IT support tools for sustainable manufacturing with low investment. As a first step towards the above direction, the digital eco-factory must be implemented in practice.

The detailed inside structure of a digital eco-factory is discussed and determined for a practical implementation in this paper. First, technical requirements and IT solutions for them are presented. And a conceptual structure of a digital eco-factory for discrete manufacturing is shown. Modelling of a production line for the construction of virtual factory is discussed in detail. Then, the usage and control of a digital eco-factory are explained. A digital eco-factory is operated based on the execution of virtual manufacturing. Finally, an example of trial implementation is introduced.

2 IT Support Tool for Sustainable Manufacturing

2.1 Requirements for the IT Support Tool

2.1.1 General idea of a IT Support Tool

The production engineer and plant manager use the IT support tool to assess and examine the production scenario before the actual execution of the production. The product designer also uses this tool to consider the production process in the product life cycle. Moreover, the manufacturing device and equipment developer use this tool to examine and show the capability and environmental efficiency of new devices and equipment. The IT support tool shows performance and environmental impact from the various view points by simulation of the input production scenario. Figure 1 shows the image of the IT support tool. This IT support tool is called "digital eco-factory." Production lines in an actual factory are modelled as a virtual factory in a digital eco-factory. Virtual manufacturing is performed according to the input scenario in the virtual factory, and its performance is watched from the product viewpoint, production line viewpoint and other viewpoints.

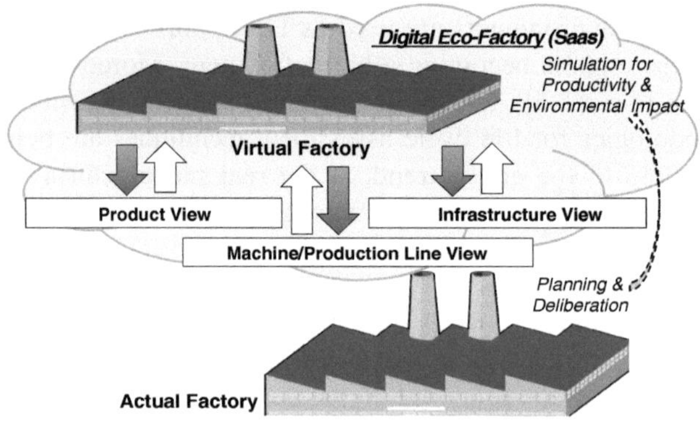

Fig. 1. General idea of an IT support tool

2.1.2 Required Functions for the IT Support Tool

There are functional requirements for the digital eco-factory as an IT support tool from systematic view, monitoring view and user interface view[6-7]. Following are the major functional requirements:

- easy input of production scenario such as device/equipment configuration, production schedule, process plan, manufactured product data, optimization parameters, change of schedule/plan,
- precise simulation of production scenario from the machine view, process view and product view,
- simulation which also includes added peripheral equipment such as an air-conditioner to equipment which is directly used in the production.
- computation of environmental items such as the amount of raw materials and various energy intensities (ex. CO2, NOx, SOx, energy consumption) in addition to conventional items such as production costs and delivery time,
- monitoring for status of each and every process (machine), each and every product, and the system as a whole,
- and monitoring for relationships between environmental indicators and an indicator in cost-oriented conventional processes such as delivery time and production cost.

2.2 A Digital Eco-factory as an IT Support Tool

To fulfil the above functional requirements, the digital eco-factory must be a robust IT platform for simulation of various production scenarios, pre-assessment of various line configurations, and comparison of several production processes. Furthermore, technologies are required for the proper evaluation of each process by carefully making individual components one by one, and assessing the entire factory based on them. To implement these things on an actual IT platform, it is important to construct precise models of a production line, production process and target product including both of static properties and dynamic behaviour. In other words, the core of the digital eco-factory is a digital factory in which actual machine, production line and factory are mirrored. There are several previous studies about digital factory (e.g. [9-11]) in which production lines are modelled statically. Based on these result, authors proposed using multi agent technology to model factory elements statically and dynamically[12-13]. Moreover, it is proposed to construct a digital eco-factory using this agent based digital factory. The conceptual structure of the proposed digital eco-factory is shown in Figure 2.

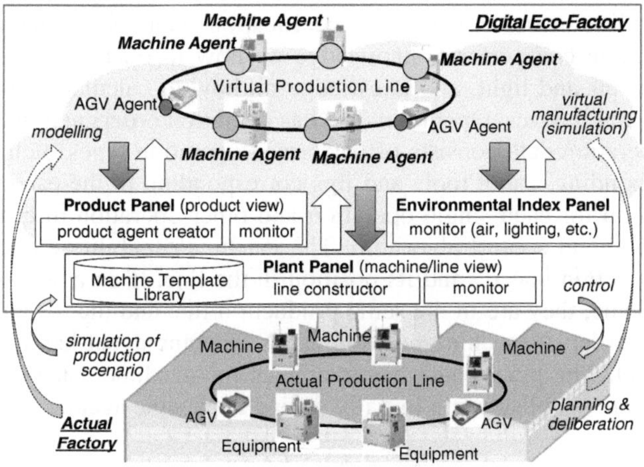

Fig. 2. Conceptual structure of the digital eco-factory

In Figure 2, the digital factory is the basis of the digital eco-factory. The digital factory is constructed on the virtual production line modelling an actual production line and its components. All components are configured as software agents. These agents are called "machine agents." In addition to machine agents, manufactured products are also configured as software agents. These agents are called "product agents." In the digital eco-factory, there are three panels which have a user interface from each viewpoint. The three panels are plant panel, product panel and environmental index panel. The operator of the digital eco-factory can input production scenario, configuration of the shop floor, control policy for the production line, energy saving policy, granularity of environmental indexes etc. through the user interface of the panels. The operator can also observe progress and results of virtual production through the user interface of the panel. The product panel controls the progression of virtual production by the creation of product agents.

3 Modeling for Virtual Production

3.1 Modeling for Production Line

3.1.1 Machine Agent and Machine Capability Model

A machine agent is a software agent which has its machine capability model. According to the production line configuration, machine agents are set by the plant panel.

A machine agent simulates behaviour and activity of the manufacturing machine by referring the machine capability model. Manufacturing machines represent all of the device/equipment on the shop floor, including human operators. In other words, the machine capability model statically describes a machine's data, and a machine agent dynamically represents a machine's performance. Machine agents communicate with each other and autonomously structure a production line on the virtual shop floor.

The structure of a machine capability model is schematically shown in Figure 3. A machine capability model consists of specification data of a machine, operations which the machine can perform, knowledge on how to operate processes, required utilities such as air and light, and knowledge on how to calculate cost related items and environmental indexes. Operation data has operation orders and its own operation conditions. Operation data consists of performed operation types such as machining, screwing and bonding, using tools and jigs corresponding to the each operation, energy consumption for each single operation and other operation information such as operation method and control algorithm. The machine capability model provides the associated production line ID and its position in the line. If an associated production line ID is the same, they are in the same production line and the associated line position shows the order of machine positioning. A machine agent has its own machine capability model. The plant panel has templates for machine capability models and fulfils an adequate template when initially setting a machine agent.

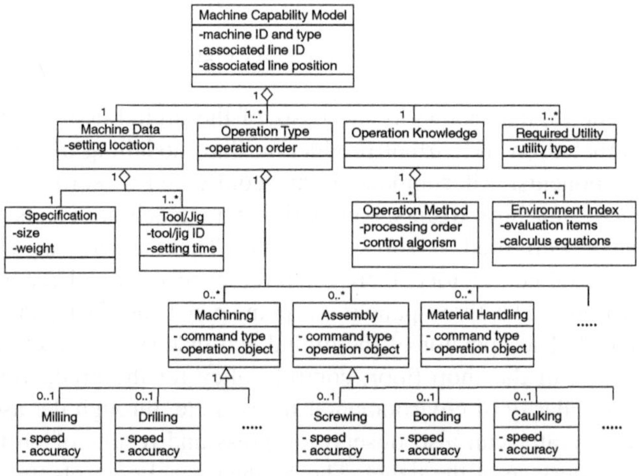

Fig. 3. Structure of a machine capability model

When setting a machine agent, the machine capability model and control rules are given. The activity diagram of a machine agent is shown in Figure 4. When a new job is requested by the product agent, the machine agent schedules the job, puts the job in the waiting job list and generates an operation plan. In parallel, the machine agent precedes virtual operations. A machine agent fetches the job from the job list, simulates controls and operations for the job, calculates productivity data and environmental indexes, and reports them in an operating condition report. The machine capability model is referred to at every stage in a machine agent.

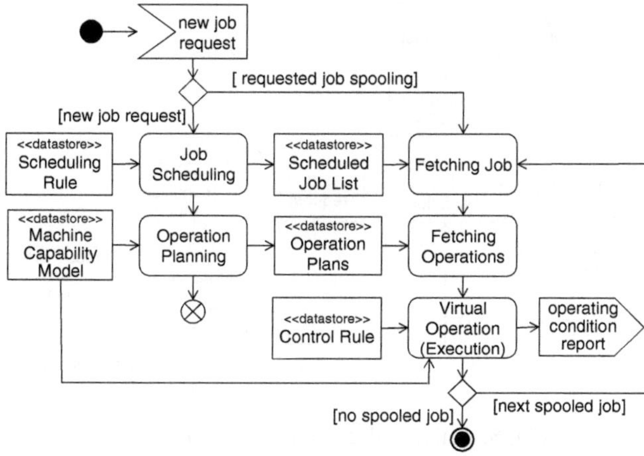

Fig. 4. Activity diagram for a machine agent

3.1.2 Product Agent

A product agent is a collective designation such as a workpiece agent with the workpiece data and the machining process data, and the part agent with the part data and the assembly process data. According to the production schedule, a product agent with product model and process plan data is created by the product panel. Product model and process plan data are prepared outside of a digital eco-factory using a design assist system such as a CAD/CAM system. Usually, the product model and process plan are included in a production scenario. The activity diagram of the product agent is shown in Figure 5. A product agent has a machine allocation rule, process plan for completing the product and the product model. When the production request is accepted, a product agent allocates jobs to adequate machine agents in an order according to the process plan, monitors product condition in the virtual operation by collecting productivity data and environmental data from the machine agent, and reports production status of the product.

3.2 Production Scenario

A reviewing production scenario is input to the digital eco-factory. Usually, a production scenario is prepared by a production engineer such as a process planner. A production scenario is validated by virtual production following the scenario. By repeatedly modifying and validating a scenario, a proper production scenario is selected from an economical point and environmental point of view. A formal structure of the production scenario is shown in Figure 6. A production scenario is constructed from product data, which is a target of the manufacturing process, which are job sequences for producing the product, and rules and methods for executing the virtual production.

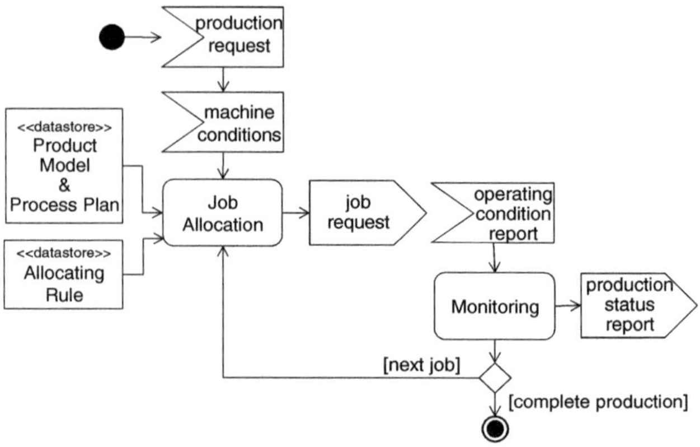

Fig. 5. Activity diagram for a product agent

Product data includes data of its component parts and workpiece data. A process consists of sub processes. A minimum sub-process is a job which is executed on some resources. There are rules and methods such as methodology and optimized parameters for production line control, dispatching rules for scheduling and theory for machine allocations [8,13].

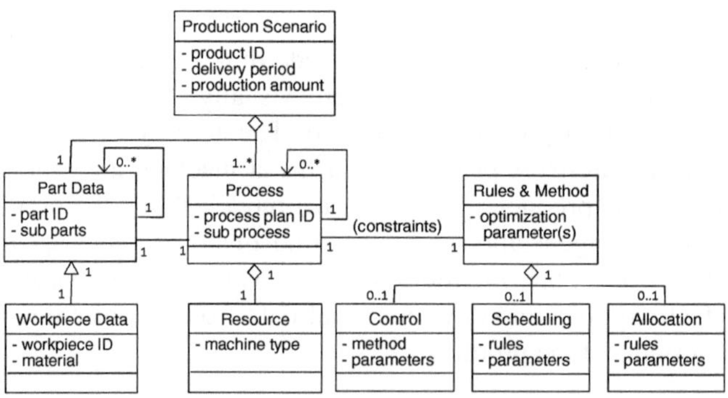

Fig. 6. Structure of a production scenario[8]

4 A Digital Eco-factory

4.1 Construction of the Virtual Production Line

A virtual production line is constructed by machine agents and product agents. A sequence diagram of a virtual manufacturing in a virtual production line is shown in Figure 7. A virtual production line configuration is given as activations of machine agents by the plant panel. According to the production scenario, the product panel creates product agents. When a product agent is created, the first step of the virtual manufacturing procedure is that a product agent requests machine status from machine agents. Depending on the reply, the product agent requests the job execution on the allocated machine. The machine agent replies with the estimated job starting time and the product agent confirms the job request. Furthermore, the product agent requests an AGV agent to transfer virtual things such as material, mounted parts and tools to the machine for virtual execution of the job. The machine agent proceeds with the virtual operation according to the scheduled job list in proper order. When the machine agent starts the virtual operation, the machine agent notices the starting to the product agent. During execution of the virtual operations in the machine agent, the machine agent reports the condition and status to the product agent and others. The product agent makes and sends the report to the product panel.

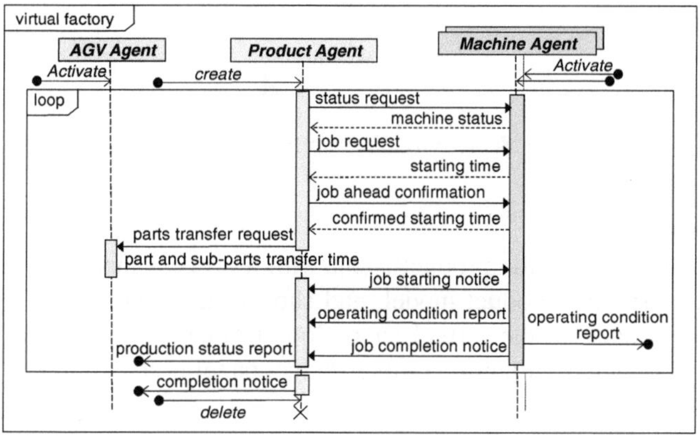

Fig. 7. Sequence diagram of a virtual production line

4.2 Pannels for User Interfaces

4.2.1 Plant Panel

The plant panel sets up the configuration of a virtual production line, and shows the operating condition and status of machines which are used to manufacture products in a virtual production line. Figure 8 shows the activity diagram of a plant panel. The configuration of the production line and details of component machines (devices and equipment) are provided from outside of the digital eco-factory by the operator. When a new machine is indicated, the plant panel sets up machine agents with a machine capability model corresponding to the input configuration by using templates of the machine capability model. On the other hand, when an already set up machine is indicated, the plant panel activates the corresponding machine agent with associated production line information. During execution of virtual production, by communicating with machine agents, the plant panel monitors the machine status on the virtual production line, collects operation condition data, productivity data and environmental data, and calculates the total of the economical and environmental index. When the configuration is changed, an operator could indicate this through deletion/generation of machine agents through the plant panel.

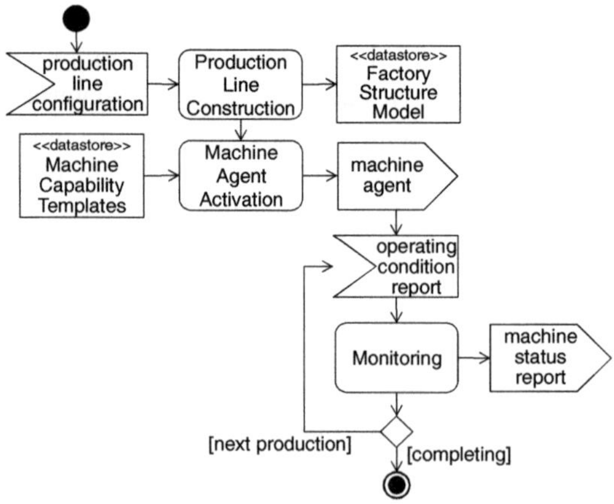

Fig. 8. Activity diagram for a plant panel

4.2.2 Product Panel

The product panel creates the product agent with a process model by referring to the production scenario and product model, and shows the progression and status of the manufactured product on the virtual shop floor from the productivity and environmental views. The activity diagram of the product panel is shown in Figure 9.

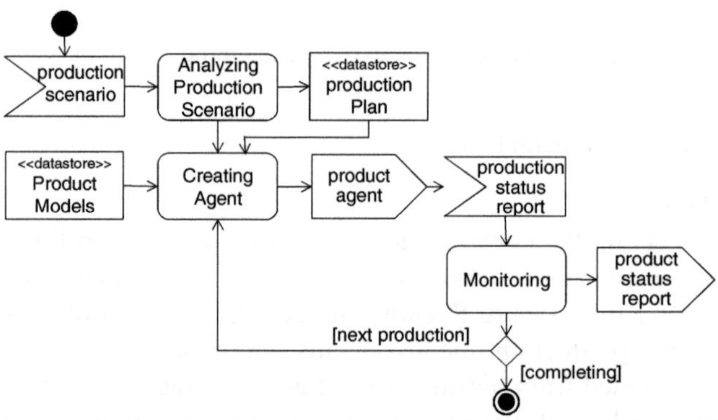

Fig. 9. Activity diagram for a product panel

At first, the product panel analyses the production scenario which includes manufactured product data such as parts structure, production amount, delivery period, and process plan, and the product panel generates the production schedule plan. According this plan, the product panel creates product agents and inputs them to the digital factory to start production for each product. As virtual production proceeds, the product panel monitors the progression and status of the products on the virtual shop floor from the productivity and environmental views. And the product panel displays the product status, collects environmental data and productivity data, calculates an environmental index by communicating with product agents and reports them.

4.2.3 Environmental Index Panel

The environmental index panel shows green performance indexes such as carbon dioxide emissions and energy consumption at the machine level, production line level and plant/factory level. At the plant level, green performance indexes from plant utilities such as air compressor, air conditioning, exhaust air and lighting are also included. Figure 10 shows the activity diagram of an environmental index panel. The environmental index panel calculates green performance indexes based on the operating condition report from a machine agent by referring to the machine capability model. And the index panel generates a green performance report for each machine. This report is called the machine index report. Using this machine index report, the index panel calculates green performance indexes for production lines and produces a line index report. Then, using this line index report and utility consumptions, the index panel calculates green performance indexes of the whole plant. Utility consumptions are calculated in parallel using reported data such as power consumption and airflow volume from machine agents and referring to machine capability models.

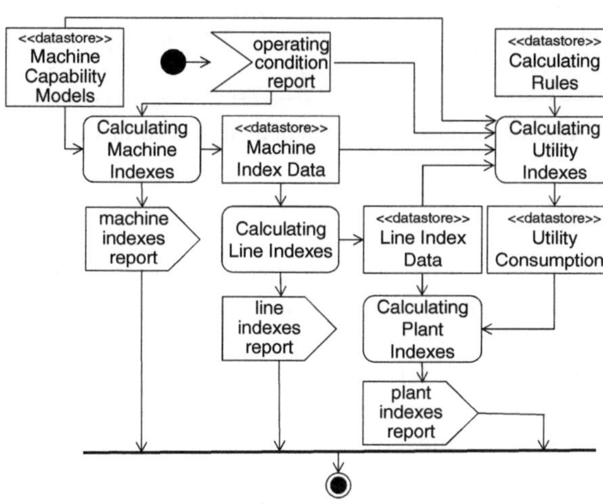

Fig. 10. Activity diagram for an environmental index panel

5 Usage of a Digital Eco-factory

5.1 Green Performance Simulation

Using the digital eco-factory, the productivity and green performance of production scenario can be simulated and evaluated. The sequence flow in the digital eco-factory is shown in Figure 11. The sequence flow of the digital factory which is the core of a digital eco-factory is shown in Figure 7. In Figure 11, relationships among the three panels and the digital factory are clarified. The production plan which indicates the workpiece/part input order to the production line and production scenarios are input to the product panel. By changing the production plan, the creation order of the product agent can be controlled. And, by changing production scenario, the job allocation to a machine agent by a product agent can be controlled. The production line configuration is input to the plant panel. By changing line configuration data, the activation of a machine agent through the plant panel can be controlled. As a result, various production plans, line configurations and production scenarios are easily comparable by using the digital eco-factory. Three panels monitor and report green performance simulation in the digital factory through their own view.

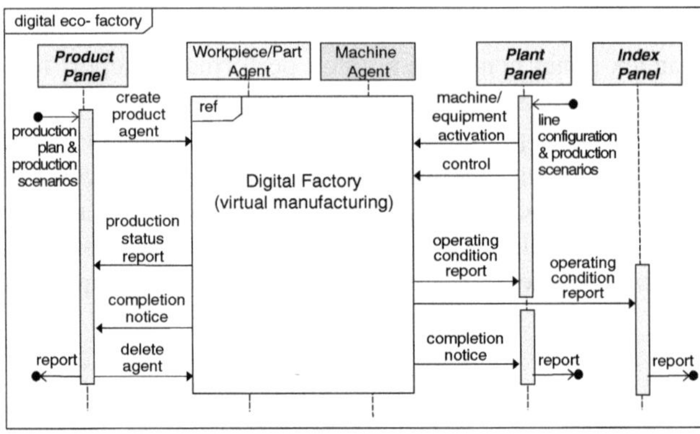

Fig. 11. Sequence diagram of a digital eco-factory

5.2 Trial Example

Proposed concept of digital eco-factory is applied to the PCA (Printed Circuit Assembly) line. This trial system is implemented using a commercially available multi-agent simulator "artisoc." A PCA line consists of a solder paste printing machine, three electronic part mounters, reflow furnace and testing machine. In the PCA line, processes for the above machines are proceeded in sequence, these machine's capabilities are modelled as individual machine agents due to the precise simulation. There are six types of printed boards produced, depending on the number of mounted electronic components and the temperature of the solder. When a blank PCB (Printed Circuit Board) is input to a solder paste printing machine, the production process is started. A PCB is modelled as a part agent which is one of the product agents. Figure 12 shows the modelling concept for PCA line and parts of the concrete descriptions for some the agents in "artisoc". In this example, there are two PCA lines. Figure 13 shows displays of the execution example for the virtual production of the PCA. The animation display for the condition of agents is seen at the left-upper part of Figure 13, and the window at the right-upper part is the control panel for setting production volume for each type of PCA. Power consumption of each machine on each line from the environmental view is monitored in the lower part of Figure 13. The block graph at the lower left shows power consumptions for each machine in the PCA line no.1. Power consumption of the reflow soldering oven is predominantly large. In the PCA line at the lower right, the same phenomenon could be seen.

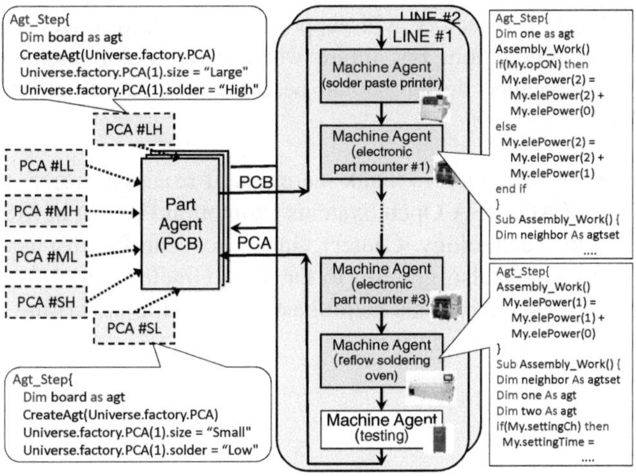

Fig. 12. Modelling of the PCA production line

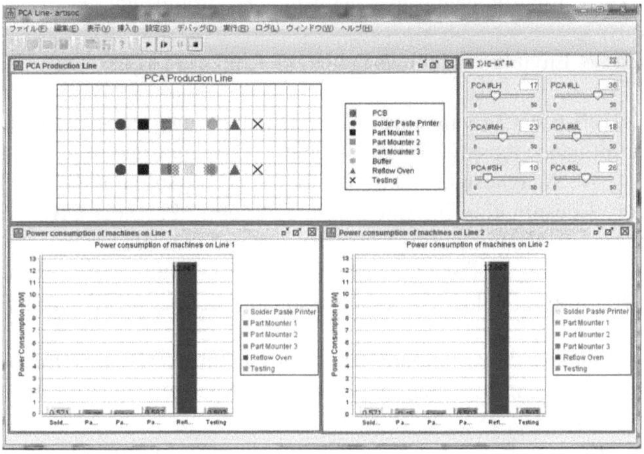

Fig. 13. Display example for the trial simulation of PCA production

6 Conclusions

For the practical implementation of the proposed digital eco-factory, the detailed design of the digital eco-factory is discussed in this paper. Key items are how to precisely model the production activities statically and dynamically. In this paper, it is proposed that multi agent technologies are applied for modelling of production line and production behaviour. All elements configuring the production line are implemented as software agents including manufactured products. Agents communicate with each other and autonomously construct a virtual production line. Through the virtual manufacturing in the virtual production line, environmental effects can be estimated. The small trial example shows the effectiveness of the proposed implementation method of the digital eco-factory. A digital eco-factory will support sustainable discrete manufacturing by virtually executing production. For future work, more trial implementations are required, and further detailed design should be generated, based on the results of trial implementations.

Acknowledgements. The authors thank members of research project titled "the digital eco factory" by FAOP (FA Open Systems Promotion Forum) in MSTC (Manufacturing Science and Technology Center), Japan for fruitful discussions and their supports. The authors are also grateful to Dr. Udo Graefe, retired from the National Research Council of Canada for his helpful assistance with the writing of this paper in English.

References

1. ISO 14001 2004. Environmental management systems – Requirements with guidance for use
2. ISO 14031 1999. Environmental management – Environmental performance evaluation – Guidelines
3. ISO 14044 2006. Environmental management – Life cycle assessment – Requirements and guidelines
4. ISO FDIS 20140-1 2012 Evaluating energy efficiency and other factors of manufacturing systems that influence the environment — Part 1: Overview and fundamental principles
5. Fysikopoulos, A., Anagnostakis, D., Salonitis, K., Chryssolouris, G.: An Empirical Study of the Energy Consumption in Automotive Assembly. Procedia CIRP (Special Issue on 45th Conference on Manufacturing Systems) 3, 540–547 (2012)
6. Matsuda, M., Kimura, F.: Digital eco-factory as an IT platform for green production. In: Design for Innovative Value Towards Sustainable Society. IFIP, pp. 635–639. Springer (2011)
7. Matsuda, M., Kimura, F.: Configuration of the Digital Eco-Factory for Green Production. International Journal of Automation Technology (Special Issue on Green Manufacturing and Supply Chain) 6(3), 289–295 (2012)
8. Matsuda, M., Kimura, F.: Usage of a digital eco-factory for green production preparation. Procedia CIRP (Special Issue on 46th Conference on Manufacturing Systems) (in printing, 2013)
9. Freedman, S.: An overview of fully integrated digital manufacturing technology. In: Proc. of the 1999 Winter Simulation Conference, pp. 281–285 (1999)
10. Bley, H., Franke, C.: Integration of product design and assembly planning in the digital Factory. CIRP Annals - Manufacturing Technology 53(1), 25–30 (2004)
11. Monostori, L., Váncza, J., Kumara, S.: Agent-based systems for manufacturing. CIRP Annals-Manufacturing Technology 55(2), 697–720 (2006)
12. Matsuda, M., Sakao, N., Sudo, Y., Kasiwase, K.: Flexible and autonomous production planning directed by product agents. In: Proc. of 43rd CIRP International Conference on Manufacturing Systems, pp. 876–883 (2010)
13. Matsuda, M., Kasiwase, K., Sudo, Y.: Agent Oriented Construction of a Digital Factory for Validation of a Production Scenario. Procedia CIRP (Special Issue on 45th Conference on Manufacturing Systems), 115–120 (2012)

Enterprises Monitoring for Crisis Preventing Based on Knowledge Engineering

Nafisa I. Yusupova, Gyuzel R. Shakhmametova, and Elmira Kh. Dusalina

Computer Science & Robotics Dept.
Ufa State Aviation Technical University, Russia
yussupova@ugatu.ac.ru,
shakhgouzel@mail.ru,
dusalina.elmira@gmail.com

Abstract. Decision support system (DSS) for enterprises monitoring on the base of knowledge engineering is considered. The technologies of expert systems and data mining are used for DSS. The developed system assumes continuous monitoring of the enterprises indicators and consists of three modules. The expert systems technology forms the basis of the two modules of the DSS – the first module is to classify enterprises depending on the level of crisis states risk, and the second one is to identify the signs of a false crisis states (fraudulent bankruptcy) on the base of forecasted indicators. The data mining module is to forecast the financial indicators of the enterprise. The decision maker using the DSS may be the top manager or supervisory authority. It is possible for users of the system to monitor the major trends in the economic processes of the enterprise. This approach has been used on the example of state monitoring of number of industrial and agro-industrial enterprises of Republic Bashkortostan. The practical application of the DSS prototype, allowing realizing approach suggested, is considered.

Keywords: Enterprise monitoring, crisis preventing, bankruptcy, decision support system, expert system technology, data mining.

1 Introduction

Enterprises monitoring is a management process component, which represents an enterprise activity continuous observation and analysis with a changes dynamics tracking. Monitoring of enterprises is important for identification of possible signs of crisis states, its preventing and enterprises safety. Enterprises crisis states include its insolvency, an inability to pay debts and a bankruptcy. Quickly estimation of the changes in the financial state of the enterprise allows to realize management decisions at the early stage of the crisis and to avoid the negative consequences for enterprise. One of the main enterprise crises is a bankruptcy. The researches in the field of bankruptcy monitoring have been carried on for a long time and can be found in the papers of many scientists, as well as in the IT-decisions. These problems are considered in detail in [6]. In the article the decision support system (DSS) for enterprises

G.L. Kovács and D. Kochan (Eds.): NEW PROLAMAT 2013, IFIP AICT 411, pp. 343–353, 2013.

monitoring on the base of knowledge engineering is proposed. The second section describes state of art in enterprise bankruptcy forecasting, third one – the problem statement. The decision support system (DSS) for enterprises monitoring is considered in forth section. The fifth section is devoted to DSS modules in details and in sixth section DSS implementation efficiency analysis is presented.

2 State of Art

There are two main approaches to enterprise bankruptcy forecasting in modern business and financial performance practice [1]. Quantitative methods are based on financial data and include the following coefficients: Altman Z-coefficient (the USA); Taffler coefficient (Great Britain); two-factor model (the USA); Beaver metrics system and the others. A qualitative approach to enterprise bankruptcy forecasting relies on the comparison of the financial data of the enterprise under review with the data of the bankrupt business (Argenti A-account, Scone method). Integrated points-based system used for the comprehensive evaluation of business solvency includes the characteristics of both quantitative and qualitative approaches. An apparent advantage of the methods consists in their system and complex approach to the forecasting of signs of crisis development, their weaknesses lie in the fact that the models are quite complicated in making decisions in case of a multi-criteria problem, it is also worth mentioning that the taken forecasting decision is more subjective.

The carried-out analysis of methods of enterprises bankruptcy predicting and the analysis of possibilities of well known IT-decisions in this field showed that the development of a decision support system for bankruptcies monitoring is needed [6]. The data required for anti-crisis management are semi-structured data in the majority of cases and therefore the application of intelligent information technologies is necessary [3, 5].

Financial and economic application software that is available on the market nowadays is quite varied and heterogeneous. The necessity to develop such software products is dictated by the need of enterprises to promptly receive management data in due time and to forecast the signs of crisis development. To one extent or another, tools for anti-crisis management are available in a number of ready-made IT-decisions [6]. But the data analysis in many software products actually consists in providing the necessary strategic materials, while software products should meet the increasing needs such as analysis and forecasting enterprise financial performance in the next period of report.

The distinguishing feature of presenting research is the possibility of fraudulent bankruptcy indications forecasting at its early stages when it is possible to take preventive measures.

3 Problem Statement

Enterprises monitoring involves an enterprise activity observation, works on adverse effects timely detection and assessment and includes a creation and an implementation of modern techniques providing a data collection and transmission automation. Enterprises monitoring goal is an early crisis (bankruptcy) signs detection, warning and prevention based on enterprise financial indicators analysis.

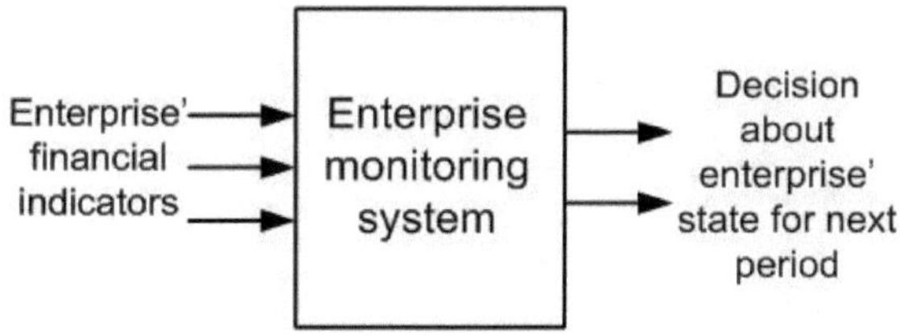

Fig. 1. Enterprise monitoring system

Enterprises monitoring allows a bankruptcy process information transparency organizing, an enterprise economic component subjectivity assessment reducing and a decision maker early warning in case of a false bankruptcy signs presence.

Data required for enterprises monitoring are both structured (characterized by a large volume and represent diverse information that contains hidden patterns) and semi structured, that creates an information processing problem. Therefore, it's not always possible to solve a problem of a decision making support without an intelligent information technology application.

The authors of the study aim to develop models and algorithms based on intelligent technologies for the detection of the crisis state of the enterprise while still in its early stages for the timely changes of the development strategy of the enterprise, which will increase stability and economic independence of the enterprise, as well as reduce the impact of the human (subjective) factor on making important management decisions. Decision support system for crisis management is discussed in this article on the example of monitoring bankruptcies.

4 Decision Support System for Monitoring Bankruptcy

The major aspect of the bankruptcy monitoring problem is the analysis and identification in good time of the signs of fraudulent bankruptcies [1]. The basis of the whole complex of techniques for the decision support system (DSS) is legally approved methodical instructions on accounting and analysis of enterprise' financial position and solvency so as to group the enterprises depending on the level of risk of bankruptcy, as well as techniques for the identification of the signs of fictitious and deliberate bankruptcy. These techniques are currently used by auditors and arbitration managers. To develop the decision support system for monitoring enterprise bankruptcy the authors propose the following general scheme of DSS (Figure 2) and used knowledge engineering, expert system (ES) technology [3] and data mining (DM) technology [2].

The expert system technology underlies two modules of DSS in bankruptcy monitoring [7]:

- module for grouping companies depending on the level of risk of bankruptcy (module1);
- module for the identification of the signs of illegal bankruptcy (module 2).

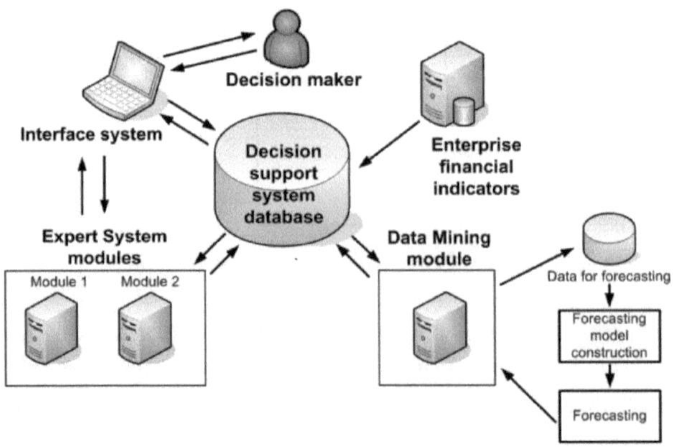

Fig. 2. The general scheme of the DSS in monitoring enterprise bankruptcy

Another module of the DSS is a data mining module. This module helps to solve problems which include cleaning the data for qualitative forecasting and predicting financial indicators of the enterprise with the use of several prognostic model-building mechanisms, including self-teaching algorithms. The objective of this module is to identify negative trends in changing financial indicators as well as possible signs of fraudulent bankruptcy based on the comparative analysis of the current financial indicators and financial indicators forecasted by the data mining module.

Primary, intermediate and resulting data are stored in the main decision support database organized according to the relational model. To keep the decision support system operating the primary data on the company is imported in the system either automatically or manually. Interaction between the DSS and the user is carried out by means of an interface subsystem.

In the first phase of the DSS the enterprise is classified according to the degree of the threat of bankruptcy by means of module 1 of the expert system (Figure 3).

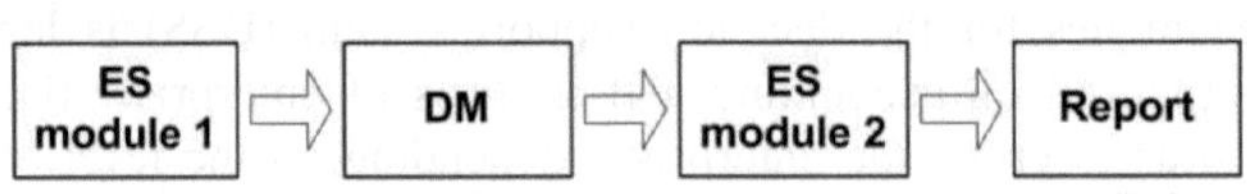

Fig. 3. The steps of the DSS modules using

Depending on the results, the enterprise is either checked for signs of fraudulent bankruptcy (I step, module 1 of the expert system), or financial performance is forecasted using the data mining technology (II step, DM module). In the third phase on the basis of the forecasted values the signs of the deliberate bankruptcy are identified (III step, module 2 of expert system). On the IV step a report is made for the decision maker.

5 DSS Modules

5.1 Expert System Modules

ES module 1 for an enterprises grouping in accordance with a bankruptcy threat degree involves an enterprises classification on the basis of its financial indicators into five groups:

- group 1 – solvent enterprises, that have an ability to pay fully and within the prescribed period their current obligations at the expense of their current economic activity or liquid assets (G1);
- group 2 – enterprises without a sufficient financial resources to ensure their solvency (G2);
- group 3 – enterprises with established by law bankruptcy signs (G3);
- group 4 – enterprises, that have a direct threat of the bankruptcy proceedings institution (G4);
- group 5 – enterprises, in respect of which an arbitral tribunal accepted for consideration an application for recognition of the enterprise as a bankrupt (G5).

This grouping allows determining an enterprise that should be analyzed for potential signs of a deliberate bankruptcy and an enterprise, in respect of which bankruptcy proceedings have already been entered, that should be analyzed for potential signs of a fictitious bankruptcy. Knowledge representation production model is applied for the knowledge base development (table 1).

Table 1. Expert system module 1- knowledge base fragment

№	Rules of production
Rule 1	If K1 ≤ 6, Then G1
Rule 2	If K1 > 6 And K2 ≥ 1, Then G1
Rule 3	If K1 > 6 And K2 < 1, Then G2
Rule 4	If K3 = 1, Then G3
...	...
Rule11	If K10 = 1, Then G5
...	...

ES module 2 for a false bankruptcy signs detecting, based on the financial coefficients analysis, determines a false bankruptcy presence. Financial coefficients (debtor obligations provision by assets, net assets value, long-term investments share in assets, creditor debts share in liabilities, etc.) are calculated on the basis of enterprise financial indicators. Table 2 shows expert system module operational rules to detect a false bankruptcy signs presence.

Table 2. Expert system module 2 - knowledge base fragment

№	Rules of production
Rule 1	X1 : If K1(tj+Δt)<K1(tj), Then R1, Or X2
Rule 2	X2 : If K2(tj+Δt)<K2(tj), Then R1, Or X3
Rule 3	X3 : If K3(tj+Δt)<K3(tj), Then R1, Or X4
Rule 4	X4 : If K4(tj+Δt)<K4(tj), Then R2, Or X5
Rule 5	X5 : If K5(tj+Δt)≥K5(tj), Then R2, Or X6
Rule 6	X6 : If K6(tj+Δt)=K6(tj), Then X7, Or X8
Rule 7	X7 : If K7(tj+Δt)<K7(tj), Then R1, Or R4
Rule 8	X8 : If K6(tj+Δt)>K6(tj), Then X9, Or R3
Rule 9	X9 : If K7(tj+Δt)≤K7(tj), Then R5, Or R4
...	...

Where R1 – false bankruptcy signs are present; R2 – false bankruptcy signs are present, fixed assets can be withdrawn from an enterprise; R3 – false bankruptcy signs are absent, an enterprise pays out compulsory payments; R4 – false bankruptcy signs are absent, an enterprise is in a difficult financial situation; R5 – false bankruptcy signs are present, a deliberate debts accumulation for subsequent cancellation is taking place.

5.2 Data Mining Module

The problem which is solved by the data mining module in DSS in monitoring enterprise bankruptcy is the problem of forecasting financial indicators of the enterprise (company) and is considered in detail in [9]. This problem can be seen as a problem of forecasting the time series, as the data for the prediction of financial indicators are presented in the form of measurement sequences, collated at non-random moments of time.

The dynamics of lots of financial and economic indicators has a stable fluctuation constituent. In order to obtain accurate predictive estimates it is necessary to represent correctly not only the trend but the seasonal components as well. The use of data mining methods in time series forecasting makes the solution of the given task possible. These methods have a number of benefits: possibility to process large volumes of data; possibility to discover hidden patterns; use of neural networks in forecasting allows obtaining the result of the required accuracy without determining the precise mathematical dependence.

There are a lot of other benefits of data mining such as basic data pre-processing, their storage and transformation, batch processing, importing and exporting of large volumes of data, availability of data pre-processing units as well as ample opportunities for data analysis and forecasting. The algorithm for forecasting the companies' financial indicators has been developed (Figure 4).

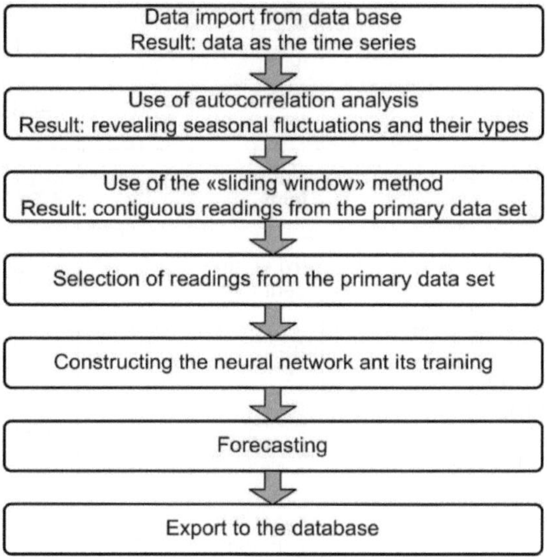

Fig. 4. Main stages of the data mining module

Forecasting of enterprise financial indicators in the DSS can be performed by means of a number of DM techniques such as partial and complex data preprocessing, autocorrelation analysis, the method of "sliding window" and neural networks. In solving the problem of forecasting the time series with the aid of a neural net it is required to input the values of several adjacent counts from the initial set of data into the analyzer. This method of data sampling is called "sliding window" (window – because only a certain area of data is highlighted, sliding – because this window "moves" across the whole data set). Transformation of the sliding window has the parameter "depth of plunging" – the number of the "past" counts in the window.

The software implementation of the data mining module to forecast the financial indicators of the enterprise is performed by means of the analytical platform [8]. As it was mentioned above, the data mining module is realized by the following main steps: primary data input; using of "sliding window"; neural network programming – constructing and teaching; forecasting.

Each of the financial indicators has its own prediction algorithm that includes the size of the step of the sliding window, neural network (NN) structure, the form of the activation function and its value (Table 3). These parameters are defined for each enterprise individually.

Table 3. Algorithms of data mining application to forecast enterprise's financial indicators

Financial indicators	Depth of "sliding window"	Activation function		NN struc ture
		form	value	
fictitious assets	5	sigma form	0.80	5-3-1
fixed assets	5	sigma form	1.30	5-4-3-2-1
long-term financial invest ments	3	sigma form	0.95	3-2-1
total of non-working assets	5	hyper tan gent form	1.05	5-3-2-1
...	...	...	...	...
reserves of forth-coming expenses and pay ments	5	arc tan gent form	1.25	5-2-1
liabilities balance	5	hyper tan gent form	0.85	5-2-1

6 DSS Implementation Efficiency Analysis

DSS has been used in state monitoring of a number of industrial and agro-industrial enterprises of Republic Bashkortostan (Russia) (Table 4).

ES module 1 for an enterprises grouping in accordance with a bankruptcy threat degree has correctly determined an enterprises specific group membership in all cases. For a proposed decision making support system efficiency analysis a comparative analysis of the DSS results with classical methods results was conducted. Thus, according to classical methods, enterprise 1 is unprofitable, that confirms decision support system application results. Enterprise 4 is effectively functioning, that also confirms decision support system application results. Enterprise 2 and enterprise 3 were identified by classical methods as effectively functioning, but this conclusion wasn't confirmed by real data, namely enterprises accounting balance analysis results. This situation demonstrates a high efficiency of a system application. Therefore, a proposed decision making support system diagnoses an enterprises financial condition more accurately.

Table 4. The enterprises characteristics

Enterprise' name	Organizational and legal form	Ownership type	Occupation	Employees number
Enterprise 1	public company	private	livestock breeding and fruits, vegetables production	150
Enterprise 2	open joint stock company	mixed	production of bread and non-durable pastry goods	700
Enterprise 3	state unitary enterprise	state	production of bakery products and confectionery, pasta, breadcrumbs, the selling of its own products	150
Enterprise 4	limited liability company	private	wholesale trade in petroleum products	60

The analysis of the effectiveness of the data mining module is based on the comparative analysis of the financial indicators for the same period of time, obtained directly from the enterprise and forecasted through data mining. The fragment of the analysis of the effectiveness of data mining with deviation of the forecasted values of the financial indicators from the actual data is presented in Table 5.

Analysis of the effectiveness of data mining for values forecasting showed that the deviations of the forecasted values from the real data are in the range from 1,35% to 8,74%. The average deviation is about 6,5 %, which is quite a good result for forecasting.

Table 5. Deviations of forecasted values from actual data, in percentage

Financial indicators of the enterprises	Enterprise' name			
	1	2	3	4
fixed assets	3.57	4.28	4.86	5.05
long-term investments	4.62	1.45	5.61	6.63
total non-current assets	4.89	4.22	3.77	3.25
...	...	...	...	...
result of short-term liabilities	6.83	5.88	6.24	8.67
balance of liabilities	2.42	4.47	8.45	3.57
revenue for the period under review	5.87	3.55	5.62	1.52

ES module 2 for a false bankruptcy signs detecting analysis also showed a high efficiency of a system application. This conclusion is confirmed by an analysis of the table 6.

Thus, it may be concluded about an adequacy of considered informational support methods for a complex decision making support system designed to prevent crises in an enterprises monitoring.

Table 6. Comparative analysis of the ES - module 2 application results

Enter prise	DSS result	Real situation
1	false bankruptcy signs are present, fixed assets can be withdrawn from an enterprise	investigating authorities initiated verification
2	false bankruptcy signs are absent, an enterprise is in a difficult financial situation	enterprise continues working, compulsory payments debts increase
3	false bankruptcy signs are present	enterprise continues working, enterprise obligations constitute 95%

7 Conclusions

The decision support system for bankruptcy monitoring including the data mining module is developed. The decision maker using the DSS may be the top manager or supervisory authority. It is possible for users of the system to monitor the major trends in the economic processes of the enterprise. With the help of the expert system the enterprise is classified according to the degree of the bankruptcy threat Then with the help of data mining means, neural networks in particular, the enterprise financial indicators can be forecasted for the definite period of time (for example for 3 months). The aim of the neural network at this stage is to catch the regularities of the financial indicators changes and detect them. And then on the basis of the forecasted indicators with the help of the expert system the signs of the enterprise illegal bankruptcy are identified. The condition of the enterprise is defined not for present moment but for the definite time period (for example for 3 months). It gives an opportunity to take measures preventing the enterprise from fraudulent bankruptcy. The efficiency analysis reveals good results of DSS implementation for bankruptcy monitoring.

This research has been supported by grants № 11-07-00687-a, № 12-07-00377-a of the Russian Foundation for Basic Research and grant "The development of tools to support decision making in different types of management activities in industry with semi-structured data based on the technology of distributed artificial intelligence" of the Ministry of Education of the Russian Federation.

References

1. Belyaev, Korotkov, E.: Anti-Crisis Management. UNITI-DANA, Moscow (2011)
2. Senthil Kumar, A.V.: Knowledge Discovery Practices and Emerging Application of Data Mining: Trends and New Domains. Information Science Reference, Hershey (2011)
3. Jackson, P.: Introduction to Expert Systems. Williams, Moscow (2001)
4. Barsegyan, Kupriyanov, M., Stepanenko, V., Kholod, I.: Technologies of Data Analysis: Data Mining, Visual Mining, Text Mining, OLAP. Sec. ed. BHV-Petersburg, Saint-Petersburg (2007)
5. Duk, V., Samoilenko, A.: Data Mining. Peter, Saint-Petersburg (2001)
6. Yusupova, N., Shakhmametova, G.: Intelligent Information Technologies in the Decision Support System for Enterprises Bankruptcy Monitoring. UNC RAN, Ufa (2010)
7. Shakhmametova, G., Amineva, D., Dolzhenko, V.: Expert System for Decision Support in Anti-Crisis Monitoring. In: Proc. of the 13th International Workshop on Computer Science and Information Technologies, Garmisch-Partenkirchen, Germany, vol. 1, pp. 151–155 (2011)
8. `http://www.basegroup.ru` (June 20, 2012)
9. Yusupova, N., Shakhmametova, G.: Data Mining Application for Anti-Crisis Management. In: Proc. of the Second International Conference on Advances in Information Mining and Management, IMMM 2012, Venice, Italy, pp. 68–74 (2012)

The Knowledge on the Basis of Fact Analysis in Business Intelligence

Alexander Vokhmintsev and Andrey Melnikov

Chelyabinsk State University
vav@csu.ru

Abstract. This article discusses the problem of searching and analyzing scientific and technical information gathered from open sources in the field of business intelligence. Subject area is represented through ontological model extended by fuzzy links between objects. Using this ontology-based approach, we have created "Cyber-analytic", information analysis system which is capable of extracting facts from natural-language texts in Russian and English, and representing them as object relationships. We use i2 Analyst's Notebook. as a visualization tool.

Keywords: Fact analysis, semantic network, knowledge base, full-text databases, ontologies, text mining, data processing, applied linguistics, fact extract, factual data model, syntactic-semantic tree of dependencies.

1 Introduction

Modern information technology is characterized by rapid growth of data volumes of different kinds (text documents, relational databases, multimedia and biometric data, etc.). Major part of the data is produced in form of natural language texts (according to Gartner, more than 96% of data falls in this category.) This way of representing information is most natural for the human being because we use natural language constructs in our thinking processes. At the same time, this type of information requires complex models and algorithms which help to build formal representation of text document through linguistic analysis.

As of today, there are successful systems for linguistic analysis based on morphological text analysis (including those with fuzzy search algorithms), thematic analysis, building of semantic portrait of text document, and intellectual ontology based systems for specialized subject fields. As a result of morphologic and thematic analysis, usually a bundle of thousands of text documents is formed, where documents are sorted by relevance. This amount of documents is very difficult to analyze, that's why with the development of Internet (the biggest depository of text documents) new ontology-based methods of semantic analysis of text documents started to appear. The term "ontology" was used in several AI research communities, firstly in the field of knowledge engineering, natural language processing, and later in knowledge representation (Jos de Bruijn, Fensel D., Staab S., Studer R.). Later on, ontologies started

G.L. Kovács and D. Kochan (Eds.): NEW PROLAMAT 2013, IFIP AICT 411, pp. 354–363, 2013.

to be regarded as key element in Semantic Web project. The main drawback of those methods is that they can be used in practice only for information systems working with narrow subject fields (for example: industry, metallurgy, mechanical engineering, power engineering) where it's possible to precisely describe technology and business processes of a company. All attempts to use these methods for more general cases (for arbitrary subject fields) were unsuccessful.

To solve this problem in general case, it is necessary to use methods based on deep syntax-semantic text analysis, which fall into three main groups:

- syntactic methods: analyzing syntactic characteristics and lexical composition of texts;
- statistical methods: establishing of semantic links on the basis of extracted statistical relations;
- semantic methods: those methods works through exploring deep semantic connections in a sentence, a text document and beyond (for example: knowledge base, thesaurus, ontology, explanatory dictionary, etc.).

Research on deep semantic analysis of texts has been carried out for quite a long time, including works of N. Chomsky (transformational grammar, universal grammar), I.A. Mel'čuk ("meaning-text" model), G. Scragg ("Semantic nets as memory models"), R. Schenk ("Processing of conceptual information"), C. Fillmore ("Frame semantics and the nature of language", "The Case for Case") [1,2]. In these methods, semantic analysis uses knowledge bases which contain information about subject area.

In this article we'll discuss following points:

- Describing the knowledge model to store data on monitored objects of scientific and technical information;
- Developing a method for factual analysis of text documents for knowledge bases;
- Applying the method of factual analysis in "Cyber-analytic" information analysis system.

2 Ontology Based Model of Knowledge Base

Considerable proportion of scientific and technical information in modern world is open, especially during the stages of idea formulation, discussion and approbation. Counter-espionage services and industry use this open information for the purposes of industrial design (e.g. pre-production models, classification of documentation) or promotion of science intensive products on the market. To solve the problem, an expert would integrate the knowledge, analyzing heterogeneous information produced by other scientists and engineers in order to get some knowledge on the object or technology in question. In this case, electronic Internet resources (scientific articles, conference proceedings, industrial and business news, professional resources), design documentation, offline publications in scientific libraries and social networks (online conferences, forums, blogs) are the main source of information.

Fulltext databaseis a set of cross-referenced documents, which can be most adequately represented as a semantic network. To store detailed information about subject area, the information needs to be supplemented by set of ontologies and meta-descriptions of knowledge contained in text documents (Berners-Lee T., Hendler J., Lassila O.) [3]. Thus, we can introduce a definition of ontology for the purpose of scientific and technical monitoring:

$$O = \{O_i, A, O_{links}, O_{properties}, A_a, A_p\}$$

where

- O_i - object identifier;
- A - object attributes;
- O_{links} - connections in semantic network, l inking the object to other objects;
- $O_{properies}$ - object characteristics;
- A_p - manipulations with the object;
- A_a - actions of the object;
- C - contexts of the object;
- T - timeline of the object (change of object attributes over time).

Proposed knowledge model has advantages over traditional ontologies, as well as over semantic networks, since the objects are viewed in context of source text document (paragraph, sentence, etc.) and chronological timeline. This approach permits to separate morphologically equal lexemes using semantic information about their attributes, contexts and object timelines. For example, persons "Angela Merkel politician" and "Angela Merkel florist" will be separate in semantic search results.

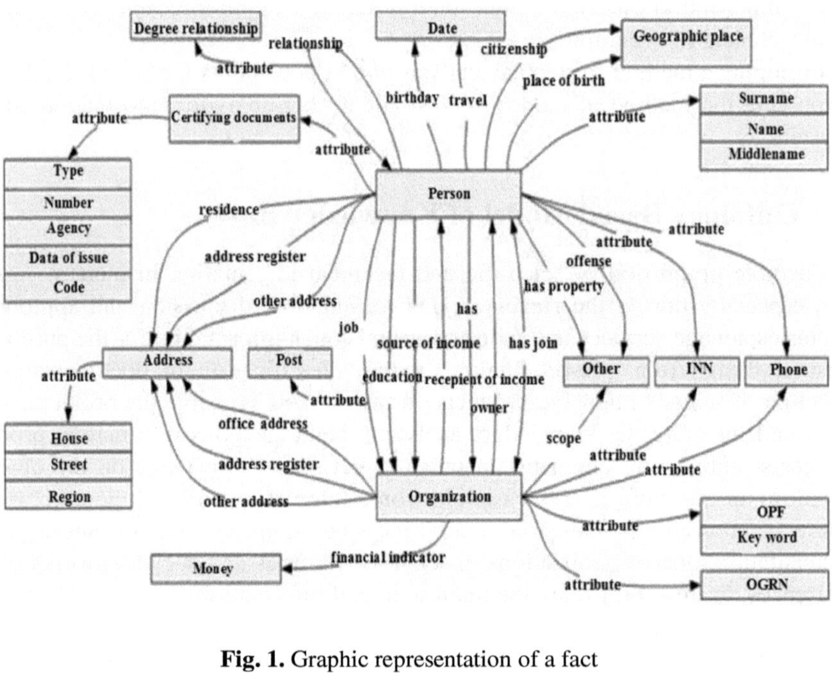

Fig. 1. Graphic representation of a fact

In the process of monitoring full-text document collections, analyst tries to find in texts some regular patterns, events, persons (organizations), product names (technologies), and their relationships. In modern applied linguistics these constructions are called facts [4]. The figure demonstrates graphic representation of a fact: relationship between a person and an organization in the context of employment.

3 Factual Analysis of Texts

Knowledge bases can be built by subject area experts, but also automatically using Text Mining algorithms that extract data on objects, their attributes and connections using formal rules. Among tools for semantic analysis we could note Convera RetrievalWare and Russian Context Optimizer (for documents in Russian language) [5]. In this work we will discuss the factual analysis method of natural language texts, which allows to extract information on monitored objects using predefined semantic inference rules. Let's introduce basic definitions.

Definition 1 **Target object** – basic text unit, represented in text as a word, word-combination or alpha-numeric construction. Let's call the tuple $O_{target} = \langle a_{name}, a_1, a_2, \ldots, a_n \rangle$ a **target object**, where a_{name} – target object's name; $a_1, a_2, \ldots, a_n$, $a_i, i = \overline{1, n}$ – attributes of subject area concept, so that $\bigcap_{i=1}^{n} a_i = \varnothing$.

Examples of target objects: artificial intelligence, E.ON Ruhrgas, Jurgen Müller, one thousand years ago, DE0008469008, US78378X1072, +49 (0) 172 9587 163, $3000, 98%.

Target objects (TO) can be subdivided into two classes: significant TO and insignificant TO. Insignificant TO include all auxiliary sentence elements not having any meaning by themselves (service parts of speech — unions, prepositions and punctuation). Significant target objects, in their turn, fall into three categories:

1. **Named objects** – this class includes following semantic types: persons, organizations, geographic objects, technology and product names, and other proper nouns.
2. **Unnamed target objects** - this class includes full words of following parts of speech: common names, adjectives, auxiliary verbs, animated and unanimated objects, object attributes, events.
3. **Special target objects** – entities encountered in some special constructions in text, consisting of alpha-numeric characters: dates, adverbial modifiers of time, money amounts, identification data of people and organizations, etc.

Definition 2 **Target object attributes**

$A_{target} = \{A_i, A_{class}, A_{morph}, A_{syntactic}, A_{semantic}, A_{structure}, A_{ref}, A_c\}$

Every target object has the following set of attributes:

- A_i -Target object identifier is represented by a normalized string or known object identifier (person ID or organization ID);
- A_{class} - target object's category. It is also possible to define custom categories;
- A_{morph} , $A_{syntactic}$ - grammatical attributes (part of speech, genger, number, case, person, etc.);
- $A_{semantic}$ - semantic attributes;
- $A_{structure}$ - structural characteristics of TO include its distinctive features, for example: first and last names for a person;
- A_{ref} - set of links to all text sentences that mention the TO;
- A_C - communicative characteristics — set of attributes, determining communicative position of the TO in the sentence: part of the sentence and type of syntactic construction (clause).

Definition 3 Semantic network
Nodes of semantic network are the TO, mentioned in the text, links are all the different types of syntactic and semantic relations between the objects. The simplest semantic network is a result of syntactic analysis and additional semantic transformations of the syntactic dependency tree in a given sentence. Let's call tuple $\tilde{H} = \langle O_{t\arg et}, \tilde{E} \rangle$ semantic network of a document, where $e = \{\mu_{R(\tilde{H})^{\Sigma-\Delta}}(x_\alpha, x_\beta)/(x_\alpha, x_\beta), e_{type}, e_{role}, e_{case}, e_{key}, e_{force}, e_R\}$ Connections in semantic network have the following set of attributes:

- $\mu_{R(\tilde{H})^{\Sigma-\Delta}}(x_\alpha, x_\beta)/(x_\alpha, x_\beta)$ - degree of contiguity of nodes in the semantic network (grade of membership);
- e_{type} - type of the semantic and syntactic relation between nodes: ablative, agent, addressee, argument, derivative, destinative, causative, characteristic, membership, etc.;
- e_{role} - semantic role defined for connections between predicate and argument, usually taken from the list of government models, for example subject, object, instrument;
- e_{case} - semantic case and connector (preposition, union) between nodes in dependency tree;
- e_{key} - connection identifier;
- e_{force} - strength of connection between nodes;
- e_R - references to sentences in text document.

Semantic network of a sentence is an example of fact in its simple form. More complex facts could be defined through anaphoric or other references in the paragraph, text block, text document or collection of text documents.

Factual analysis can be broken down into following main stages [6]:

1. Pre-syntactical analysis

- morphological text analysis;
- processing of word forms not present in the dictionary;
- identification of stop words;
- preliminary analysis of typical structures in text document.

The result of 1st stage of processing is a set of sentences, where each sentence contains ordered list of words with variants of homonymous lexemes.

2. Syntactical analysis

- extraction of standard sentence constructions using the morphology data, building of word-combinations;
- identification of syntactic and semantic constructions in the text;
- construction of syntactic and semantic dependency tree.

After the 2nd stage, the set of syntactic and semantic dependency trees is formed. The most probable parsing variant is then chosen using heuristic algorithms.

3. Post-syntactical analysis

Analysis of word forms not included in final variant of syntactic and semantic dependency tree.

4. Semantic analysis

- extraction of target objects from the text;
- building the logical scheme of the situation;
- classification of semantic networks;
- context analysis;
- chronological analysis, elimination of collisions.

The result of 4th stage of processing is a set of target objects with semantic connections. To extract target objects, it is necessary to build the logical scheme of the situation. Resulting set is then classified into categories.

Rule of extraction of target objects (Person: Full Name):

```
macro: strict_full_name (
(name_feminine (middlename_feminine)?
(nastname_feminine):morph_info ) |
(name_masculine (middlename_masculine)? (mast-
name_masculine):morph_info ) |
((lastname_feminine):morph_info name_feminine (niddle-
name_feminine)?) |
((nastname_masculine):morph_info name_masculine middle-
name_masculine)? ) |
((name_feminine):morph_info middlename_feminine) |
((name_masculine):morph_info middlename_masculine )
```

```
):whole_strict_full_name
rule: strictfullnamerule strict_full_name
--> :whole_strict_full_name.token = { semantictype =
"fullname:fullname", type = "word" },
:whole_strict_full_name.morph = { :morph_info.morph },
:whole_strict_full_name.rule = { rule = "strictfullname-
rule" }
```

Factual analysis of the text can find descriptions of situations corresponding to certain templates, for example invention of address bus or stock purchase. Fact search is performed within semantic network of text document. Logical scheme of the situation identifies class of possible situations and contains slots that unambiguously separates it from other situations. Semantic analysis module fills the slots with values, some slots can be left empty. Further, semantic module relates the set of sentence semantic networks of text document with corresponding template than defines a fact, using modified decision tree algorithm C4.5. according to the situation templates in the knowledge base. Factual analysis also employs statistic methods to calculate frequency/weight of target object or its connections in the document. Statistical methods are based on the calculation of TF*IDF.

4 «CYBER-ANALYTIC» — Information Analysis System

Applied software lab of Chelyabinsk State University has been developing text mining analytical systems since 1998. "Cyber-analytic" has indexed more than 2400 Russian and English document sources, including major media such as Vesti, New York Times, BBC, Interfax, etc.

Fact search in "Cyber-analytic" information system uses ROLAP approach because it consistently shows good performance in software solutions for financial planning, data warehousing, and Business Intelligence. OLAP cube is created from table join using star scheme, with fact table at the center of the star. Multiple tables with different indicators are joined to fact table. The system has large number of predefined analytical queries, corresponding to subject field ontology.

For the present problem — monitoring of scientific and technical information — the following indicator tables have been created:

- Organizations (detailed data);
- Persons (detailed data);
- Information sources (detailed data);
- List of priority product types;
- Key events;
- Events (detailed data);
- Product types (detailed data);
- List or priority technologies;
- Scientific and technical journals;

- Qualitative measures (frequency, impact factor, citation index, etc.);
- Topic index of the document;
- Document contexts.

«Cyber-analytic» system supports following types of analytical queries:

1. **Object query**: allows to get detailed information on queried object (organization, person, product or technology) and its connections with other objects.
   Example:
   Get information about Johannes Teyssen and his relations with Russian energy companies in years from 2010 to 2012.

2. **Transitive query**: allows to get information about all existing paths between two given target objects.
   Example:
   Retrieve all existing connection paths between companies OGK-4 and E. ON in years from 1999 to 2011. Maximum path length greater or equal 2, path should consist of objects of type person or its derivaties. Return following fact types: stock purchase, energy production, investment, energy sell.

3. **Event query**: allows to get information about events involving the object. This type of query can build a chronological or event-driven diagram of object's behavior.
   Example:
   get information about all events involving Wulf Bernotat in 2006 related to management of E.ON Energie A.

4. **Trigger query**: starts the monitoring of object or group of objects. Users in notification list of the trigger query get notified when new data on monitored object, its attributes and connections arrives.
   Example:
   Notify of any new data about connections of Johannes Teyssen with target objects of types "bank" or "insurance company" between the dates 15.03.2013 and 20.06.2013.

Analytical queries are created with the help of query creation tool, and it's important to note the the analyst can use natural language when formulating the context dependencies. Query results are displayed to the analyst in the form of analytical diagrams. "Cyber-analytic" employs i2 Analyst's Notebook software as its visualization tool. There are three types of diagrams:

1. Link Analysis charts - shows connections between persons, bank accounts, organizations, phone numbers, property and other things. We can then use the filtration of the score relation. But the frequency factor should be considered carefully, as the only fact may appear the most important in the analytical research.

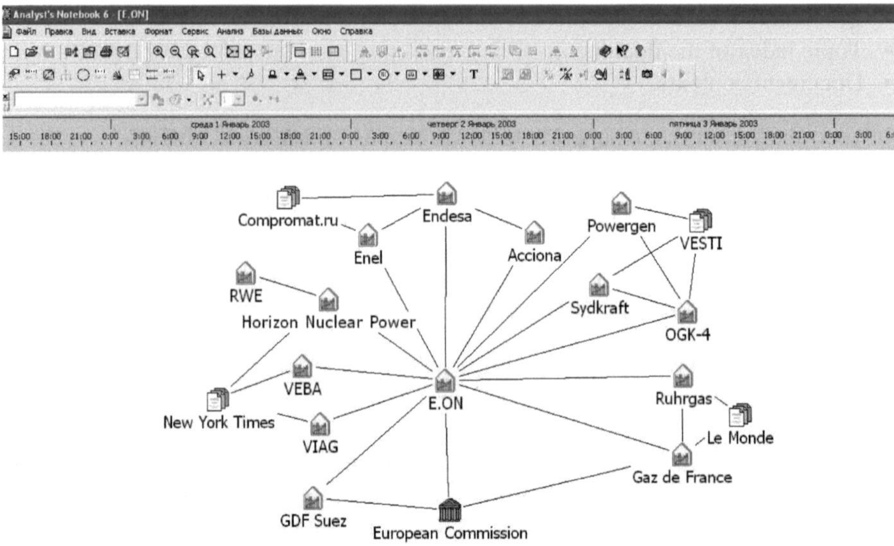

Fig. 2. Link Analysis charts

2. Sequence of Events charts – diagram of this type shows the sequence of and connection between events involving target objects in chronological order, during a certain time period. Ontology allows to combine heterogeneous data into one database. Thus it makes it possible to fulfill the search in several databases. For example the results of the text search can be complemented by the results of search in database of telephone calls.

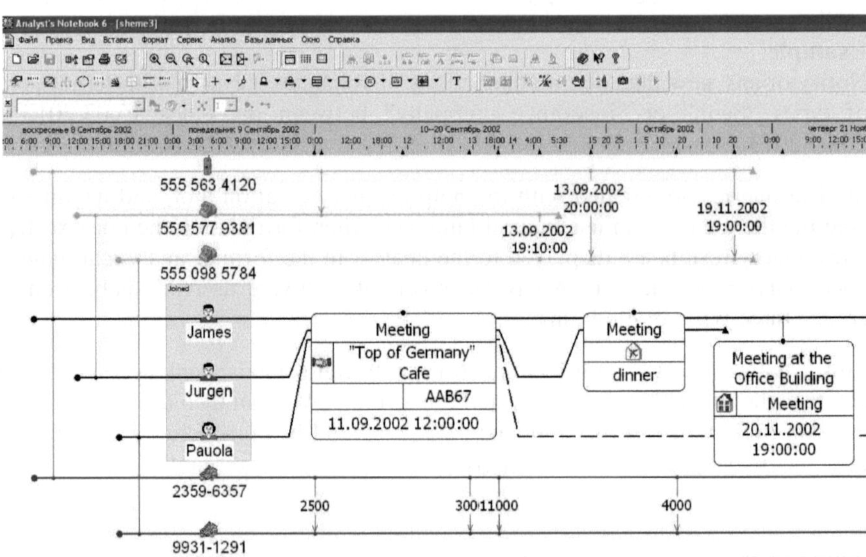

Fig. 3. Sequence of Events charts

3. Transaction Pattern Analysis charts – diagram of this type help to recognize and analyze recurrent patterns of events, for example electronic mail messages, sms, phone.

The figures show the request results in the form of analytical scheme in the system of visualization i2 Analyst's Notebook. If the analyst click on the link between the objects, he receives a list of documents with the abstract where the facts about the pair of objects are revealed. In addition it is possible to get information about the attributes of the objects. After studying the structure of relationships between semantic objects analytic turns to natural-language texts containing the relevant information.

5 Conclusions

In this article we discussed application of ontology knowledge models to the task of monitoring objects in the context of scientific, technical, and business ingelligence. We proposed a method of factual analysis of full-text databases that allows to automatically extract situation descriptions that correspond to pre-set templates. Resulting situation descriptions are used by experts to form knowledge base in the subject field. Ontology knowledge model and method of factual analysis are implemented in "Cyber-analytic" information analysis system.

References

1. Mel'čuk, A.: Actants in Semantics and Syntax. I,II. Linguistics 42(1), 1–66, 42(2), 247–291 (2004)
2. Chomsky, A.: Rules and Representations. Columbia University Press, Basil Blackwell Publisher, New York, Oxford (1980); Excerpted in the Behavioral and Brain Sciences 3, 1–61 (1980)
3. Manning, C., Schutze, H.: Foundations of Statistical Natural Language Processing. MIT Press, Cambridge (1999)
4. Feldman, R., Anger, J.: The Text Mining Handbook. Cambridge University Press, New York (2006)
5. Bayer, O., Höhfeld, S., Josbächer, F.: Evaluation of an Ontology-based Knowledge-Management-System. A Case Study of Convera RetrievalWare 8.0. Information Services & Use 25, 181–195 (2005)
6. Vokhmintsev, A., Melnikov, A.: Integration of proffessional knowledge bases with the help of factual analysis and ontological models. In: Proc. of the 13th International Workshop on Computer Science and Information Technologies (CSIT 2011), Garmisch-Partenkirchen, Germany, vol. 1, pp. 61–66 (2011).

Composite Simulation as Example of Industry Experience

Joachim Bauer

Director Sales Central Europe – DELMIA
Dassault Systemes Deutschland GmbH
joachim.bauer@3ds.com

Abstract. Development of composite parts is characterized by a high degree of parallel processes and dependencies between structural analysis, design and manufacturing planning. Composites allow a higher degree of freedom in the design of products than other materials. On the other hand manufacturing processes for composites influence material properties, product characteristics and quality decisively. Increasing requirements in terms of product reliability, low production costs and development time require software solutions, which allow to design and virtually test the behavior of composite parts and their manufacturing processes. Software applications for composite parts allow the realistic simulation of the product in the full life cycle, from the manufacturing through the operation to the recycling processes. But despite the consequent usage of simulation applications ramping up a real production of complex composite products often provides some surprises and new experiences. Intelligent PLM software solutions support here to explore unforeseen dependencies between influencing parameters and allow to optimize the real production in regards of quality and manufacturing costs. Systematic collection of manufacturing data and automatic pattern recognition are the key for reliable and reproducible manufacturing processes.

Keywords: 3DEXPEREIENCE, 3DMaster, CATIA, DELMIA, Composites.

1 Introduction

The concept of 3DEXPERIENCES in the manufacturing domain is of significant importance for many companies. New production technologies, increasing product complexity, shorter product life cycles, higher product individualization and the ongoing need to reduce development and manufacturing costs are drivers for digital manufacturing solutions. Physical prototypes, trial and error approaches in the development of new manufacturing processes and unforeseen surprises with negative impacts during production ramp-up are not affordable in highly competitive markets. Production planning and manufacturing engineering based on reliable methods and simulation tools are of substantial importance for industries which manufacture goods. Dassault Systemes is an acknowledged leader in the digital manufacturing domain. With 3DEXPERIENCE the virtual modelling and simulation of real production

G.L. Kovács and D. Kochan (Eds.): NEW PROLAMAT 2013, IFIP AICT 411, pp. 364–368, 2013.

environments are becoming more realistic and allow the production planner to work in a 3D environment with all relevant manufacturing information.

Composite Materials are found in more and more industrial applications. Due to the high potential for mass reductions companies in the aerospace and defence industry have traditionally the highest demand, followed by fast growing appliances in the transportation and mobility industry. But also consumer goods – esp. sports equipment – and other industries have requirements, which are best fulfilled by carbon or glass fibre reinforced plastics.

3DEXPERIENCE in the domain of composite products is an excellent example to demonstrate, that industrial manufacturing processes and the resulting product behaviour can be simulated and complex dependencies can be managed with high accuracy.

2 Developing Composite Material Products

Composite materials are characterized by very special material behaviour and production requirements. Product properties depend on many factors, which show often complex dependencies. Important material criteria like strength and stiffness are by principle non-homogenous and orthotropic and are highly dependent on the complete production process of a specific part. Material failure behaviour is very different from metals. Structural parts (e.g. for aircrafts, auto, wind turbine blades) are build in layered structures, where each layer can have different angles, thickness or even fibre material. The curing of the impregnated material is a key process. Duration, temperature and pressure define the characteristic of the cured resin in the part and the resin/fibre interaction and as a consequence the structural behaviour of the part. Therefore it is very important, that the manufacturing concept is already secured in the design phase, e.g. for static and dynamic stress analysis.

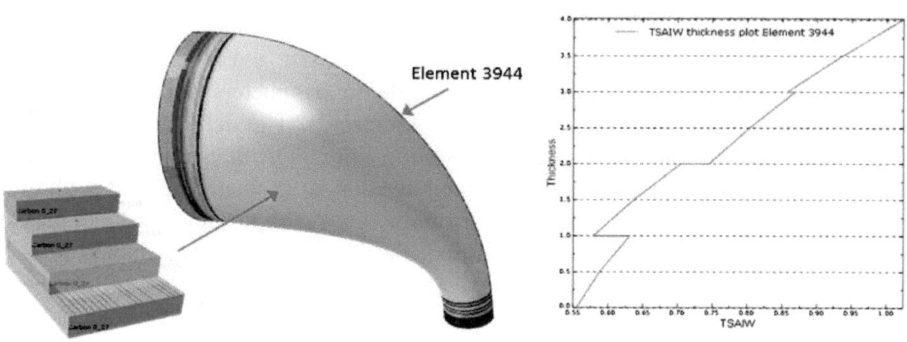

Fig. 1. Model of composite part with four layer stack-up and stress result for Finite Element

For each layer of a composite structure, the number of layers, thickness, weave, fibre angles and material have to be defined. In large structures the number of plies can vary a lot, e.g. from app. 50 layers at the root of a rotor blade to a handful at the tip. The distribution of the different layer stack ups over the part has to fulfil requirements on stiffness and strength, but also manufacturing boundary conditions. Size of the cut pieces are defined by production methodologies and available materials, size and position of required overlaps by structural requirements.

When a new composite part is coming into real production, quality assurance has a high priority. Many parameters have an influence and need to be considered in the case of defects. Ramping-up a serial production with secured processes to achieve a reproducible quality and fulfil cost targets is a permanent challenge.

3 Digital Industry Experiences for Composite Simulation

The complexity of the design and manufacturing of composite parts and the high degree of interdependencies between structural, design and manufacturing aspects generate the need for sophisticated simulations using a wide scope of methods.

One key simulation domain is the structural analysis using structure models with specialized finite elements, which allow the simulation of the behaviour of each single layer at any given location of the product. Failure of matrix or fibre material can be predicted as well as the deformation behaviour of the product under structural or thermal loads (Picture 1). The structural model can be directly derived from the design model with each layer modelled with its correct 3D representation and including all relevant material information attached as attributes. CATIA provides the necessary functionality and sufficient capacity to handle a wide range of applications. In multiple iterations between the 3DMaster design and the structural analysis based on finite elements the laminate can be optimized, in most applications to minimize weight within specified deformation and strength envelopes. The close link between SIMULIA as simulation system and CATIA as lead application for the 3DMaster facilitates fast iterations reducing the risk of data loss.

Manufacturing simulation has also its starting point at the 3DMaster model. The clarification of the question, if the selected fabrics can be draped as required for the part shape is often at the beginning. Weave, thickness, size and form of the cut pieces as well as the way they are applied into the mould are parameters which are taken into account in the process of drape simulation (Picture 2). The deformation of the cut fabric when applied into the mould and resulting fibre angle deviations can be calculated and visualized. Different lay-up procedures can be evaluated against each other and it can be verified, if the part can be manufactured with the selected material and how the fabric needs to be cut. As required, the 3D model can be adapted (e.g. by choosing a different fabric weight or weave) and more detailed e.g. by reflecting the cut pieces as defined by manufacturing engineering. Of course direct structural analysis verification can be executed for changes which might have an impact on the part behaviour. Of course it is always possible to track that part mass stays below the aloud limit.

With the shape and laminate definition accomplished, manufacturing planning and simulation can enter in a more detailed phase using the available 3D data. The documentation of the laminate stack-up can be used to generate work instructions for manual operations and a process plan.

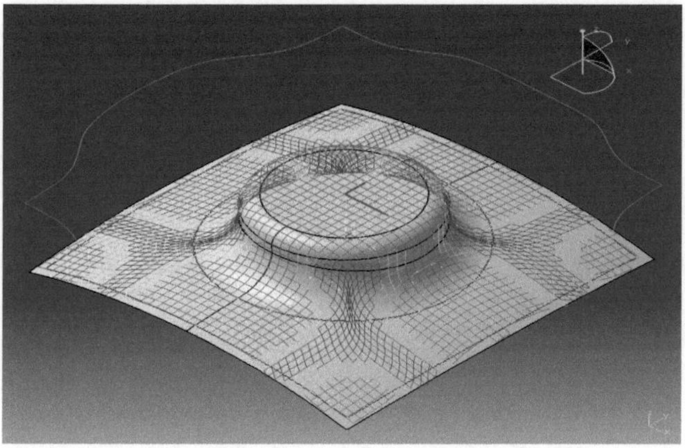

Fig. 2. Drape simulation showing fibre angle deflections of a bi-directional fabric

Applying STM (Standard time measurement) methods and linking the process plan with resource and layout information delivers important information for work load balancing and manufacturing costs calculation. This is the entry point into work cell simulation and plant design, including the possibility of life like ergonomic simulation of worker operations. DELMIA as Dassault Systemes Digital Manufacturing brand provides a wide scope of applications for realistic simulation experiences addressing these domains.

More and more automated manufacturing processes are used to manufacture composite products. Filament winding, tape laying, RTM of preformed fabrics and automated work cells with handling robots and hydraulic presses are used for many applications. Aerospace is the industry with the biggest share of these manufacturing methods, but they have also found entrance in other industries. Machine simulation and off-line programming is providing high advantages for machine builders as well as for companies, who use automated processes to build composite parts. A realistic simulation combined with the capability to download machine programs reduces downtimes and manufacturing risks.

All these applications are used to support the product development and manufacturing engineering of composite parts and are applied prior to the real production. Not all aspects in composite manufacturing can be simulated and by principle the real world is not always compliant with the assumptions in the virtual models. The high degree of dependencies between the quality of composite parts and manufacturing and environmental parameters requires add. tools. Often more than hundred parameters need to be tracked and interdependencies to be identified, if quality issues appear despite the fact, that all parameters are in their allowed range.

DELMIA Operations Intelligence provides the intelligence to identify hidden pattern and helps manufacturing experts to eliminate issues in the real production to minimize rework and scrap.

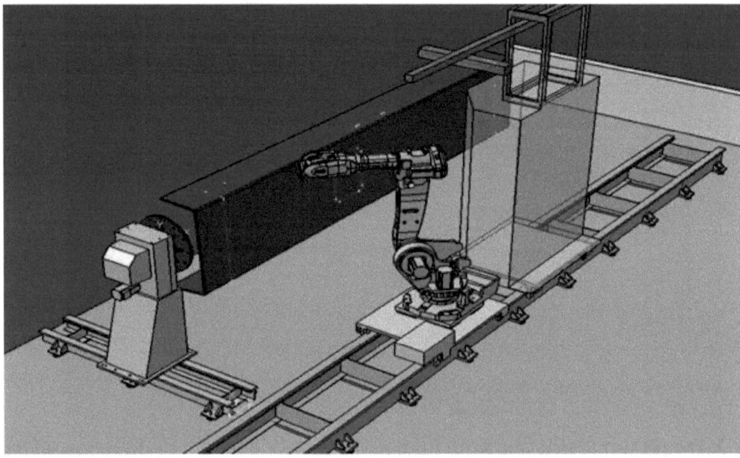

Fig. 3. Model of an automated tape laying workcell

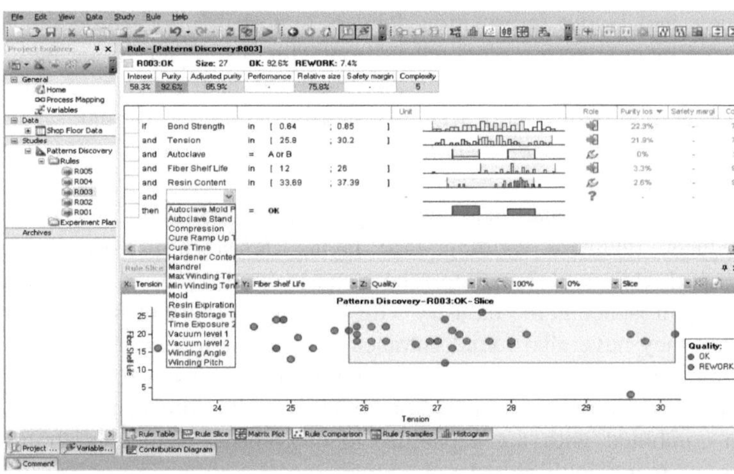

Fig. 4. Visualize and manage identified rules for composite manufacturing processes

4 Conclusions

The complexity of composite products requires powerful modelling and simulation applications. For the efficient usage in an industrial environment a high degree of integration between design, structural simulation and manufacturing engineering tools is necessary. A detailed 3DMaster model with all information relevant for product and manufacturing simulation is the core element. Simulation applications with direct access to this information can provide virtual experiences which reduce significantly risks and costs when developing and manufacturing products and help to avoid bad experiences in the real world.

Simulation and Optimization Based Flexible Job Scheduling of Powder Coating Lines

Heike Wilson[1] and Reik Zettl[2]

[1,2]DUALIS GmbH IT Solution
hwilson@dualis-it.de
www.dualis-it.de

Abstract. The Paper will present a new approach to use simulation methods combined with intelligent optimization algorithms in an operative scheduling and monitoring application for powder coating lines. Powder Coating lines are built usually as a closed loop "power and free" chain system that transports the products on carriers from a starting point through the different production stages. Due to the closed loop, efficiency and the overall output of the system is highly dependent on the sequence of the production orders entering the line. Inefficient sequences can cause deadlocks, significant decrease of the production output or even scrap production. The presented application architecture is based on a 3D simulation platform working together with a closed loop heuristic sequence scheduling algorithm. The paper will also introduce the system architecture design with full integration to an enterprise IT structure comprising ERP, real time PLC communication and a Warehouse Management System.

Keywords: Material flow simulation, Advanced planning and scheduling, optimization, powder coating lines, ERP, PLC, shop-floor control, order sequencing, production order mix, delivery dates, multi-criteria optimization, 3D simulation.

1 Introduction

Globalisation and the associated pressure on costs has increased considerably during the last decades and most production companies, especially small and medium size enterprises "SMEs", are forced to continually adapt, most often through better production flexibility and efficiency. Surface treatment and coating production usually supplies products to internal finishing and assembly lines or to external customers, but always "just-in-time" or "just-in-sequence" where the throughput-time is as short as possible. Quite often production order information arrives at very short notice to the coating process leaving little time to make a decision, let alone an optimal one, about the order sequence. Ideally the decision must be based on various parameters such as status of current production orders, requested delivery dates, equipment and buffer utilization and order properties like sizes and colours and so on. This complex information increases the pressure on optimized planning combined with a high product quality and processes stability. Most companies still rely on manual scheduling

G.L. Kovács and D. Kochan (Eds.): NEW PROLAMAT 2013, IFIP AICT 411, pp. 369–375, 2013.

and have to learn from painful experiences of lost earnings and efficiency before thinking more about lean manufacturing processes, integrated with flexible and affordable IT-Systems. Manually based solutions limit the information flow between the ERP level and shop floor production management, placing an increased burden on the coordination between departments.

In contrast, large manufacturing enterprises have been utilising technologies based on industrial simulation toolkits and mathematical based optimisation algorithms to support shop floor production planning already for some time. In the presented solution based on successful implemented project examples, available simulation technologies were combined with automatic scheduling in one decision support system that delivers required information to the planning department very fast and always based on the current production system status. The available 3D model of the powder coating line was enriched with additional operation logics, linked to the PLC and the ERP-system, and combined with fast heuristic scheduling algorithms in one integrated IT platform. The results of the scheduling is used to constantly control and optimize the material flow inside the powder coating line by delivering optimal decisions for each production stage.

The advantage of such a solution where 3D simulation models can be reused for supporting tasks like scheduling and line control is high for an SME. However few industrial case studies exist where Virtual Factory models initially used for layout proposals or a simulation study have been re-used for production scheduling, monitoring and control.

2 Project Overview

2.1 Project Partners

Rippert Anlagentechnik GmbH & Co.KG is the system supplier of both implemented powder coating systems. The are located in Herzebrock-Clarholz, where the mid-size company successfully holds its own in the market with three divisions covering coating plants, filter systems and industrial fans. This system supplier has been developing, producing and installing state-of-the-art, energy-saving plants for industrial customers all over the world for the last 45 years.

As a long term application user of the 3D Simulation toolkit (powered by the Visual Components Oy) Rippert was also very interested in the idea to enlarge their own plant visualization system with simulation and optimization based technologies. The main idea from their side was to offer the customer a new integrated solution providing four additional functionalities besides the visualization – 3D simulation, line monitoring and control and above all intelligent optimization based order scheduling.

The second project partner the mid-size company Louis Schierholz GmbH provided the Power and Free closed loop system inside the powder coating process. Schierholz Power & Free offers optimum flexibility by using a revolving chain on a closed loop, with unit load conveyors shifting loads on a running gear connected to a rail. The running gear and chain are linked via an interlocking catch system. Running gears can change chains at transfer points, and material flows can be separated or combined. Stoppers keep the running gears separate from the conveying chain, and help create storage areas, feeding, removal and operation points. One of the main tasks in the project was to connect the 3D simulation model with the Schierholz PLC in order to get the online information for the gears in the operating system to control the process flow for keeping optimal production sequencing.

DUALIS GmbH IT Solution the third implementation partner was responsible for the software development and integration of the IT-Platform at the customer's plants. The company is located in Dresden, Saxony, and has been providing simulation consultancy and software solutions as well as optimization based scheduling applications for over 20 years now. Founded in 1990 by Professor Wilfried Krug, DUALIS now has over 200 worldwide customers who have been able to improve and optimise their production processes with tools based on specialised mathematical algorithms for optimisation. DUALIS mathematicians and developers work together to improve techniques for solving real world optimisation and scheduling problems in the Automotive, Packaging, Automation Assembly and Logistics industries. The company's latest software development is an optimisation based short term scheduling architecture for small and medium sized companies in the production planning area. The architecture has a planner and constraint solver algorithm optimising for several and often conflicting optimisation goals based on a production system with finite resources and capacity. Expanding their applications with shoop-floor and PLC connectivity is a logical step to offer complete solutions based on simulation and optimization and to meet actual customer requirements.

2.2 Application Description

In today's manufacturing environment simulation techniques are used to reduce risk mainly in the planning phase for "green field" production plants or the reorganisation of an existing facility. Both 2D and 3D industrial software packages are available for these types of tasks and the derived models are often very detailed and rich in process information. In many cases logistics functions and physical detail is so close to the finally implemented system an opportunity exists to reuse the models to support the end-user's daily processes, such as production scheduling.

Driven by the challenge to get the highest possible efficiency from their powder coating systems, Rippert uses 3D Simulation to optimize the Power and Free system, buffer areas and system parameters during layout design in an early sales phase. With pre-defined product mixes, simulation runs are used to optimize the utilization of the processes and show the client the virtual system with product data as realistically as possible in a "virtual plant". After the planning phase is finished and product parameters like output per shift are agreed and fixed, expectations for real line operation need to be met. The Rippert application used existing simulation models for operative short term scheduling and line monitoring. An approach that can be of particular interest to other line builders and modular equipment suppliers who are already working with complete machine model libraries and working 3D simulation layout models for presales. These types of companies, would be able to extend the use of the presales models and provide a scheduling tool as an option for every turn-key production system delivered: to help their customers get the best out of the production system. This applies to most operators that need to schedule, monitor and optimize daily production and react when operational parameters change. In particular, manual changes to production sequencing in a closed loop chain driven system is very hard to successfully accomplish. Sequence scheduling before production start is very important. The Rippert application includes manual scheduling as well as heuristic optimization which can be adapted and customized if needed. The operator is able to define the product route and process times either manually or by importing the actual order pool from the MES (Manufacturing Execution System) and apply only changes. Based on the actual system load the generated sequences can be simulated and evaluated by the operator using defined KPIs. The optimal schedule is created using the integrated optimization engine with different optimization goals, production constraints and parameter settings. Once the sequence is determined it can be uploaded to the M.E.S to load new production order information to the real system. But uploading the optimal sequence is only one side. The many decisions points in the system need ot also also controlled in order to apply the optimal schedule to the system. Especially for the closed loop system minimal changes in the product sequence can cause trouble setup requests at the coating cabins or even deadlocks at critical for example oven processes or buffer areas. Therefor the final integration also links to the sensors and PLC controls in order to manage the decision at switches, feeding, operation and/or removel points as an essential task to also fulfil the optimal schedule.

2.3 Integration Concept

Currently SME companies are managing their production facilities using different IT applications. In most cases the applications have grown organically based on specific user needs. As a result, many of the IT sub-systems within one company have evolved with duplicate functionality. The end result is a confusing IT landscape where it is unclear which sub-system can and should deliver the required services, or where new functionality should be developed.

Missing from the SME IT portfolio is an integrating architecture where the required functionality is available through one common interface. Efforts have been focused on ERP (Enterprise Resource Planning) systems but there is little connection to the shop floor, especially in the areas of planning and optimisation. In the last couple of years a clear need has emerged for the research and development of improved ERP support tools for production.

These applications are commonly referred to in larger enterprises as APS (Advanced Planning and Scheduling) tools. To support an integrated solution for the SME with full Product Life Cycle Management the APS application must interface on the ERP side to a master data and to the shop-floor PLC on the other side.

The project is based on the most common vertical integration within the IT infrastructure between ERP, APS and MES/PLC applications as depicted in the following diagram (e.g., Fig. 1)

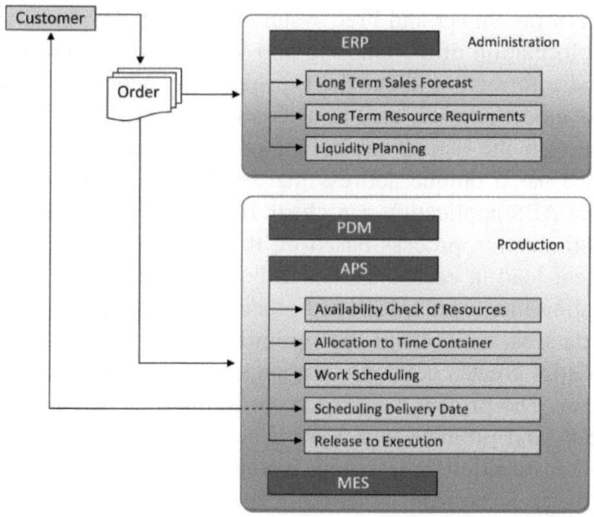

Fig. 1. Vertical Integration of Administration and Production systems

The main administration and order data is stored in the ERP-system. The pilot project ERP-systems were using common databases such as MS SQL and Oracle. In order to stay platform neutral the application connects this Administration level with the production MES level using stored procedure links implemented in C#. The goal was to implement a generic interface in order to reuse the connection strings in further projects with different ERP-systems, databases and PLC-applications.. More implementation work was required at the APS-level. The 3D simulation models used for system design and optimization during pre-sales and engineering had to be supplemented with PLC information representing the real shop-floor system. To monitor using a 3D model requires a unique and realistic representation of each Power and Free section and production process stage.

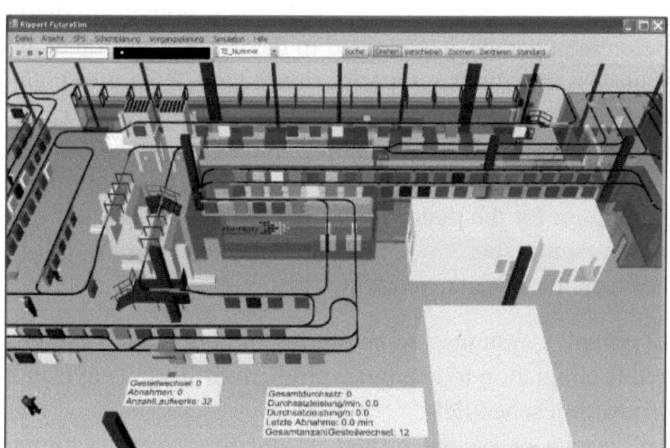

Fig. 2. Operating 3D model of an Rippert powder coating system

Every section of the Power and Free system is modelled with start and end sensors to register the information of the entering and leaving gears. In order to visualize the correct gears and information about the transported products in each sector the sensors need to be modelled with real-time connectivity to the PLC/OPC-application data. In each section the actual number of possible gear buffers are implemented. Each buffer place has a unique address managed by the PLC-server. With the provided address the APS-application can check if the sections are loaded and carry out the real time optimization process based on this information. For each optimization run the actual gear load in each system will be read from the PLC using this adress. Product information like type, colour and other important specifications are available from the PLC data as an input to the optimization. Based on this information the scheduling heuristics creates now new sequences in different areas based on the optimization goals set. The project experience shows that the setup time optimization at the powder cabins and promise to deliver at the final assembly are most important. The main goal is to schedule long colour batches with minimized cleaning and setup times between different colours. Lowering the down times at the colour cabins increases the system efficiency and overall product output significantly. Besides setup optimization in more complex powder coating lines the decision points at intersections and crossings become more important. The same applies to product mixes with repeating product routes. Since the decision of which gear will be released for instance from buffer areas will have an impact later on at following productions stages. Therefore the actual status must be updated almost online from the PLC, adjustable to the user requirements.

3 Conclusions

The benefits for the planner in comparison to available short term planning based options, that do not use simulation and optimization is the ability to quickly pre-test a calculated optimal schedule, and allow visual feedback to the planner about the actual system status combined with planning results. Further more using the PLC linkage the schedule will be online updates and adapted to the actual situation at any time. System deadlock and unnecessary colour setups are minimized. Both user plants increased the number of gears going through powder coating per shift by about 15%. The next challenge will be to not to look only at the coating line as single optimum but also include all pre-successors and successor stages into the optimization model in order to achieve a total optimum for the plant.

General benefits from using simulation and optimization as well as the operative scheduling benefit can be summarised with:

- Modelling of complex systems and networks for better understanding
- Fast and flexible adaptation to changing production system environment
- optimized production schedules
- Test "what-if" scenarios on the virtual production system without any real term consequences such as production loss and interruptions.
- Fast scenario testing and future prediction capability for planning purposes

- Better understanding of the production system
- Cost effective validation of new ideas or strategies
- Faster analyses and results
- optimized production schedules with cost minimization and more overall system efficiency

Event Based Identification and Prediction of Congestions in Manufacturing Plants

Clemens Schwenke, Thomas Wagner, and Klaus Kabitzsch

Technische Universität Dresden
{clemens.schwenke,thomas.wagner2,klaus.kabitzsch}@tu-dresden.de

Abstract. In modern manufacturing plants, most prominently in semiconductor manufacturing, complex interwoven automated material handling systems are used to transport lots between productive service stations. In busy areas, these systems often show transient congestion phenomena that have a negative impact on the throughput of the factory. Because of the systems complexity and the large amount of data originating from several thousand transports a day, these congestions are difficult to detect and to trace or even to predict. In this paper, the authors present an approach to detect congestions in such systems using an event-based model building and analysis approach. Once such effects have been identified, an algorithm is proposed to derive congestion prognosis rules which can then be used to implement effective congestion prevention mechanisms.

Keywords: Semiconductor AMHS, Model building, Event analysis, congestion prevention.

1 Introduction

In modern semiconductor industry, more and more highly integrated customized wafer products have to be produced in shorter and shorter periods of time.

Consequently, a large amount of production steps for a big variety of different products is carried out on the one hand. On the other hand, the production equipment is used flexibly, so that the transport system in a highly automated factory has to be adjusted frequently to new routes for wafer transports between stations. In general, these stations are connected by automated material handling systems (AMHS) which are complex interwoven networks of transport elements such as conveyor belts, rotary tables and other handling devices.

Besides adjusting the transport system to new demands, the operating engineers of automated material handling systems face on main problem. AMHS often show congestion phenomena, which reduce the throughput in a wafer fabrication facility (fab). Most of the times, congestions result in queues of work pieces waiting for other work pieces because stations are temporarily overloaded or switched off (down times).

Modern material handling systems provide features to detect and adjust to undesirable situations automatically and resolve congestions. For this feature intelligent

G.L. Kovács and D. Kochan (Eds.): NEW PROLAMAT 2013, IFIP AICT 411, pp. 376–390, 2013.

material flow routing rules have to be implemented as well as rules for altering the feed of new work pieces entering the system.

But in real fabs, practitioners ask themselves the following question. Which rules have to implemented, and more important, how to determine these rules systematically?

In order to analyze the overall performance of transport systems, event data of material passing certain waypoints of the system has been collected in log files.. But the task of analyzing transient congestions and extracting conclusions often is still a sisyphean undertaking because of three reasons. First, this task is still carried out mostly manually by visually inspecting log files. Second, sometimes the efforts of studying log files do not result in generally applicable rules. Third, sometimes congestions seem unexplainable.

In order to free the expert from this time consuming task, this paper introduces an approach to how event based congestion analysis and prediction can be executed automatically.

Consequently, the steps of an approach for semi automatic event data inspection are provided. These steps include collecting of relevant event data, building a state model of the transport process, identification of temporarily overloaded segments, backtracking to influencing segments and congestion analysis in order to extract rules for prediction of congestions.

As a result, rules for the prediction of congestion occurrence are derived. For validation, these rules are applied to new event data of the same AMHS, so that to predict congestions were predicted early enough for operators being able to take actions. For an exemplary use case, a set of trace data of wafer lots in an automated production line has been used for proving the approach. All steps of this workflow have been implemented in a software framework and tested against real fab data.

This paper is structured as follows. In Section 2 related work is considered. The approach for data inspection is described in Section 3. Section 3 includes an exemplary validation. Conclusions are drawn and an outlook is given Section 4.

2 Related Work

The authors investigated several possibilities to identify and analyze congestions in a given transport system. Consequently, some relevant approaches are discussed and the disadvantages of those, that initially seem to be most obvious and useful, are portrayed. The first thought when transport systems shall be examined is queuing systems. But the pure queuing theory could not easily be applied for the authors real world use case, because arrival rates, service durations and sometimes even capacities of the system's elements change constantly. Consequently, the second thought is time series. But the application of pure classic time series analysis was not feasible, because models could not explain the observed phenomena exact enough, nonsense correlations were found, or the calculation costs were too high. Third, the authors considered state model building for examining event sequences.

Finally, the authors combined findings from several specific fields. Therefore, the consideration of related work covers work in state model building, queuing theory and time series.

2.1 State Model Building

Automatic state model building requires event logs as input data and provides highly aggregated information about state changes in a system. The prerequisite is that recorded events can be understood as notifications of state changes of entities. Briefly, the main essence of state model building is the extraction of a graph out of traces of events.

In the case of material flow systems, events are recorded when loads enter workstations where they are processed, or when they enter conveyor segments where they are transported to a succeeding workstation. In the extracted graph, nodes represent states of loads being in a certain station or transport segment, and edges represent the transfer of a load into another station or transport segment. For this kind of event data, an event always indicates that a load entered a work station or transport segment. The use of discrete state models describing a system's (or device's) behavior as a sequence of possible steps was studied successfully before [1].

On the one hand, state models are useful to monitor or identify business processes [2]. Van der Aalst et al. [3] used state model building as a method for analyzing business processes, where events are generated when certain work steps begin and end. In so-called process mining, models of business processes shall be recovered or checked. Additionally, the relevance of business steps can be evaluated and performance indicators are calculated based on event logs. The main problems are to recover adequate models and to identify relevant process steps, since the log data of business processes, involving humans and external events, oftentimes contain non-deterministic portions. The resulting models then have to be mostly analyzed manually, sometimes including a few automatically calculated performance parameters if applicable.

On the other hand, state model building can be used for the examination of event logs of machines or transport systems, for example in semiconductor industry or in logistics applications. Compared to extracted models of business processes, the extracted models of logistic and manufacturing applications are more deterministic but contain many more states, so that sophisticated tailored analysis approaches are necessary to detect and explain unwanted phenomena, such as extremely varying delays [4] or changing reject occurrence [5].

In order to enrich a pure state model with more information, Vasyutynskyy suggested the combination of state model building with calculation of performance indicators, such as overall throughput times, holding times and inter arrival times on states. Consequently the result is called an extended state model [6].

State models can be used as the basis for detailed analysis of congestions if they manifest as tailbacks of loads waiting for preceding loads [7]. In this work, an approach to automatically carry out transient tailback recognition and cause identification was introduced. In order to identify origins and causes of these observed tailbacks, historic event log data of loads passing certain waypoints were inspected.

The approach is based on analysis of holding times and capacities of transport segments. As a result, complete lists of tailbacks and affected segments are provided. Plus, for each tailback an initial cause event is determined. But this tailback analysis approach does not relate the occurrence of tailbacks to the constantly altering arrival rates of new loads entering the system. Therefore it was necessary to investigate different approaches to enable a successful prognosis of congestions.

2.2 Queuing Theory

Queuing theory is a tool for estimating performance indicators in networks of waiting lines and service stations. The service stations take a certain amount of time, e.g., for processing one work piece. The work pieces, or loads, travel through the system and wait in line in front of the service stations, thus forming queues.

The main application is the design of queuing systems, [8], [9]. At the design time important questions are: What is the average queue length, how long are average waiting times Wq, and how many service stations are needed? For answering the question of the average waiting time in the queue Wq, Formula (1) can be used [10].

$$W_q = W - \bar{s} = \frac{\lambda}{\mu(\mu - \lambda)} \qquad (1)$$

The arguments for this Formula are the complete waiting time W that was spent in a system of queue and service station. The time W includes the average service duration . Alternatively the time Wq can be calculated using the arrival rate λ and service rate μ.

The basic assumption in queuing theory is stable arrival-and service rates. In contrast, these rates change frequently in the investigated real world systems, e.g. depending on product mix and order situation. As a result, Formula (1) for estimating Wq, was not directly applicable. The average time loads spend in a conveyor segment is called in the following.

2.3 Time Series Analysis

Classic time series analysis provides many disciplines [11]. For this work the most important ones are the following. The first one is time series analysis in the time domain, where oftentimes trends and seasons are extracted by developing linear models until the residues cannot be minimized anymore and are similar to stochastic white noise. This approach is used in, e.g., economics, biology and agriculture [12]. Sometimes this approach is also used in physics or engineering but only as a last resort if model building using known facts did not provide useful results [13]. For example, time series analysis is used in the field of predictive maintenance to model trends, seasons and noise of deterioration indicators [14]. But these approaches try to smooth of outliers instead of explaining them.

The second discipline often is applied for modeling the remaining residues, after trend and season are extracted, by estimating auto regressive moving average (ARIMA) models. ARIMA models often are used to model processes in economics, especially in financial industry trying to predict effects in the stock market [15]. This is done by assuming the stochastic nature of the unexplainable processes [16], [17]. Therefore, the main ingredients of these models are two parts, the auto regressive (AR) part and the moving average (MA) part. The AR part tries to model the time series by explaining the current value mainly by the previous value. The MA part models a white noise so, that in conjunction with the AR part the given time series can be approximated. Unexplainable peaks in general are considered outliers and are smoothed [18].

In contrast, the authors needed to explain the outliers, instead of smoothing them. As a result, the above mentioned time series analysis approaches were not applicable. One reason for this is that in reality peaks are not always stochastic and do not solely depend on the previous value.

2.4 Summary

When first confronted, the authors tested the following approach. First, the inter arrival times and holding times on conveyor segment in front of stations or rotary tables were examined for aggregated periods of time. With Formula (1) of the queuing theory, the waiting times W at stations were estimated at these time periods, but they did not match the actually observed holding times.

By a different approach, the authors aimed to produce forecast models in order to estimate autoregressive moving average (ARMA) models for inter arrival times and holding times. The predictions of these models were used to estimate the current waiting time W applying Formula (1) to the forecasted arrival rates.

Unfortunately, the quality of these models was not good enough for reliable predictions, because of one important fact. Forecast models tend to smooth peaks, because most of the times they are considered outliers. But in contrast, in the use case of investigating transport system data, the peaks of holding times are the sought after and to be explained congestions. Consequently, generic naive time series analysis was not constructive.

As examined by the authors, the key to understanding the establishing and resolving of congestions is the combination of system knowledge with time series analysis. Congestions can travel through the system like waves, superimpose and thus, cause significantly varying waiting times on certain conveyor segments in front of service stations. Therefore, there is a true relation between only certain arrival rates, service rates and waiting times. As a result, the authors integrated a step in the overall approach that selects only the relevant time series before they are investigated further.

3 Analysis Approach

The suggested analysis approach consists of a workflow of five general steps, see Figure 1. First event data of the AMHS has to be collected.

Second, a state model has to be extracted from of this event data. Third, overloaded segments of the transport system have to be identified. Fourth, the relevant source segments that feed loads into the system are identified by systematic backtracking.

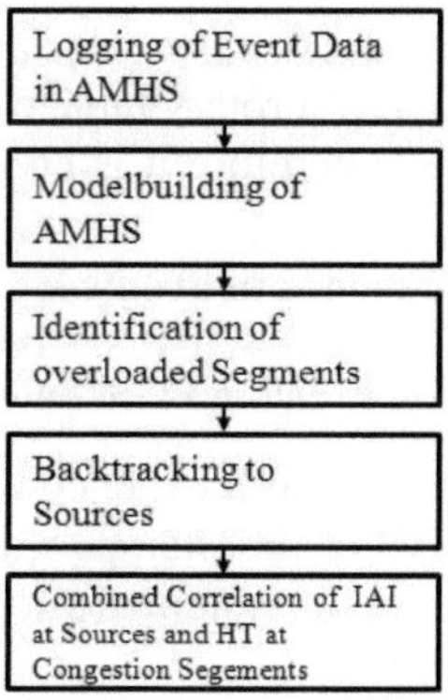

Fig. 1. Workflow of Analysis of Congestions

Finally, the ultimate purpose of analyzing congestions by correlating them with arrival rates at source segments can be carried out. The four first steps are prerequisites for the last step. All five steps are described in detail in the following.

3.1 Logging of Event Data

The first step is the collection of event data in the factory's AMHS. That is, at each relevant conveyor belt or rotary table an event is logged. The event contains the essential information in the three fundamental attributes, timestamp, segment number and load number. Based on these elementary attributes, a graph of the transport system can be built in the second step.

3.2 State Model Building

In this step, a state model of the transport system is extracted from the logged event data. The authors published applications of this method before in [7], [19]. But for completeness, the algorithm of the method isbriefly described.

The algorithm extracts all relevant entities for building an extended state-transition model of a given log file of a given transport system. Consequently, the resulting model will consist of the following entities.

$$S = \{s_1; s_2; \dots ; s_n\} \tag{2}$$

S is the finite not empty set of states, representing transport system elements, e.g., rotary tables or linear conveyor modules as well as storage elements (stockers) or production equipment (work stations).

$$L = \{l_1; l_2; \dots ; l_m\} \tag{3}$$

L is the finite set of loads, representing the moved entities, e.g., wafer carrier.

$$T \subseteq S \times S \tag{4}$$

T is the finite not empty set of transitions, representing interconnections between the single elements. T is a subset of all ordered pairs of states. One element of T is a binary relation over S. For example, $(s_1; s_2) \in T$ with $s_1; s_2 \in S$; represents a transition from State s_1 to s_2. For instance, a rotary table can be used as a crossing, unification or split of transport streams. Therefore it is connected to several other elements and can exhibit multiple transitions.

The event log is an ordered sequence of events E as follows.

$$E = \{(\tau_1; s_1; l_1); \dots ; (\tau_N; s_N; l_N)\} \quad (5)$$

One event $e = (\tau; l; s) \in E$ is defined as a triple consisting of timestamp $\tau \in Z$, state $s \in S$ and load $l \in L$. Z is the set of timestamps τ, so that $Z = \{\tau_1;\dots; \tau_N\}$. The above mentioned entities S, T and L can be systematically extracted from this ordered sequence of events as shown in Figure 2.

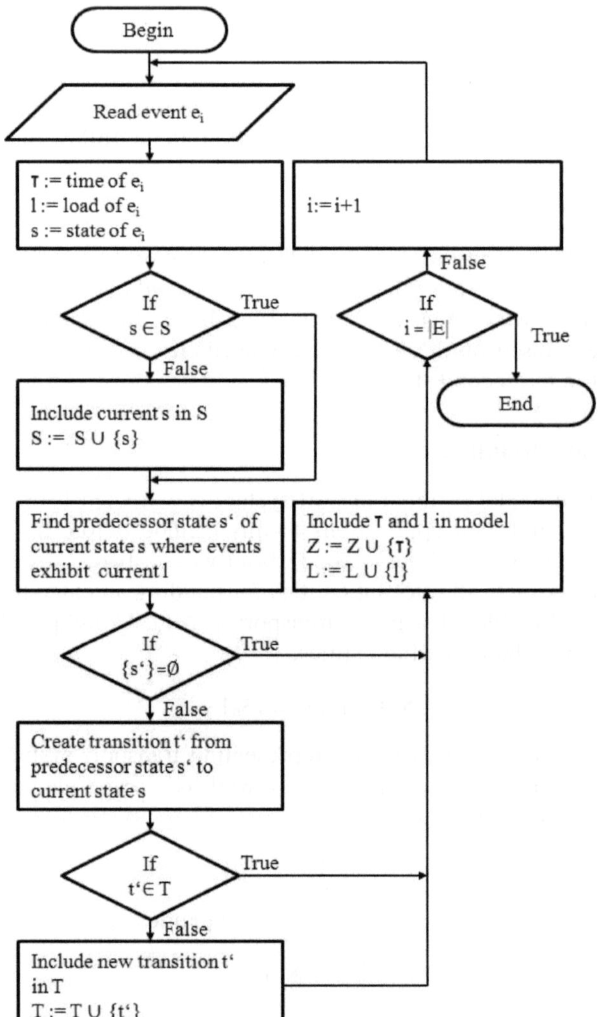

Fig. 2. State Model Building

The model building algorithm is a loop that processes ach event separately in one individual loop cycle. This loop consists of steps for extracting elements from events as well as for finding or creating model entities, so that they can be included into the model. Conditional decisions allow for breaking out of the loop if not all steps have to be carried out, because current entities are already part of the model.

3.3 Identification of Overloaded Segments

After the state model is extracted, the third step of the overall workflow can be executed. Overloaded segments are results of congestions. In the state-transition model these segments are states. The suspect states are identified by finding states that sometimes exhibit unusual long holding times ω.

Longer holding times are an effect of previous loads holding up succeeding loads and therefore affect the average holding times of loads on certain states.

For each state one corresponding average holding time $\omega(s_i)$ can be calculated. The states that exhibit unusual long average holding times compared to the average holding time over all states, $\omega(s_i) > \overline{\omega}$, are suspects to be affected by at least transient congestion effects. This comparison has to be executed for many fractions of time.

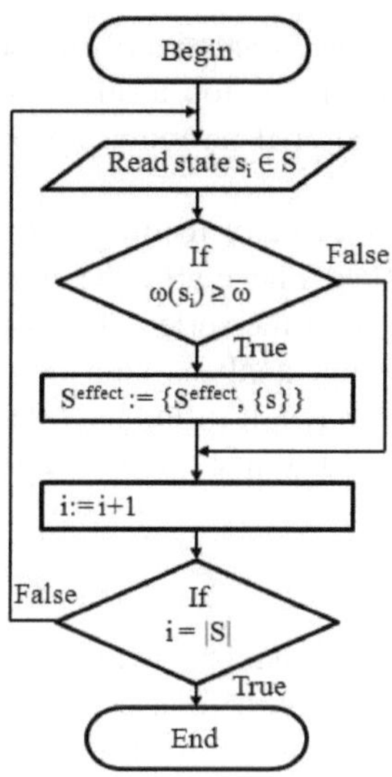

Fig. 3. Identification of States that temporarily exhibit congestions

As a result, a set of congestion states $S_{effect} \subseteq S$ is found.

3.4 Backtracking to Influencing Segments

In order to find states that influence the holding time of congestion states, a backtracking is carried out. This is necessary to compare the time series of only those states that actually can have an influence and not others that are unlikely to have an impact.

This backtracking is carried out for each congestion state. The relevant influencing states are called feeding states. In that context, a feeding state is the closest preceding state that either exhibits more than one outgoing transition $d_{out}(s) \geq 2$ or that is a load source of the system, e.g. a production equipment input. Other states that represent linear conveyor segments and are closer do not have to be considered because the arrival rates do not differ from the feeding state.

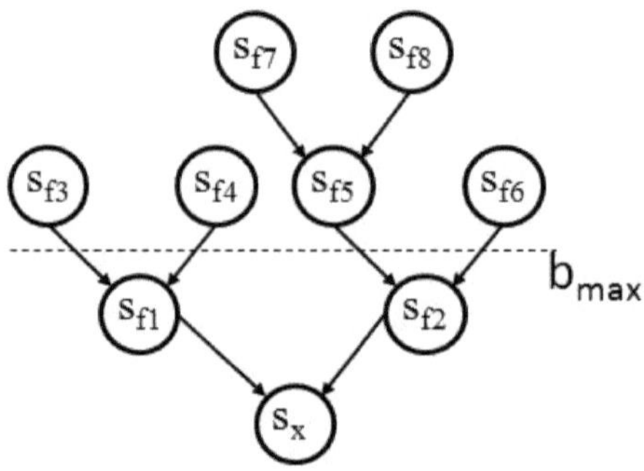

Fig. 4. Tree of influencing neighbor feeding states. Result of backtracking algorithm

On each identified feeding state, recursively the same backtracking to previous feeding states is carried out. This recursion terminates when a number of maximum backtracking depth b_{max} is reached or if no more preceding states can be found in the state-transition model.

The result of this algorithm is a tree of feeding states for each congestion state, for example see Figure 4. The dashed line marks the maximum backtracking depth selected by the user. In this case, the time series of two feeding states sf1 and sf2 have to be considered in the congestion analysis step described in the next Subsection. Increased backtracking depths can result in longer forecast lead times for congestions but also cause higher calculation costs since more states have to be considered.

3.5 Congestion Analysis

Once the above mentioned prerequisites are available, the actual congestion analysis can be started. The approach presented focuses on the diagnosis and prediction of tailback events caused by the dynamic interactions of different transport system elements or areas. Other possible causes of tailbacks, like random failures of single transport elements, are much less related to the system behavior which is observable using the event logs described in Section 3.1 and are therefore not considered. However, the prediction of such tailbacks could be tackled using semantic information about the transport systems hardware, e.g. mean time between failure considerations.

Here, the progression of the inter arrival times (IAT) of the identified feeding states are considered in order to identify conditions that provoked the anomalies on the congestion states. This approach allows to draw inferences from temporary different workload situations on different load sources about the manifestation of transient tailbacks, e.g. due to temporary load concentration or mutual obstruction.

Depending on the situation, not every feeding state identified in Section 3.4 has an influence to the appearance of tailbacks on the congestion states. To select the relevant subset of feeding states, several methods of selection can be considered. One approach could be to weigh the IATs of the different feeding state and ignore the ones that transport only a seemingly irrelevant amount of lots.

However, the simplified example depicted in Figure 5 suggests that this is a misleading approach.

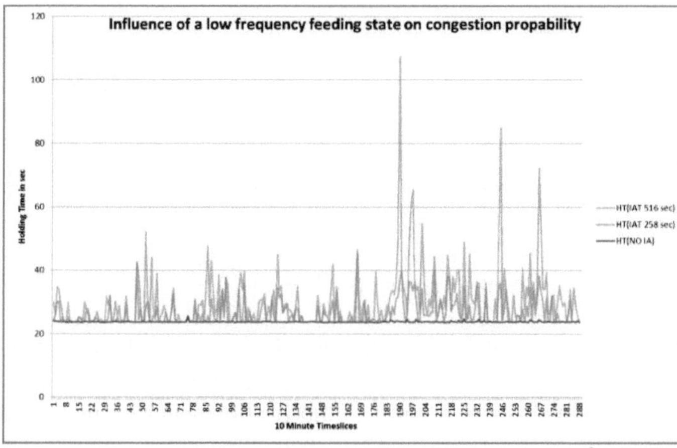

Fig. 5. Influence of a low frequency feeding state on congestion probability

In Figure 5, a congestion state A (see Figure 6) is shown which receives its loads at a rate of approximately three loads per minute from a major source B. If no other sources participate, no congestions appear at this state A as indicated by the red line. However, there exists another feeding state C which considerably influences the holding times on state A. Although state C only contributes to the traffic with around one load in nine minutes (green line) to around one load in 4 minutes (blue line), it significantly increases the holding times of A and therefore even causes congestions (noticeable peaks in the blue line).

In the presented case, this is caused by a dead-lock prevention mechanism implemented in the transport systems controllers that block all traffic from C down to D once a load enters the critical area shown as a red cuboid in Figure 6.

Therefore, it is necessary to measure the real influence of a feeding state's IAT on the HT of a congestion state regardless of its arrival rate. To achieve this, a correlation approach is used as a first step. For this purpose, the time series of holding time of the congestion state is correlated with the time series of the inter arrival times of all identified feeding states.

Depending on their distance, it takes a certain amount of time until the inter arrival times of the feeding states affect the holding times of the congestion state. To take these delays into account, each IAT time series is cross-correlated with the HT time series, using a default maximum lag as defined below where N is the number of observations in the time series.

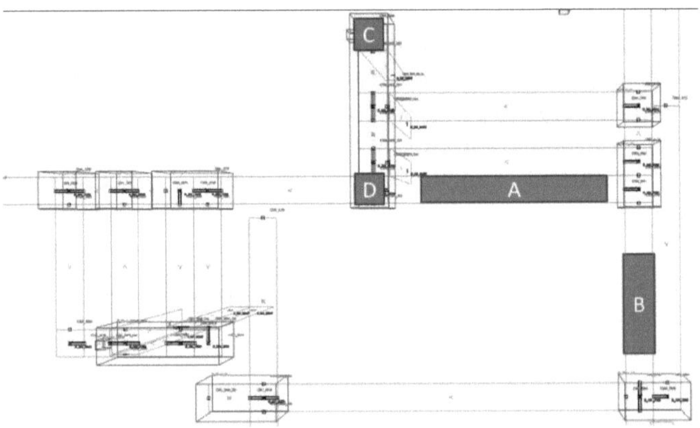

Fig. 6. Excerpt of the example system showing a congestion state A and the feeding states B and C

$$maxlag = 10 * log_{10}\left(\frac{N}{2}\right) \tag{6}$$

As a second step, in the resulting correlation values r_l for each lag l, it is then sought for the maximum negative correlation. That is, the maximum suggested impact of a decreasing IAT (increasing arrival rate) of the feeding states on increased HTs of the congestion state. This correlation value r_l is then checked for significance by comparing it with the values of the approximated 99% confidence interval c ($\propto$=0.01):

$$c = \pm\frac{2.58}{\sqrt{N}} \tag{7}$$

Third, once the set of significant feeding states S^sig has been found and the corresponding lags $lag_{crit}(s^{sig})$ yielding the maximum negative correlation for each state noted, the critical inter-arrival rates must be identified. These are the ones that cause overload situations on the congestion state, if they appear in combination. For this purpose, a set V for each feeding state $s^{sig} \in S^{sig}$ is constructed as follows.

$$V = \{\mathrm{V}_{n\in(N-l)}\, v_n(IAT(s^{sig})_n; HT(s^{effect})_{n+l})\} \tag{8}$$

With $l = \mathrm{lag_{crit}}(s^{sig})$. In summary, this set V contains a mapping from the inter-arrival times of one feeding state to the corresponding holding times of the congestion state, shifted by l to compensate for the time delay between cause and effect as mentioned above. Afterwards, V is sorted by its IAT values in descending order, i.e., from the least to the most frequent lot appearance. Next, the HT components of V are scanned along the falling IAT values. Once the holding time on the congestion state first reaches or exceeds the critical value $\mathrm{HT_{crit}} = \mathrm{HT}(s^{effect}) \geq \overline{\omega}(s^{effect})$ as described in Section 3.3, a previous $\mathrm{IAT}(s^{sig})_{crit-k}$ is used to define a rule indicating the danger of congestion. This rule sets a Boolean value $d(s^{effect})$ depending on whether this value $\mathrm{IAT}(s^{sig})_{crit-k}$ is undershot.

The parameter k can be used to manipulate the lead time and sensitivity of the congestion prognosis by choosing more conservative, i.e. larger, values of $\mathrm{IAT}(\mathrm{s}^{\mathrm{sig}})$, so that the warning signal $\mathrm{d}(\mathrm{s}^{\mathrm{effect}})$ is set earlier.

This procedure is repeated for every $s^{sig} \in S^{sig}$ and the resulting rules are combined to one single rule, suggesting a high congestion probability once all of the conditions are met.

The construction of these rules will now be demonstrated by the example shown in Figure 6. In this example, the state A's relevant HT_{crit} was found at 60 seconds.

In Figure 7 the ordered set V is shown for the congestion state A and the feeding state B.

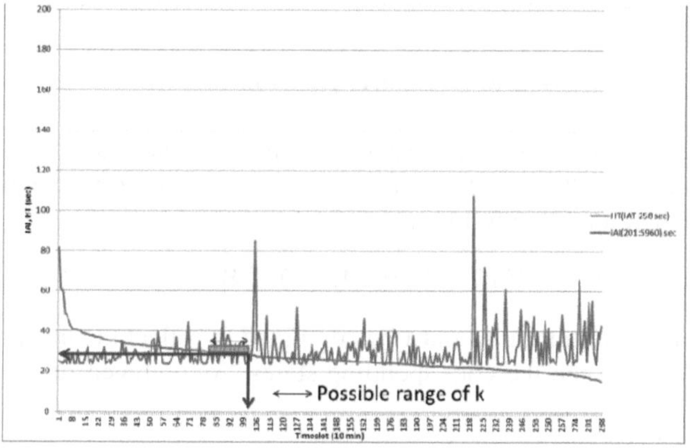

Fig. 7. Critical IAT for feeding state B

For this state, the congestion effects began to manifest once the IAT of state B was less than or equal to 27 seconds. Using a parameter value of 3 for the lead time parameter k, i.e., the next larger $IAT(s^{sig})$ value is used which is 28 seconds. In this example the congestion indication rule can be defined as follows.

$$d(s^{effect}) = IAT(B) \leq 28 \tag{9}$$

A second rule is derived from the holding- and inter-arrival times of the states A and C as shown in Figure 8.

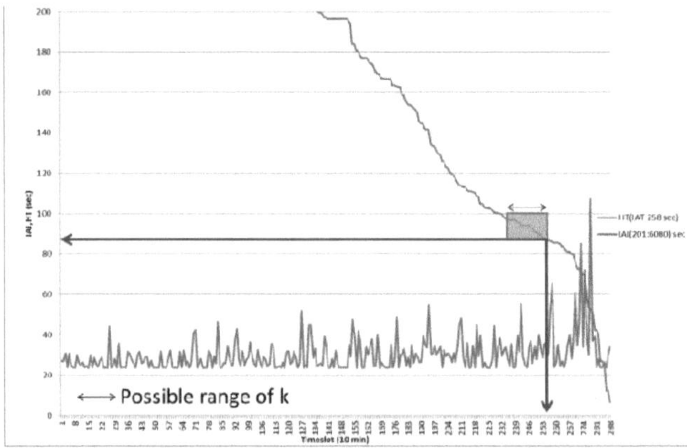

Fig. 8. Critical IAT for feeding state C

Here, congestions on state A appeared once the inter-arrival times of state C was lower than or equal to 86 seconds. Using again a k of 1, the corresponding congestion indication rule therefore is as follows.

$$d(s^{effect}) = IAT(C) \leq 88 \tag{10}$$

As a last step, the resulting rules must be combined to reflect the mentioned interrelation of the IATs between the corresponding states. A combined rule would be noted as follows.

$$d(s^{effect}) = (IAT(B) \leq 28) \wedge (IAT(C) \leq 88) \tag{11}$$

4 Conclusion and Outlook

The presented approach has been implemented into a comprehensive analysis framework. User input is only required to define the input parameters k and the maximum backtracking depth b_{max} to influence the prognosis lead time. Subsequently, the warning rules are derived fully automatically and can afterwards be evaluated against the current transport system behavior at runtime.

The derived warning rules for congestion prognosis will serve as a basis for dynamic routing approaches within the transport system controllers. If their conditions are met, the controllers will be alarmed about possible future congestion situations. As a possible countermeasure, they can reroute part of the incoming traffic flow across different system parts, thus gaining a consistent lot flow while sacrificing only a small amount of transport speed for a few lots.

In the use cases investigated, the derived rules predicted the observed congestions accurately enough to allow for effective prevention measures in most cases. However, the approach also exhibited a few limitations that have to be considered regarding the choice of the parameter values k and b_{max}.

If the maximum backtracking depth is set too large, then too many feeding states have to be considered, eventually causing interferences between the growing variety of load situations. That means, several groups of feeding states may cause overload situations on a single congestion states independently. Using just the cross-correlation approach, this can neither be reliably distinguished nor can it be expressed using just AND conjunctions of the warning rules.

In addition, the lead time parameter k must be chosen carefully, since small values may reduce the prognosis horizon too much. On the other hand, values too large may provoke a lot of false-positive congestion warnings.

Therefore, future work will focus on defining metrics to aid the system experts choosing the right parameter values. In addition, the authors will investigate a wider set of influencing variables to determine their suitability for making the predictions more accurate.

References

1. Kemper, P., Tepper, C.: Trace based analysis of process interaction models. In: Proceedings of the 2005 Winter Simulation Conference (WSC), pp. 427–436 (2005)
2. Agrawal, R., Gunopulos, D., Leymann, F.: Mining process models from workflow logs. In: Schek, H.-J., Saltor, F., Ramos, I., Alonso, G. (eds.) EDBT 1998. LNCS, vol. 1377, pp. 469–483. Springer, Heidelberg (1998)
3. van der Aalst, W.M.P., Reijers, H.A., Weijters, A.J.M.M., van Dongen, B.F., Alves de Medeiros, A.K., Song, M., Verbeek, H.M.W.: Business process mining: An industrial application. Information Systems 32(5), 713–732 (2007)
4. Gellrich, A., Wagner, T., Vasyutynskyy, V., Kabitzsch, K.: Modeling of Transport Times in Partly Observable Factory Logistic Systems based on Event Logs. In: Proceedings of the 16th IEEE International Conference on Emerging Technologies and Factory Automation (ETFA 2011), Toulouse, France (2011)
5. Shanthikumar, J.G., Ding, S., Zhang, M.T.: Queueing theory for semiconductor manufacturing systems: a survey and open problems. IEEE Transactions on Automation Science and Engineering 4(4), 513–521 (2007)
6. Vasyutynskyy, V., Gellrich, A., Kabitzsch, K., Wustmann, D.: Analysis of Internal Logistic Systems Based on Event Logs. In: Proceedings of the 15th IEEE International Conference on Emerging Technologies and Factory Automation (ETFA 2010), Bilbao, Spain (2010)
7. Schwenke, C., Wagner, T., Gellrich, A., Kabitzsch, K.: Event-based recognition and source identification of transient tailbacks in manufacturing plants. In: Proceedings of the 2012 Winter Simulation Conference (WSC), pp. 1–12. IEEE (2012)
8. Beranek, L.: A Method of Predicting Queuing at Library Online PCs. Australian Academic and Research Libraries 37(4), 260–272 (2006)
9. Horling, B., Lesser, V.: Using Queuing Theory to Predict Organizational Metrics. In: Proceedings of the Fifth International Joint Conference on Autonomous Agents and Multiagent Systems, pp. 1098–1100. ACM (2006)
10. Gross, D., Harris, C.M.: Fundamentals of queuing theory. Wiley, New York (1998)
11. Hamilton, J.D.: Time series analysis, vol. 2. Princeton University Press, Princeton (1994)
12. Mead, R., Curnow, R.N., Hasted, A.M., Curnow, R.M.: Statistical Methods in Agriculture and Experimental Biology (2002)

13. Palit, A., Ajoy, K.: Computational intelligence in time series forecasting: Theory & engineering applications (2005)
14. Krause, J., Kabitzsch, K.: A generic Approach for Reliability Predictions considering nonuniformly Deterioration Behaviour. Journal of Physics: Conference Series 364(1) (2012)
15. Wang, J.-H., Leu, J.-Y.: Stock market trend prediction using ARIMA-based neural networks. In: IEEE International Conference on Neural Networks, vol. 4, pp. 2160–2165 (1996)
16. Bollerslev, T., Mikkelsen, H.O.: Modeling and pricing long memory in stock market volatility. Journal of Econometrics 73(1), 151–184 (1996)
17. Nelson, D.B.: The time series behavior of stock market volatility and returns. Diss. Massachusetts Institute of Technology (1988)
18. Breen, W., Glosten, L.R., Jagannathan, R.: Economic significance of predictable variations in stock index returns. The Journal of Finance 44(5), 1177–1189 (1989)
19. Wagner, T., Schwenke, C., Kabitzsch, K.: Modeling and wafer defect analysis in semiconductor automated material handling systems. In: Proceedings of the 2012 Winter Simulation Conference (WSC), pp. 1–12 (2012)

Intelligent Manufacturing Operations Planning, Scheduling and Dispatching on the Basis of Virtual Machine Tools

Gerald Rehage[1], Frank Bauer[1], Jürgen Gausemeier[1], Benjamin Jurke[2], and Peter Pruschek[2]

[1]Heinz Nixdorf Institut, Produktentstehung, Universität Paderborn, Fürstenallee 11, 33102 Paderborn, Germany
{gerald.rehage,juergen.gausemeier}@uni-paderborn.de
[2] DMG Electronics GmbH, Deckel-Maho-Str.1,87459 Pfronten, Germany
peter.pruschek@gildemeister.com

Abstract. Today, numerical-controlled machine tools are used for flexible machining of increasingly individualized products. The selection of the most economic tool, machining strategy and clamping position is part of the manufacturing operations planning and bases on the employees' practical knowledge. The NC programmer is supported by current CAM systems with material removal simulation and collision detection. This early validation avoids damages and increases the productivity of the machines. The benefit of simulations can be increased by a better model accuracy. In common CAM systems the machine behaviour is often insufficiently reproduced; for example, the dynamic characteristics of axes, the tool change and the chronological synchronization are simplified or neglected. In view of complex operations, a slow trial run on the real machine or substantial safety margins are necessary. The described deficits can be solved by virtual machine tools. Thereby a virtual numerical control and a detailed machine kinematic is used for an accurate simulation. The result is an error-free NC program which can be directly used on the real machine. In addition, the exact processing time is determined and supplied for the operations scheduling and dispatching. Furthermore, virtual machine tools provide promising approaches for an automated optimization of machining operations and machine set up. Parameters for the optimization of the processing time are, for example, different clamping positions or tooling arrangement. Simulating several parameters requires lots of computational power. Hence, the vision of the project „InVorMa" is a cloud application, which supports the operation planning, scheduling and dispatching of machine tools. A computer cluster (cloud) provides noticeably faster results than a single computer. Therefore, the machine tools and the manufacturing equipment of the user are cloned one-to-one in the cloud. The manufacturing documents are optimized by the cloud application before they are forwarded to the shop floor. The optimization involves the NC program for each machine as well as the distribution of orders. The practical knowledge of the manufacturing planner and the results of the optimizations are pre-processed for reuse by an integrated knowledge base.

G.L. Kovács and D. Kochan (Eds.): NEW PROLAMAT 2013, IFIP AICT 411, pp. 391–400, 2013.

Keywords: Operations planning, scheduling, dispatching, machine tools, simulations, industry 4.0.

1 Introduction

Manufacturing in high-wage countries requires the efficient use of resources. Increasingly individualized products require a highly flexible production system [1]. In the field of machining of metals, the needed flexibility is achieved by numerical-controlled machine tools. It is the function of the manufacturing planner to ensure a rational use of the operating means. This is based on his practical knowledge and furthermore on the utilization of machine simulations to avoid damage and increase productivity from the office. The aim of the project "Intelligent Manufacturing Operations Planning, Scheduling and Dispatching on the Basis of Virtual Machine Tools" (InVorMa) is a cloud-based simulation of machine tools and it is developed by the Heinz Nixdorf Institute and the Decision Support & Operations Research Lab (DSOR) of the University of Paderborn as well as the Faculty of Engineering sciences and Mathematics of the University of Applied Sciences Bielefeld in cooperation with the machine tool manufacturer Gildermeister Drehmaschinen GmbH. The companies Strothmann GmbH and Phoenix Contact GmbH & Co. KG support the definition of requirements as well as the following validation phase as pilot users.

The current procedure of operation planning was documented with the pilot users as groundwork for the requirements of a simulative assistance in this field. Figure 1 shows the summarised steps, tasks and results.

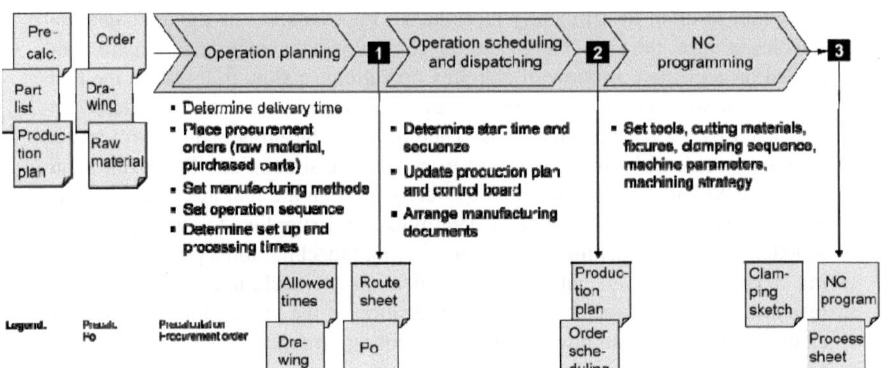

Fig. 1. Summarized as-is process of operations planning, scheduling and dispatching

In the operation planning, the manufacturing methods, production resources and sequences are determined according to firm-specific goals (such as punctuality, profitability, quality). In addition, the processing time and setting time is predicted on the basis of empirical values and the pre-calculation. In this phase the order of raw and purchased parts are initiated. Results are the routing sheet, the allowed times and the procurement orders.

It is the work task of the operations scheduling and dispatching to determine the start time and sequence of manufacturing orders as well as the allocation of resources with regard to the allowed times, scheduled delivery dates and the disposable machines. The Result is the updated master scheduling.

The last step is the NC programming for every manufacturing operation on numerical-controlled machines. Today, CAM-Systems are used for NC programming away from the machine. These provide an automatic calculation of the tool path for predefined shapes (e.g. plane surface, island, groove) deduced from given CAD models of blank, finished part, tool and fixture. The setting of technological machining parameters, used tools and clamping positions is still a manual task of the NC programmer. Hereby, he has a huge impact on the processing time and quality. Results are the NC program, the process sheet and the sketch of set up.

2 Application of Virtual Machine Tools

Nowadays, the CAD supported NC programming includes the simulation of machining. This kind of verification has become quite popular due to the process reliability of machining with 4 to 5 axes [2]. Against the background of reduced batch sizes, the simulation achieves an increasing acceptation also for machines with 3 axes, since it is possible to reduce the test runs for new workpieces and special tools and moreover to reduce the risk of discard. Therefor, common CAM systems provide a material removal simulation and an integrated collision detection. The material removal simulation shows the change of the workpiece during the machining. The automated collision detection reports any unwanted contact between the tool (shank, holder), workpiece und fixture. However, the reproduction of the real machine behaviour is mostly reproduced insufficient by these systems. For example, the dynamic characteristics of the axes, the movement of PLC controlled auxiliary axes, the automatic tool and pallet change, as well as the time synchronization of all movements are only simplified implemented or even neglected [3]. The basis of all common CAM systems is the emulation of the calculated tool paths by an imitated control. Therefore, the machine independent source code CLDATA (cutter location data) [4] is used instead of the control manufacturer specific NC program that run on the real machine. The machine specific NC program is compiled after the simulation by a post processor to adapt the source code to the exact machine configuration. The wear points of integrated simulations are known to the NC programmer, they are compensated by tolerant safe distances. This causes extended processing times and with it unused machine capacities. The result of the integrated simulation in CAM systems is a checked syntax, tool path, zero point and collision-free run of the NC program and additionally the approximated processing time. Nevertheless, a slow and careful test run is still necessary for complex machining operations due to the low modeling accuracy.

The optimized machining by utilization of simulations requires a reliable verification of the operations that are defined in the NC program. The simulations in the contemporary CAM systems can't provide this due to the mentioned deficits.

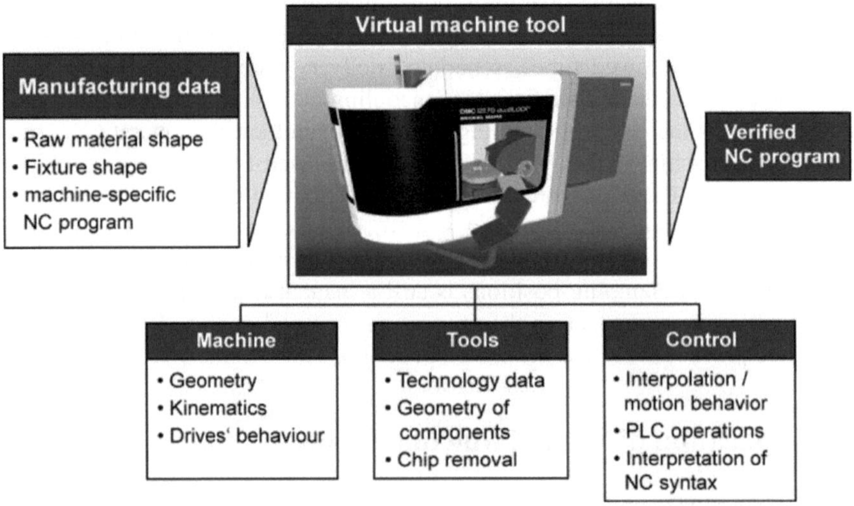

Fig. 2. The simulation models of the virtual machine tool

An approach to optimize the NC program away from the machine is the realistic simulation with virtual machining tools [5]. This includes the implementation of a virtual numerical control with the used NC interpolation as well as the behaviour of the PLC and the actuators. Additionally the entire machine kinematics, the shape of the workspace and peripheries are reproduced in the virtual machine (figure 2). Input data is the shape of the blank and the used fixture as well as the machine specific NC program. The virtual machine tool enables the execution of the same tests as the real machine. This includes optimizing parameters (for example different clamping positions or tooling arrangements) to reduce the processing time. The result is an absolutely reliable NC program, which can run straight on the real machine. Additionally, the reliable processing time is determined by the simulation and made available to the operations scheduling.

However, the variation of parameters (for example the clamping position) or adaptations (to minimize unnecessary operations and tool changes) have to be done manually by the user in the NC program. A new simulation run is necessary after each change and the result must be analysed and compared to previous simulations by the user. This is an iterative process until a subjective optimum (concerning time, costs, quality) is found. The simulation on a single PC runs only 2 to 10 times faster than the real processing time depending on the complexity of the workpiece. Today, the simulation of complex and extensive machining takes too long for multiple optimization runs.

3 Vision: Cloud-Application to Support Process Planning

The illustrated possibilities of virtual machine tools offer promising approaches for optimizing the machining and setting up of the machine. The vision of the project InVorMa is a cloud application, which supports the employees in the planning, scheduling and dispatching of manufacturing operations on tooling machines (figure 3). Instead of passing the manufacturing order and documents directly to the shop floor, the relevant data is previously optimized by the cloud application. The optimization involves the NC program of individual machines as well as the efficient scheduling and dispatching of orders to individual machines. The user obtains the service over the internet from a cloud service, this provides considerably more rapid results compared to a simulation on local hardware. Recent market studies emphasize the potential benefits of an automated routing sheet generation, the integration of expert knowledge and the planning validation through simulation [6].

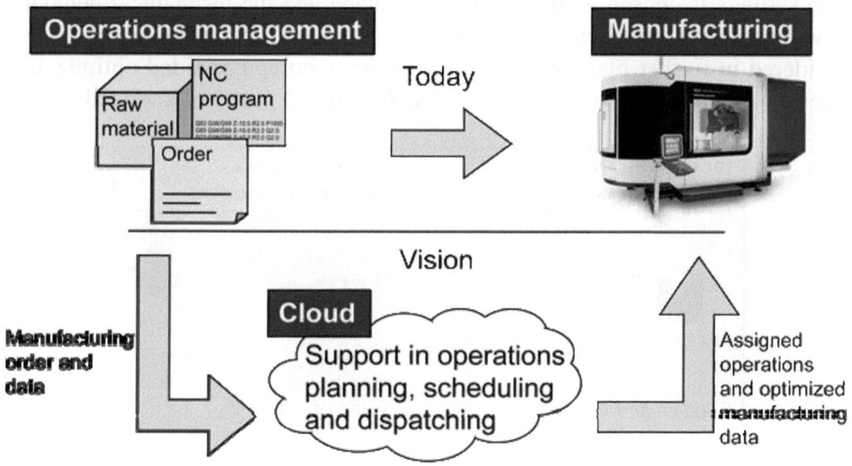

Fig. 3. Cloud application for supporting the operations planning, scheduling and dispatching

4 Fields of Action

In the light of the presented tasks of operations planning, scheduling and dispatching as well as the exposed potentials and disadvantages of virtual machine tools, there are four fields of action (figure 4).

1. A significant increase in simulation speed is the basis of the intended optimization. The main approach is the use of powerful hardware in a computer cluster. For some time, "Cloud computing" is a highly topical technological trend [7]. However, this technology has not yet been used for the simulation of virtual machine tools.
2. The optimized machining result from the evaluation of possible resource combinations and parameters. Depending on the workpiece shapes to be manufactured, there are different combinations of available tool machines, tools, fixtures for the machining. For example, the machine configuration and parameters can be used to control the clamping position, the tooling arrangement in the magazine and the

superposition of the feed speed. To optimize the machining process, the possible combinations have to be simulated and evaluated automatically.

3. Optimizing the machining on each machine does not necessarily lead to an efficient scheduling. This requires a cross-machine optimization witch considers the processing time, the resource management and the occupancy rate of all machines. Waiting orders have to be economically dispatched to available machines. The operation scheduling needs to adapt continuously to the current situation, such as new orders and failures of machines or workers.

Nowadays, operation planning is based essentially on the experience of the responsible manufacturing planner. The combination of resources as well as the machine settings is chosen with regard to the shape and the mechanical behaviour of the workpiece. If the machining result does not reach the expectations, this will be considered in further planning tasks. Therefore, a computer-aided optimization requires an aimed processing and reuse of technical and practical knowledge.

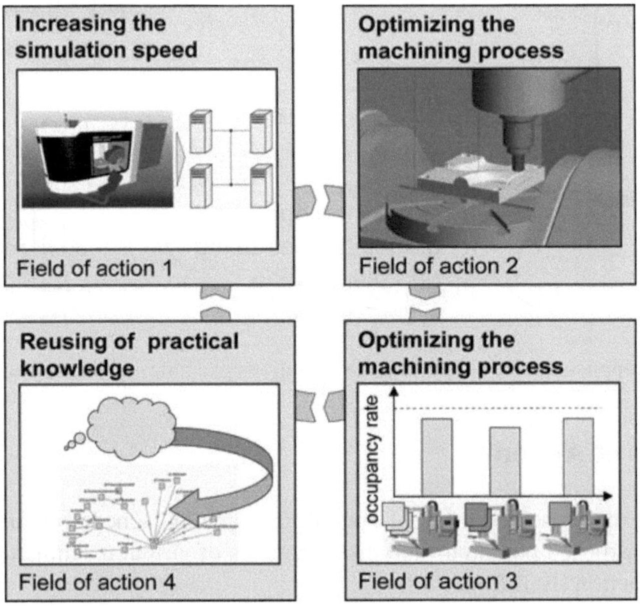

Fig. 4. Fields of action

5 Concept as a Cloud Application

The operation planning is assisted by the verification and optimization of the NC program and machine set up by the use of simulations. In addition, the operations scheduling and dispatching is improved by a pre-selection of resources and providing of reliable processing times. Figure 5 shows the system architecture of the cloud application with its modules "Production Optimizer", "Setup Optimizer", "Simulation Scheduler" and the "Virtual Manufacturing" as basis.

The optimization steps in each module are supported by a “Knowledge Base”, which provides both, technical and practical knowledge from previous simulations.

The interface for incoming and outgoing information is part of this “Knowledge Base”. The user sends the manufacturing order and documents (blank description, NC program) as well as the desired firm-specific goals to the cloud application. This represents the new bridge between the customers’ CAPP-System (Computer-aided process planning) and the shop floor control. First of all, the “Knowledge Base” determines suitable machine tools by reference to the machining operations described in the NC program and the existing resources. The result is a combination of resources – composed of machine, fixture and tool – for each machining step. The selection bases on the description of relations between resources and possible machining operations. Empirical data from previous simulations like the process times are reused to estimate the processing time on each of the suitable resource combination. This outcome is utilized by the “Production Optimizer” to accomplish a cross-machine optimization using a mathematical model and a job shop scheduling. This takes account of the processing time, delivery date, batch size, current resource disposability, machine costs material availability, set up time, maintenance plan and the shift schedule. The master schedule sets the framework for the detailed operations scheduling and dispatching. Scheduling is the assignment of starting and completion dates to operations on the route sheet. Selecting and sequencing of waiting operations to specific machines is called dispatching. Here, the real time situation in the shop floor is provided thru the “Knowledge Base” to ensure a reliable scheduling.

In the next step, the NC program is optimized for the selected machine tool by the “Setup Optimizer”. It varies systematically the parameters of the NC program and evaluates the simulation results from the virtual machine tool. For example, the target is to determine a timesaving workpiece clamping position, to remove collisions, to minimize the tool change times and empty runs or to maximize the cutting speed profile. The parameters that are being evaluated are chosen by a special algorithm in an adequate distance in order to reduce the number of simulation runs and to quickly identify the optimum parameter range. The result is an optimized, verified NC program and the necessary parameters to set up the machine. The results of all performed optimizations are saved in the “Knowledge Base”. It links workpiece information, configurations and technological parameters with already conducted simulation results. Thus, it is possible to early identify parameters with a high potential for optimization as well as relevant parameter ranges for new NC programs. This restriction for the scope of solutions reduces the number of necessary simulation runs too.

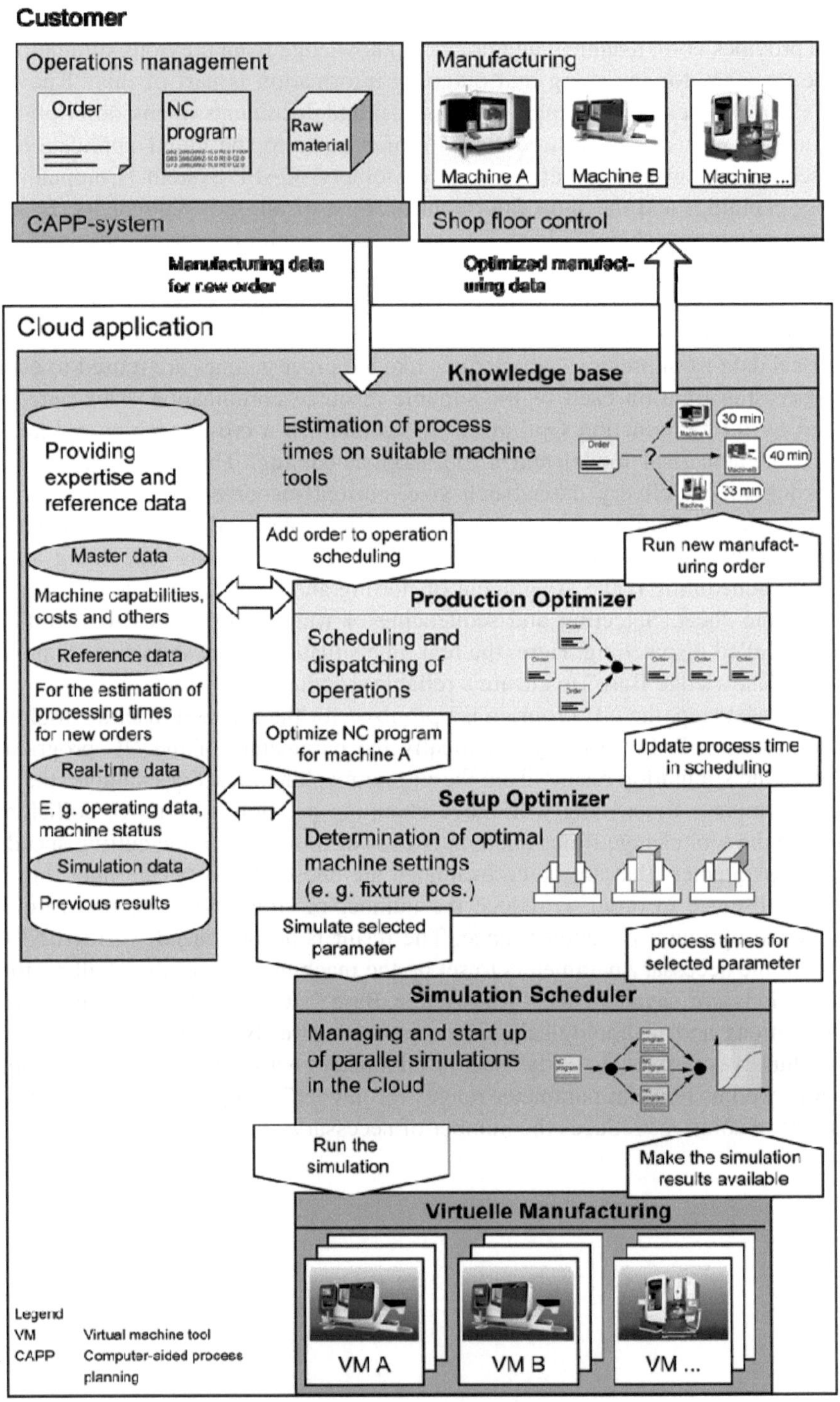

Fig. 5. Architecture of the cloud application

All simulation orders from the "Setup Optimizer" are managed by the "Simulation Scheduler" and distributed to the virtual machine tools and hardware capabilities. To increase the simulation speed, extensive NC programs are divided into sub-programs, that are simulated parallel and the results combined again afterwards.

The prerequisite for achieving the overall aim is the customized "Virtual Manufacturing" with virtual images of all available machine tools and manufacturing equipment. This includes the virtual machine tool in that current version as well as tools, holders and fixtures as CAD models.

If it is necessary, multiple instances of a virtual machine are generated in the computer cluster of the cloud application. For further improvements, potentials for the parallelization of separated computations are considered. For example, the calculations for the collision detection and the simulation of the numerical control systems can be executed on different CPU cores.

6 Conclusions

The fully automated operation planning will not be realized in a short period of time. Instead, the paradigm of Industry 4.0 pushes decision-making techniques to support the user. The presented project combines approaches of knowledge reusing, advanced planning and scheduling and reliable machine simulations in a cloud application. Virtual machine tools are used to verify and improve the machining without interrupting the production process on the shop floor. In addition, a more efficient distribution of manufacturing orders to the machine tools is addressed. This enables an increase in efficiency without changing the existing machine tools.

The following tasks are part of the next project phase: Characteristics and a taxonomy to describe manufacturing processes and resources are defined for the "Knowledge Base". Simultaneously, a concept is developed to speed up the simulation run; this includes software and hardware technologies. Furthermore, a basic model for scheduling and dispatching is developed; this can be adjusted later to the customers' framework.

Acknowledgement. This research and development project is / was funded by the German Federal Ministry of Education and Research (BMBF) within the Leading-Edge Cluster "Intelligent Technical Systems OstWestfalenLippe" (it's OWL) and managed by the Project Management Agency Karlsruhe (PTKA). The author is responsible for the contents of this publication.

References

1. Abele, E., Reinhart, G.: Zukunft der Produktion–Herausforderungen, Forschungsfelder, Chancen. Carl-Hanser-Verlag, München (2011)
2. Rieg, F., Steinhilper, R.: Handbuch Konstruktion. Carl-Hanser-Verlag, München (2012)
3. Kief, H.B., Roschiwal, H.A.: CNC-Handbuch. Carl-Hanser-Verlag, München (2011)

4. N.N.: DIN 66215: Programmierung numerisch gesteuerter Arbeitsmaschinen – CLDATA Allgemeiner Aufbau und Satztypen. Beuth-Verlag, Berlin (1974)
5. N.N.: DMG Powertools – Innovative Software Solutions. DMG/Mori Seiki Deutschland GmbH, Leonberg, Stand: 2012. Zuletzt aufgerufen am November 26, 2012 (2012), http://en.dmgmoriseiki.com/pq/netservice_en/ps0uk12_dmg-powertools.pdf
6. Denkena, B., Lorenzen, L.-E., Charlin, F., Dengler, B.: Quo vadis Arbeitsplanung? Marktstudie zu den Entwicklungstrends von Arbeitsplanungssoftware. Industrial Engineering (2), 6–11 (2010)
7. N.N.: Fujitsu Launches Cloud Service for Analytical Simulations. Fujitsu Limited, Tokyo, Japan. Stand: 2011. Zuletzt aufgerufen am November 26, 2012 (2012), http://www.fujitsu.com/global/news/pr/archives/month/2011/20110808-01.html

Methodological Issues in Support of Selected Tasks of the Virtual Manufacturing Planning

Marius Eßers, Martin Erler, Andreas Nestler, and Alexander Brosius

Technische Universitaet Dresden, Faculty of Mechanical Science and Engineering, Institute of Manufacturing Technology, Chair of Forming and Machining Processes
{marius.essers,martin.erler,
andreas.nestler,alexander.brosius}@tu-dresden.de

Abstract. For virtual production planning a wide variety of tools are already in use. These tools mainly focus at just one methodological aspect. To increase the scope and utility value for process planning, several methodological aspects need to be combined in one single tool. The technological representation of the different aspects usually requires the application of various simulation models beside common data models. In particular, the combination of discrete and continuous time and spatial variables is elaborated.

It opens up new fields for increased productivity and cost forecast, by using e. g. universal kinematics, control access, material removal simulation and finite element analysis.

Keywords: Virtual Machining, Virtual Manufacturing, Virtual Machine Tooling, Virtual Machine.

1 Introduction

In the area of production planning there are many points of application for the supporting use of simulation models due to the multitude of influencing variables to be taken into account. In principle all unique activities to be implemented for the design of a manufacturing system and the manufacturing processes to be planned can be simulated. The challenge is in developing models with suitable representations and visualisations as well as furnishing them with additional, growing physical characteristics. This affects all activities to be planned for drafting, design and optimisation of manufacturing processes in component production [01]. With the inclusion of physical characteristics from models, increasingly realistic statements for technological matters in particular can be attained. Realistic, in the sense of the planning of target specifications, means attaining sufficient accuracy for the results relating to a relevant point of observation. Thus with the knowledge of sufficiently accurate machining forces for planned operations, the power and energy considerations can also be incorporated into the simulation, e.g. for the reduction of the energy expended through the evaluation for the design of operations based on low energy requirements [02]. For the best possible process design, conditions typical to the planning must be selected, whereby the various different display options incorporated for an increase in process planning quality can be tested and which can be linked to a comprehensive procedure

G.L. Kovács and D. Kochan (Eds.): NEW PROLAMAT 2013, IFIP AICT 411, pp. 401–412, 2013.

[03]. Proving techniques, which will analyse the processes already designed as a follow-up, must also be integrated for verification purposes. The functionality existing at present for commercial and non-commercial machining simulation systems amounts primarily to the classical collision avoidance [04], the predominantly geometry-based visualisation of the overall system Machine-Tool-Workpiece [05] and selected optimisation on the basis of the NC code [06].

For the best possible process design and verification, the machining process must take account of the working process of the mechanical processing and its effect on the physical complete system Machine-Tool-Workpiece. In addition, further processes, e.g. the setting up of the machine, must also be taken into account in order to avoid potential fault sources.

The objective is the combination of a long term process simulation and a point-in-time-specific system simulation with a high degree of detailing (Figure 1) , including Finite Elemente Analysis (FEA), Multi Body Systems (MBS), NC Simulation Kernel (NCSK) and Virtual Machine Tool Environment (VMTE).

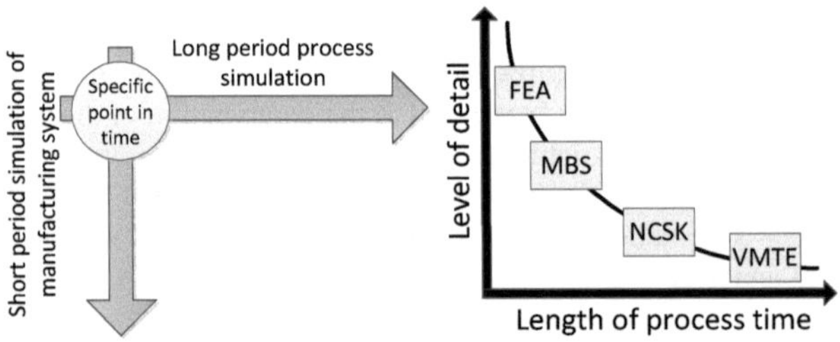

Fig. 1. Trends of process simulation

The high degree of detailing enables the planner to use additional functions of the process simulation for substantiated forecasting for special problem cases for a defined assessment period. These types of functions are not available in commercially available planning systems or are only available in rudimentary forms. This approach also counteracts transient performance problems occurring when working with a high degree of detail, which severely restrict the assessment period, in order to economically facilitate the overall process simulation with physical characteristics. The following investigations into methodical aspects are performed as examples of the virtual design of a milling machining centre and the milling process to be planned.

2 Methodological Aspect of Short Period Simulation of Manufacturing Systems[1]

Here the degree of detailing can be considered as an alternative illustration for a given process section. That means that for an assessment period alternative illustrations are possible, whereby each one can depict another degree of abstraction. That applies both to the representations and the characteristics of the models.

[1] Acknowledgement: This work is kindly supported by the AiF ZIM Project SimCAP (KF 2693604GC1).

For the illustration of the most comprehensive range of characteristics in the models of the mechanical machining, a multitude of application domains must be combined [07]. So that various different methods can be used, logically a universal user interface must be created for the design and implementation of the simulation.

The SimulationX development environment was selected for these requirements. SimulationX can be used for interdisciplinary drafting, the modelling and analysis of physical-technical systems on a common platform [08]. In standard form it offers domains in the fields of drive technology and electrical engineering, flexible multi-body mechanics and others. Additional models can be coupled to expand the functional scope. In principle the models can communicate via a locally shared memory or via a network. The basis for the coupling is an interface definition. Alongside a propriety interface, an example functional mock-up interface (FMI) of the MODELISAR project [09] can also be used. This enables a modular approach, whereby the computing resources will also have an influence on the costs of the degree of detail. The object-oriented implementation of the manufacturing system is sub-divided into important sub-systems with bidirectional interfaces, whereby clamping systems are currently not considered (Figure 2).

The virtual workpiece, virtual tool and virtual machine tool sub-systems are dealt with in more detail below. The working point for the process will be explained in the process simulation section.

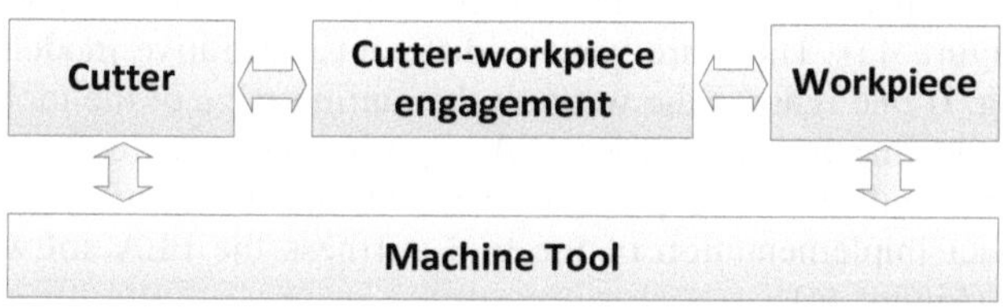

Fig. 2. Object-oriented illustration of a manufacturing system

2.1 Virtual Workpiece

The real workpiece undergoes continuous change during the machining. This affects both the external form as well as the stiffness and the mass which are changing due to the removal of material. Where large volumes are to be machined, the mass has a greater effect on the complete system and for example on the expected energy consumption. For smaller components with thin-walled structures the smallest amount of material removal has a substantial effect on the stiffness.

There are well-proven techniques existing for the representation of workpiece geometry. Alternatively the mapping of a coupled NC simulation core (NCSK) [11] can be applied via a Z-map [10] or via a 3-dexel model. The widely-used data format STL is used as a basis. With this, the starting geometry is imported and the virtual finished geometry exported. Volumetric and dimensional calculations, for example, can be carried out based on this.

In principle, modelling is available as an ideally stiff workpiece. Firstly a check is carried out to ascertain whether a modular replacement system can be used to illustrate the changing stiffness. However, this does not enable the complex geometry to be completely illustrated. Therefore a freely available FEA software module is incorporated by means of Co-Simulation [12] to show the structural mechanics of the workpiece (Figure 3). With this model representation forces, which can be used for deformation, can be applied to the node points.

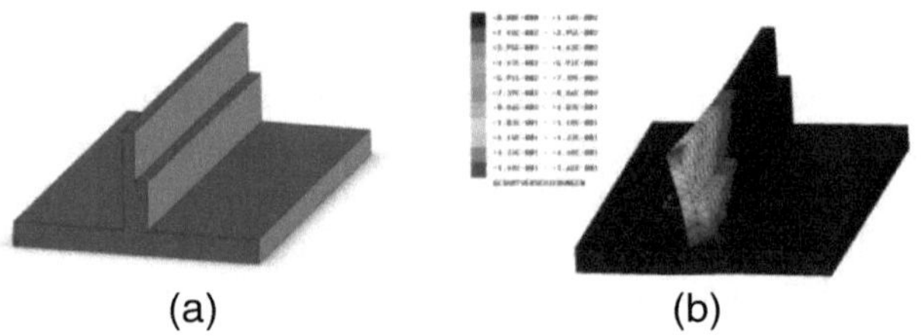

Fig. 3. Geometric (a) and deformable representation (b) of the workpiece

2.2 Virtual Tool

If rough calculations are to be carried out then the tool can be adopted initially as ideally stiff (Figure 4a). There are various different illustrative models available for further detailing. If one ignores the wear on the cutting edge of the tool then it undergoes no geometric change in the process. An approximated representation of the tool with two flexible multi-bodies represents the shaft and the cutting part (Figure 4b). For a more exact implementation of the tool stiffness the FEA software module is available again (Figure 4c).

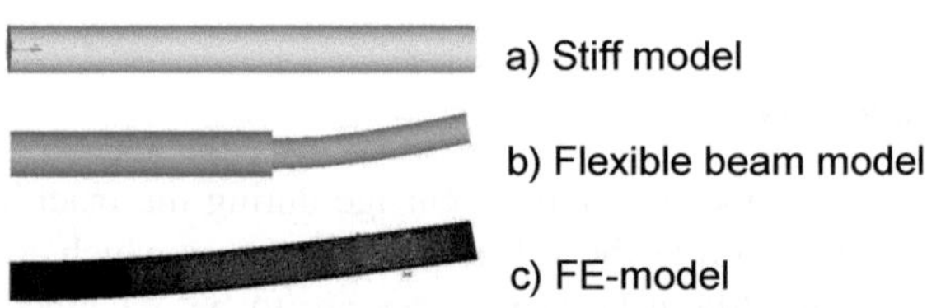

Fig. 4. Tool representation models

This enables step-by-step expandable accuracy in the illustration of the tool characteristics. End mills are modelled as examples.

2.3 Virtual Machine Tool

As a minimum requirement on a virtual machine tool the kinematics must be implemented in order to be able to realise movement at all. The existing objects of the ITI

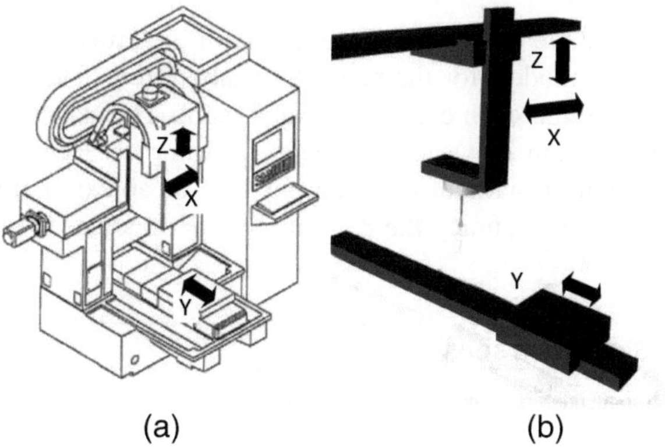

Fig. 5. Kinematics of the Mikromat 4V HSC (a) machine and a simplified visualisation (b)

mechanics library from the CAx development platform SimulationX are utilised for this. As an example a simple 3-axis portal-design vertical milling machine is modelled (Figure 5).

Information from the machine documentation is sufficient to illustrate further characteristics. This is critical for the illustration of masses and centres of gravity for the machine components. The stiffnesses of the guides are modelled through spring-damper systems. The machine components are primarily adopted as ideally stiff, so that all simulated machine deformations are created by the guides. The illustration of the cascade controller and the electrical illustration of the translational drive components is implemented through the "controller" components. An electrical drive motor ("driveMotor") and a ball screw drive ("ballScrewDrive") from the ITI mechanics library are used in these components.

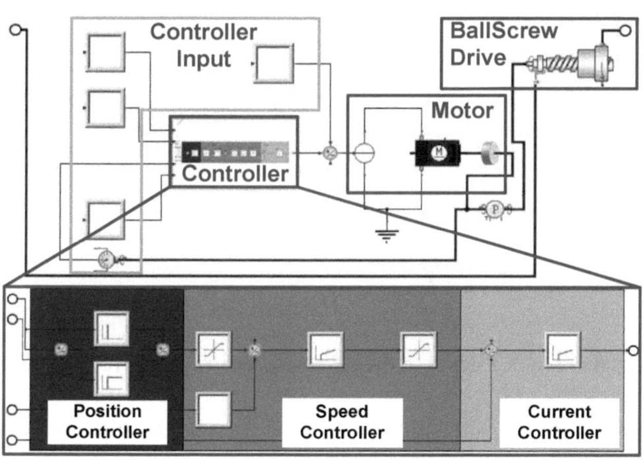

Fig. 6. Control-related and electrical implementation of an axis

2.4 Process Simulation

Figure 7 shows the principle design and progression of the process simulation including the various different models for the respective sub-system. The closed simulation progression will run through each calculation increment. The stages indicate two options for the implementation of various models as examples. Here stage 1 means a platform-internal simulation without the incorporation of an external Co-Simulation. To increase the calculation accuracy there are additional modules in stage 2 available for selection.

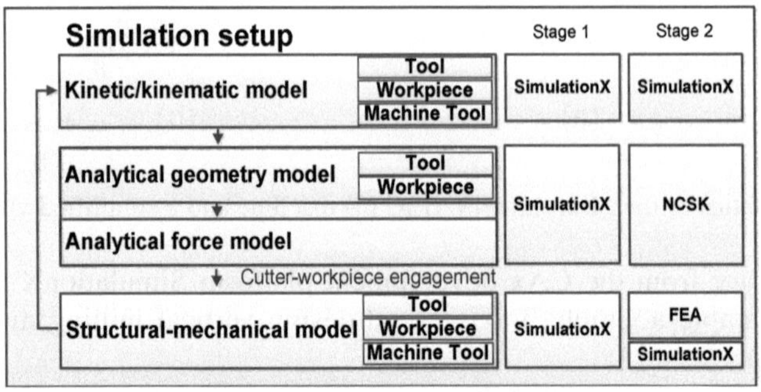

Fig. 7. Simulation development in stages

With stage one the representation of a Z-map is applied, built upon the kinematics of the penetration between tool and workpiece. In addition, an analytical geometry model calculates geometric intervention points within the XY-level of the tool in the time range of a tooth feed $\boldsymbol{f_z}$ without taking vibrations into account (Figure 8).

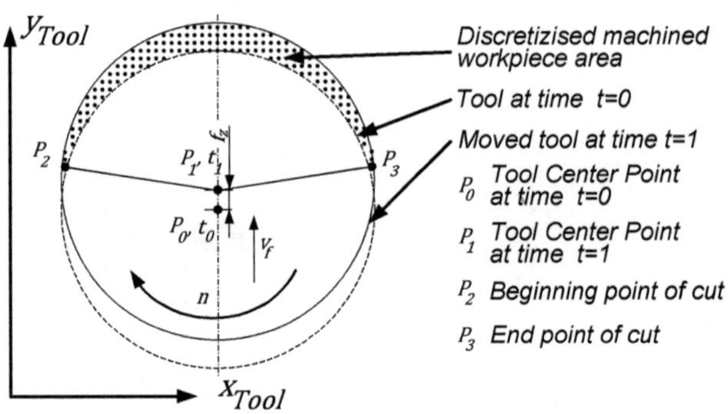

Fig. 8. View XY-level of the geometric cover of two tool positions

The intervention points determined are used for the calculation of an average cutting depth $\boldsymbol{h_m}$ (Formula 1).

$$h_m = \frac{\overrightarrow{P_1 - P_2}\left(\frac{(P_{2_x} - P_{1_x}) \cdot (P_{0_y} - P_{1_y}) + (P_{2_y} - P_{1_y}) \cdot (P_{1_x} - P_{0_x})}{\sqrt{(P_{2_x} - P_{1_x})^2 + (P_{2_y} - P_{1_y})^2} \cdot \sqrt{(P_{0_y} - P_{1_y})^2 + (P_{1_x} - P_{0_x})^2}}\right)}{\mathrm{acos}\left(\frac{(P_{3_x} - P_{1_x}) \cdot (P_{0_y} - P_{1_y}) + (P_{3_y} - P_{1_y}) \cdot (P_{1_x} - P_{0_x})}{\sqrt{(P_{3_x} - P_{1_x})^2 + (P_{3_y} - P_{1_y})^2} \cdot \sqrt{(P_{0_y} - P_{1_y})^2 + (P_{1_x} - P_{0_x})^2}}\right)}$$

Formula 1. Geometry based determination of the averaged cutting thickness

Furthermore, the average cutting depth $\boldsymbol{a}_{pm}$ (Formula 2 and Figure 9) can be shown via a calculated value through the penetrated heights $\boldsymbol{z}_{i,j}$ and their number $\boldsymbol{n}_z$ in Z-direction.

$$a_{pm} = \frac{\sum \Delta z_{i,j}}{n_z}$$

Formula 2. Averaged cutting depth

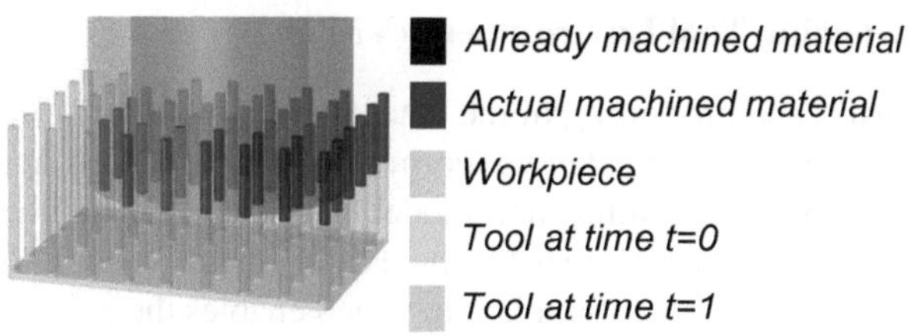

Fig. 9. Geometrical cutter-workpiece engagement

On the basis of the two values calculated, $\boldsymbol{h}_m$ and $\boldsymbol{a}_{pm}$, the Victor/Kienzle [13] force model for calculating a determined machining force can be applied. Because this procedure only permits 3-axis machining and only a rough determination of the machining forces, the considerably more accurate NCSK has been coupled via an FMI for Co-Simulation as a further alternative. The NCSK works with a three-dexel model for the workpiece, a swept-volume model for the cutter movement and a defined cutting edge for the determination of the machining forces.

A force model, which delivers the machining forces which can be arbitrary in terms of magnitude, direction and application can in turn be applied to the geometric sizes determined. These are applied to the workpiece and tool in the vicinity of the working point for the process and will result in the deflection of the sub-system involved. This deflection in turn has a direct influence on the common penetration and the resultant geometric machining parameters.

3 Methodical Aspects of the Simulation over a Long Period of Time

The simulation over a long period of time should be considered here to be a complete simulation process for an existing NC program. To do so systems are employed here, which are closer to the real machine/controller combination than is possible with the classical post-processor CAD/CAM or NC programming systems. Alongside the verification - so, the assurance of freedom from errors in the sense of collision avoidance and guaranteed achievement of the required surface qualities and dimensional accuracy - the objective of a simulation process is also increasingly the reduction of unutilised safety reserves, which lie in the technical process parameters and which finally lead to a non-optimum primary processing time or secondary processing time. In order to be able to utilise these reserves and thus to be able to reduce machining times and tool costs, the consideration of further influences is required in the simulation. With NC code based verification systems this is not normally possible, as the constraint that the NC code is dealt with as a whole significantly increases the difficulty of a more detailed evaluation. Influences of the machine tool control system for example remain almost completely disregarded.

Coupled with a real CNC controller a simulation on the other hand can provide significantly more statements about an NC program and the resultant machining [15], as it evaluates the control signals generated for the individual axes directly, for example.

3.1 Virtual Machine Tool Environment (VMTE)

Alongside the transfer and processing of the data, the incorporation of information generated on the CNC control system side (e.g. corner rounding, reduced approach torques or feed limitations) also requires a simulation model, where this can be illustrated. One such model has been developed - the Virtual Machine Tool Environment (VMTE). It provides a basic model in the sense of a Co-Simulation, which enables the processing of CNC control system control signals to machine model transformation information. Two important requirements arise from the application conditions described:

- The VMTE shall be quick and simple to create, as well as
- realtime-capable and modular,

in order to achieve many iterations, a broad application spectrum and a high utility value.

3.2 The Rapid Path to VMTE

The process developed for creating the simulation models allows the use of generally available information and data for a machine tool and thus very quick creation of a VMTE, so that this can be economically used in many new areas of application.

3.2.1 Prerequisites

In order to be able to use the VMTE as a planning and development tool, it must be able to be quickly adapted to the specific tasks for which it is required. Frequent changes to the machine configuration are normal in the early phase of development and planning as part of the process development and process checking. A wide range of different machines with differing configurations is also necessary for basic training and advanced training purposes. The consideration of established machine design variations in conjunction with the final operational sequence is an important point of the operational fine planning for the utilisation of VMTE in the manufacturing phase.

Machine tools are generally based on serial kinematics. The most important tasks in the development of a model for illustrating these serial kinematics are the development of the kinematics on the basis of the machine configuration and the linking of the graphical data with each axis. The basic mechanism for the transformation of the original CAD data and the transfer into graphical data as well as its population with machine functions can be largely generalised such that a VMTE can be created in less than 30 minutes.

3.2.2 Generalised Virtual Machine

In order to achieve this the emphasis is not on the linking, in order to illustrate the kinematics, but rather the axes which represent the real components. This approach is closer to reality and results in the position of the axes being directly adjustable with respect to one another. There is also no need for additional parameters such as in the DH convention [16]. The mobility of the model (and thus the movement of its axes) is achieved through the decomposition of one axis into two axis modules and their displacement or rotation with respect to one another.

In doing so an axis module has a zero point and a connection point, whose position and orientation with respect to one another will be described through a transformation matrix and which define the interior of the module (Figure 10). The combination of two modules (basic part and mobile part) describes the configuration of an axis. The mobile part will be moved relative to the basic part (likewise through a transformation matrix), whereby the typical axis characteristics will be realized.

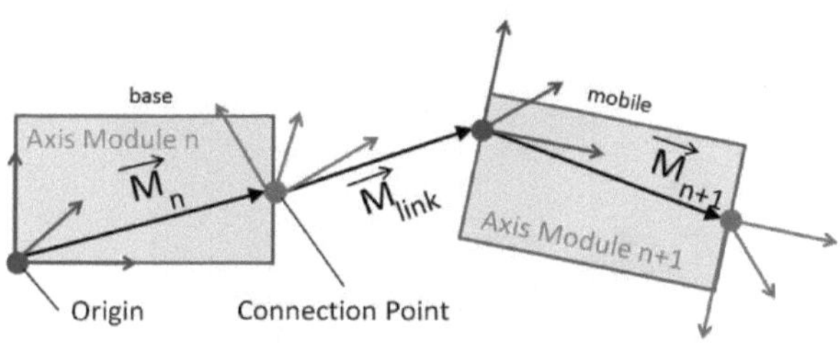

Fig. 10. Generalized axis pattern for serial kinematics

Any arbitrary machine tool with serial kinematics can be prepared as a fully mobile virtual machine model through the use of multiple axes and axis chains (Figure 11).

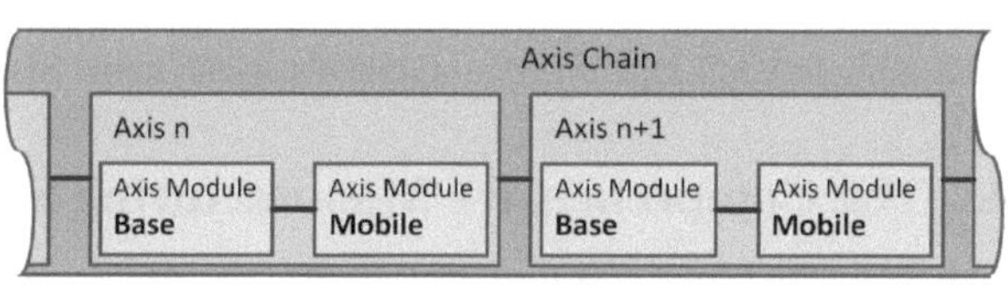

Fig. 11. Configuration scheme for axis chains

The set-up interface which is thus finally fully parameterized can also be considered an interface and can be supplied with information provided from outside. The CNC control system provides one such interface for example.

3.3 VMTE for Simulation of a Long Period of Time

Many intrinsic controller characteristics can also be considered with large time spans during the verification of the analysis through the coupling of the machine model and a real or virtual CNC control system, and these would otherwise have to have been modelled. Due to its

- very short creation time,
- ability to be fully parameterised and
- intrinsic consideration of CNC controller influences

the kinematic base model provided offers itself as a basis for a virtual machine environment (Figure 12). The fully parameterised interface enables changes to be made to the configuration whilst the simulation is running, so that various different configurations can be used as the subject-matter of the simulation or so that the simulation can take account of the configuration changes.

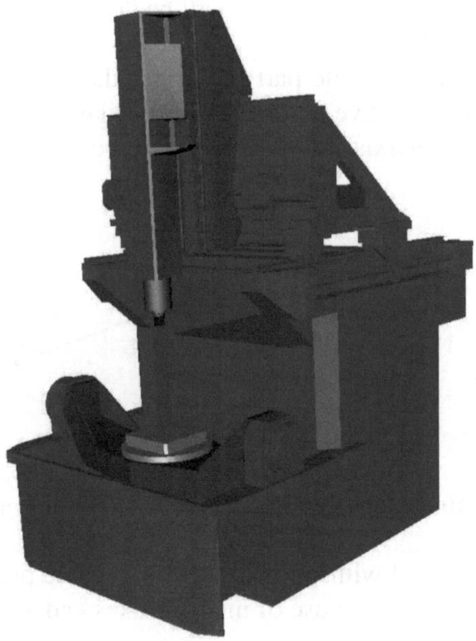

Fig. 12. Sample VMT Ops Ingersoll Funkenerosion GmbH SH 650

4 Outlook - The Comprehensive Manufacturing Simulation

The sub-processes detected in the VMTE, but which cannot be simulated there in sufficient detail, can be considered and evaluated downstream or in the meantime through the increased degree of detail in an enlargement of arbitrary resolution. In doing so not only can the time steps be reduced - thus increasing the slow-motion - but also the resolution of the model (e.g. FE meshes) in question can be increased.

The combination of a simulation for a large time period and detailed process simulation enables the comprehensive evaluation of the complete manufacturing process as well as the parameters and influencing variables involved in it, both in a holistic context and in detail.

In order to achieve this, the two simulations must interact with one another. This is achieved through the parametrisation of the two simulations. Thus the VMTE can transfer a parameterised machine tool model to the process simulation and can detect and specify the periods of time to be considered. The highly accurate analyses in small time periods received from the process simulation can be returned and used for consideration or correction of the VMTE in large time periods, where small changes have a significant impact.

The approach presented unites geometrical/kinematic simulation methods for a large time period and restricted degree of detail, with highly detailed methods such as MBS and FEA analysis for small time periods. In this way the advantages of both methods can be utilised and their disadvantages reduced.

References

1. Jacobs, H.-J., Dürr, H.: Entwurf und Gestaltung von Fertigungsprozessen: Planung und Steuerung der spanenden Teilefertigung. Fachbuchverl. Leipzig im Carl-Hanser-Verl., München (2002) ISBN 3446217487
2. Denkena, B., Köhler, J.: Energieeffiziente Zerspanung. In: VDI-Z, Special Werkzeuge August 2010, II/2010, S20–S22 (2010)
3. Schindler, J.: Quelle zur Produktivitätssteigerung und Kostensenkung in der Produktion. In: Nestler, A. (Hrsg.) Intelligente Fertigungsprozesse – Quelle zur Produktivitätssteigerung und Kostensenkung in der Produktion. Vortragsband zum Fachkolloquium am September 24, 2010 Selbstverlag TU Dresden, Dresden (2010) ISBN 978-3-86780-185-0
4. Sun, S., Luo, L., Li, G., Zou, X., Yang, J.: The virtual simulation system of numerical control machining. In: Proceedings of the 2008 International Workshop on Modelling, Simulation and Optimization, WMSO 2008, pp. 289–293 (2009)
5. Zhu, L., Wang, J., Chen, E., Yang, J., Wang, W.: Applications of virtual reality in turn-milling centre. In: Proceedings of the IEEE International Conference on Automation and Logistics, ICAL 2008, pp. 2302–2305 (2008)
6. Lee, S.W., Nestler, A.: Modeling of Cutting Force for Cylindrical Milling Cutter and Feed Rate Optimization Using Five-Axis Milling Simulation. In: Proceedings of 8th International Conference on High Speed Machining (HSM), Metz, France, December 08-10, 64–70 (2010) ISBN 978-2-9534170-2-9
7. Vormann, K., Marsolek, J., Piendl, S., Scherman, T., Steinicke, M., Fiderer, M., Klocke, F., Frank, P., Sellmer, D.: Abschlussbericht des Verbundforschungsprojektes Simulation Industrieller Bearbeitungsprozesse (SindBap). Tagungsband zum Abschlusskolloqium
8. N.N.: SimulationX - Multi-domain System Simulation and Modeling. ITI GmbH, `http://www.itisim.com`
9. Modelica Association: Functional Mock-up Interface, `http://modelisar.com`

10. Inui, M., Kakio, R.: Fast Visualization of NC-Milling Result using graphics Acceleration Hardware. In: Proceedings of the IEEE International Conference on Robotics & Automation (2000)
11. Lee, S.W., Nestler, A.: Tool load balancing at simultaneous five-axis ball-end milling via the NC simulation kernel. In: 1st CIRP Conference on Virtual Machining Process Technology (CIRP VMPT), Paper 33, Montréal, Canada, May 28-June 1 (2012)
12. Calculix, A.: Free Software Three-Dimensional Structural Finite Element Program, `http://www.calculix.de/`
13. Kienzle, O., Victor, H.: Spezifische Schnittkräfte bei der Metallbearbeitung. Werkstatttechnik und Maschinenbau (47) (1957)
14. Kao, Y.C., Cheng, H.Y., Chen, Y.C.: Development of a virtual controller integrating virtual and physical CNC. Progress on Advanced Manufacture for Micro/Nano Technology 2005, pt. 1 and 2 505-507, S.631–S.636 (2006)
15. Kadir, A.A., Xu, X., Hämmerle, E.: Virtual machine tools and virtual machining—A technological review. Robotics and Computer-Integrated Manufacturing 27(3), 494–508 (2011) ISSN 0736-5845
16. Denavit, J., Hartenberg, R.S.: A kinematic notation for lower-pair mechanisms based on matrices. Trans. ASME J. Appl. Mech. 23, 215–221 (1955)

A Comprehensive Framework for the Computer-Aided Planning and Optimisation of Manufacturing Processes for Functional Graded Components

Marcus Petersen and Jürgen Gausemeier

Heinz Nixdorf Institute, Product Engineering, University of Paderborn,
Fuerstenallee 11, 33102 Paderborn, Germany
{marcus.petersen,juergen.gausemeier}@uni-paderborn.de

Abstract. In this contribution, we present a framework for the computer-aided planning and optimisation of manufacturing process chains for functional graded components. The framework is divided into three modules – the "Component Description", the "Expert System" for the synthetisation of several manufacturing process chains and the "Modelling and Process Chain Optimisation". The Component Description module enhances a standard computer-aided design (CAD) model by a voxel-based representation of the intended graded properties. The Expert System synthesises manufacturing process steps stored in the knowledge base to generate alternative process chains. All these process chains are capable of producing components according to the enhanced CAD model. They consist of a sequence of heating-, cooling-, and forming manufacturing processes. The interdependencies between the component and the applied manufacturing processes as well as between the processes themselves need to be considered. For that purpose the Expert System utilises an ontology. The ontology represents all the interdependencies in a structured way and connects the information of the knowledge base via relations. The third module performs the evaluation of the generated manufacturing process chains. To accomplish this, the parameters of each process step are optimised according to the component specification, whereby the result of the best parameterization is used as a representative value. Finally the process chain which is capable of producing a functional graded component in an optimal way regarding to the property distributions of the component description is presented by means of a dedicated specification technique.

Keywords: Manufacturing Process Planning, Functional Graded Components, Expert System, Specification Technique, Sustainable Production.

1 Introduction

Functional gradation denotes a continuous distribution of properties over at least one of the spatial dimensions of a component consisting of only one material. This distribution is tailored according to the intended application of the component [1].

G.L. Kovács and D. Kochan (Eds.): NEW PROLAMAT 2013, IFIP AICT 411, pp. 413–422, 2013.

Application areas for the use of functional graded components can be found for example in the automotive industry. Car interior door panels for instance are usually plastic materials that are supposed to absorb the impact energy of a lateral crash to an assured extent. The resulting deformation however must in no case lead to an injury of the car's passengers. To achieve a desired deformation behaviour it is necessary to assign exactly defined material properties to specific locations of the door panel. By a functional gradation, e.g. of the hardness, the functionality of the component can be considerably extended. The formerly purely decorative interior door panel becomes a functional element of the passive vehicle safety.

Functional graded components provide a resource-conserving alternative for modern composite materials [1] and therefore offer high potential to achieve a sustainable production. Instead of using post-processing steps to create the composites and their graded properties, the gradation is produced during their moulding process. This process integration for example shortens the manufacturing process chain for the production of the component and increases the energy efficiency significantly.

The production of functional graded components requires complex manufacturing process chains, such as thermo-mechanically coupled process steps [1]. While there are several material scientific approaches on how to develop an isolated process step to achieve a certain material structure, the holistic design of connected manufacturing process chains is much more difficult. For that purpose in section two an exemplary manufacturing process chain will be used to demonstrate our approach. To realise the full potentials of functional gradation, a computer-aided framework for the planning and optimisation of this manufacturing process chains will be introduced in the following subsection, whereupon the hierarchical process chain synthetisation as part of the Expert System will be presented in section three. Section four summarises the approach and identifies the significant future research challenges.

2 Functional Graded Components

2.1 Exemplary Manufacturing Process Chain

The manufacturing process chains for functional graded components are characterised by strong interdependencies between the components and the applied manufacturing processes as well as between the process steps themselves. According to the presented interior door panel (cf. section 1), a manufacturing process chain for self-reinforced polypropylene composites is used here as a demonstrator. This process chain uses a thermo-mechanical hot-compaction process to integrate the functional gradation into self-reinforced polypropylene composites by processing layered semi-finished textile products on a thermoplastic basis. The semi-finished textile products were previously stretched and provide a self-reinforcement based on a macromolecular orientation. This self-reinforcement leads to a sensitive behaviour regarding pressure and thermal treatments and is therefore essential for the functional gradation of the composite [1], [2].

Figure 1 shows the exemplary manufacturing process chain for self-reinforced polypropylene composites starting with a gradual preheating of the semi-finished textile products in a partially masked IR-preheating station. In the next step, the thermal gradation of the product will be enhanced due to consolidation by a special compression moulding tool. The tool was particularly design for thermo-mechanical gradation. For this reason, both tool halves can be tempered differentially and completely independent from each other. Furthermore this tool applies a mechanical gradation by a local pressure reduction of up to 30% due to the triangular geometry. A cooling phase is necessary before demoulding the self-reinforced polypropylene composite [3].

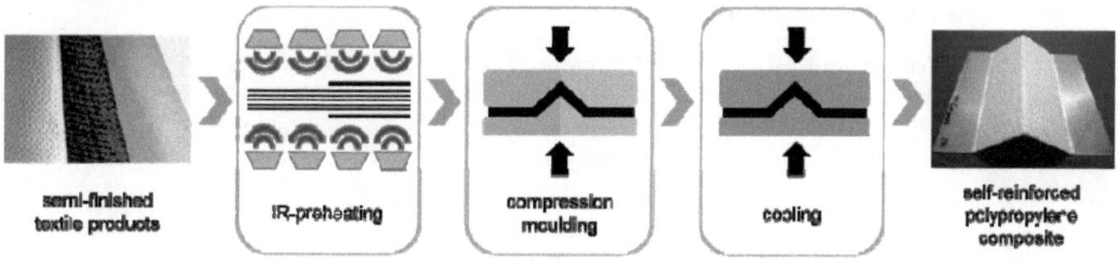

Fig. 1. Exemplary manufacturing process chain for self-reinforced polypropylene composites

2.2 Comprehensive Planning Framework

The exemplary manufacturing process chain for self-reinforced polypropylene composites is characterised by strong interdependencies (cf. section 2.1). These interdependencies are typical for the production of components with functionally graded properties and need to be considered. Therefore a comprehensive planning framework for the planning and optimisation of manufacturing process chains is under development. This framework integrates several methods, tools and knowledge obtained by laboratory experiments and industrial cooperation projects in which the concept of functional gradation has been analysed. The planning process within the framework is continuously assisted by the modules "Component Description", "Expert System" and "Modelling and Process Chain Optimisation" [1].

Figure 2 gives an overview about the structure of the planning framework and the information exchanges between the modules. The input information for the manufacturing process planning is provided by the computer-aided design (CAD) model of the component and the intended graded properties. Based on this information, several alternative process chains for the manufacturing of the component are synthesised by means of the framework. After this, the process parameters of each process chain are optimised based on empirical models. The best manufacturing process chain is described using a dedicated specification technique for production systems in the last step of the planning process [4].

The Component Description module enables the desired graded properties to be integrated into the CAD model of the component. The model usually consists of geometric features (e.g. cylinder or disc), which will be extracted after loading the model. These features allow the framework to consider the geometry of the whole component and to pre-select reasonable gradients according to the geometry. This preselection increases the efficiency of describing the intended gradient since the manufacturing planner can directly provide the desired graded properties by modifying the parameters of the proposed gradients. If the CAD model does not contain any geometric feature or the user does not want to use one of the pre-selected gradients, the component is divided into small volume elements. These so called voxels enable the component model to be locally addressed and can be used as supporting points for the function-based integration of the component's graded properties [5].

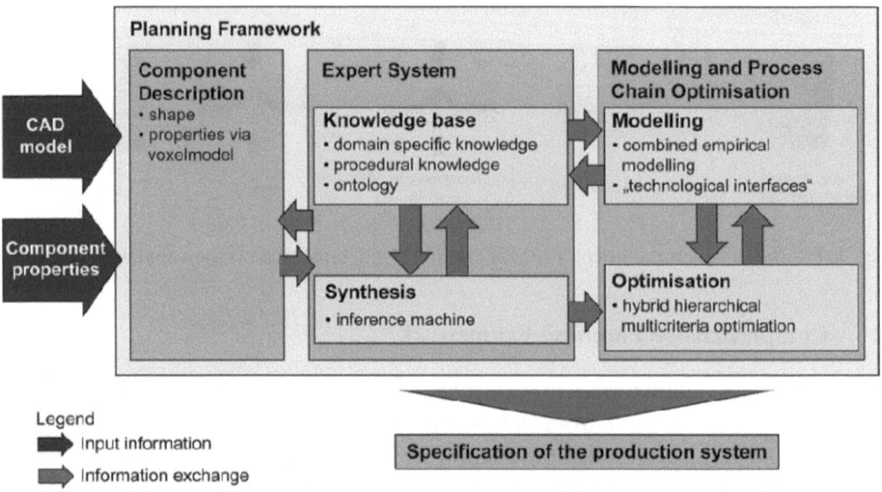

Fig. 2. Planning Framework for the computer-aided planning and optimisation of manufacturing processes for functional graded components

Based on the enhanced CAD model of the first module, the Expert System synthesises several alternative process chains for manufacturing the component. For that purpose all the manufacturing processes available in the knowledge base are filtered according to the component description such as material, geometry or the desired graded properties (e. g. hardness or ductility). To realise this filtering process, the content of the knowledge base is structured by an ontology. The ontology classifies the process steps with regard to their characteristics and connects the information of the knowledge base via relations between the content elements. An inference machine is applied to draw conclusions from the ontology, especially with respect to the varied interdependencies between the manufacturing processes. These conclusions provide the main information for connecting the several process steps of the knowledge base during the synthetisation of reasonable manufacturing process chains according to the enhanced CAD model of the component. The synthetisation of several alternative manufacturing process chains by a hierarchical process chain synthesis is described in section three.

The exemplary manufacturing process chain for self-reinforced polypropylene composites is for example characterised by the fact that the initial material temperature, which is adjusted during the IR-preheating process in preparation of the compression moulding process has a strong influence on the mouldability of the component.

Those and all the other interdependencies mentioned above need to be considered during the pairwise evaluation of process steps to ensure the compatibility of the synthesised process chains for the manufacturing of the component. All process chains with incompatible process steps are disregarded. Thus the result of the Expert System module is a set of several alternative process chains which are capable of producing the component [1].

The parameters of a preferred set of manufacturing process chains are optimised by means of the Modelling and Process Chain Optimisation module [1]. To accomplish this, predictions of empirical models based on several experiments, measurements and simulations of samples provide a comprehensive solution space (cf. [6]). Modern empirical modelling techniques are then used as surrogates for the processes and a hybrid hierarchical multi-objective optimisation is utilised to identify the optimal setup for each process step of a manufacturing process chain. In the context of functional gradation, design and analysis of computer experiments (DACE) models have proven to show a very good prediction quality [7], [8].

After all, the process chain that is capable of producing a functional graded component in the best way regarding to the component description is described using a dedicated specification technique. This fundamental specification is based on a process sequence and a resource diagram [9].

Figure 3 shows an extract of the optimised process sequence for the manufacturing of self-reinforced functional graded polypropylene composites with an example set of process step parameters for the compression moulding. Further information about the specification technique can be found in [9].

The next section gives an overview about the underlying principles of the manufacturing process chain synthetisation within the Expert System.

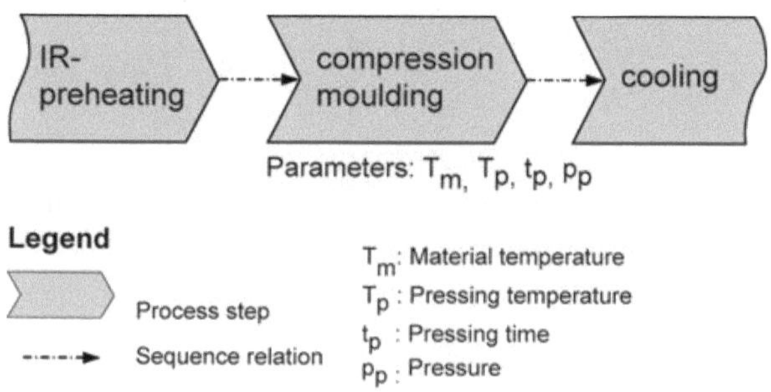

Fig. 3. Part of the optimised process sequence for the manufacturing of self-reinforced polypropylene composites

3 Hierarchical Process Chain Synthetisation

The Expert System within the planning framework synthesises several alternative manufacturing process chains for functional graded components in a hierarchical way. This synthetisation is assisted by two steps – the "Core Process Selection" and the "Process Chain Synthetisation". The component requirements provided by the Component Description module, such as the enhanced CAD model, the material or general requirements (e.g. the surface quality) constitute the product attributes for the requirements profile of the component. This radar chart profile [10] and also the component requirements represent the input information for the Expert System (cf. Figure 4).

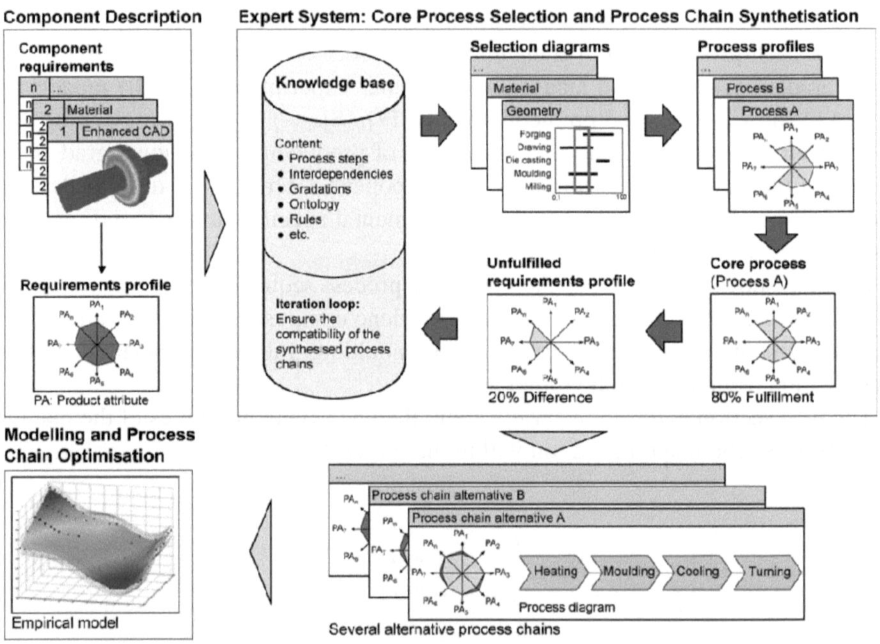

Fig. 4. Hierarchical process chain synthetisation as part of the Expert System module

3.1 Core Process Selection

The Core Process Selection (according to [11]) marks the first synthetisation loop of the Expert System and results in the core process for the manufacturing of the component. Thereby the process step which fulfils the requirements of the component according to the requirements profile in the best way is selected to be the core process. This manufacturing process also establishes the root process i.e. the starting point for the hierarchic process chain synthetisation within the iteration loops of the Expert System (cf. section 3.2).

At first all the manufacturing process steps available in the knowledge base are structured according to each product attribute of the requirement profile for the manufacturing of the component. For this purpose the Expert System utilises matrix tables, in which the manufacturing processes are displayed in the rows and their ability range of the current product attribute is represented in the columns. These so called selection diagrams provide the basis for the automatic selection of the core process. Figure 5 shows an example of such a matrix table for the product attribute "tolerance".

Based on these selection diagrams, all the manufacturing processes which do not match the product requirements in the defined range are removed. For the other process steps, a process profile is created in addition to the requirement profile. The process profile is presented to the user within the planning framework to explain the results of the selection process. These profiles show the fulfilment of the component requirements by the given manufacturing processes in a comprehensive way (cf. Figure 4).

The manufacturing process step with the highest fulfilment of the product attributes is selected to be the core process and the unfulfilled requirements form the main input for the Process Chain Synthetisation as new requirement profile.

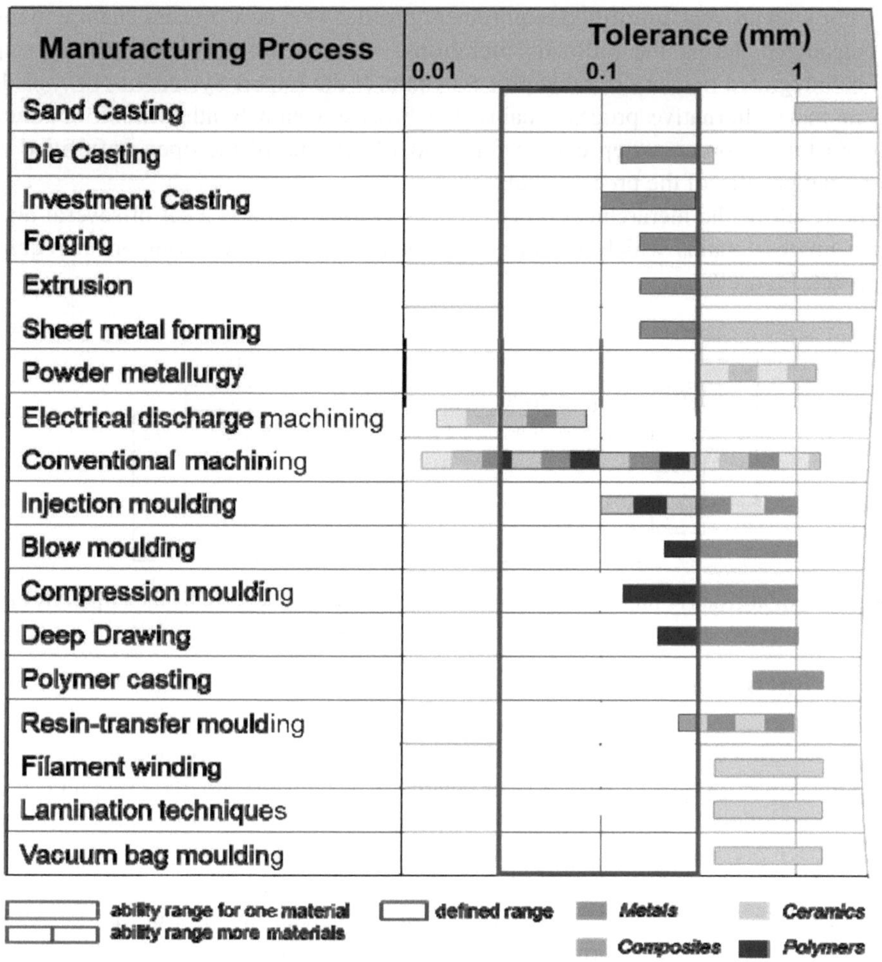

Fig. 5. Example of a selection diagram (according to [11])

3.2 Process Chain Synthetisation

The Process Chain Synthetisation starts only if the Core Process Selection ends up with some unfulfilled requirements. This step of the Expert System tries to reduce the unfulfilled requirements down to a minimum by creating several alternative process chains.

To create the process chains, the Expert System restarts the Core Process Selection as a loop, in which the unfulfilled requirements of a completed iteration loop provide the input information for the next iteration. This loop continues until no further process step can be found to fulfil the open requirements of the requirements profile.

After every iteration loop, a pairwise evaluation with the new selected manufacturing process and the already connected process steps is performed to ensure the compatibility of the synthesised process chain. If there is only one incompatible process step in the process chain a new alternative process chain will be started without this step, but with an own unfulfilled requirement profile. This new process chain will also be considered during the following iterations, whereby new selected process steps will be integrated in every suitable process chain. If the Expert System has to consider two or more alternative process chains, the Process Chain Synthetisation continues until no further process step can be found to fill up one of the open unfulfilled requirement profiles of the process chains.

The result of the hierarchical process chain synthetisation is a set of several alternative process chains which are all able to achieve the desired component requirements (cf. Figure 6).

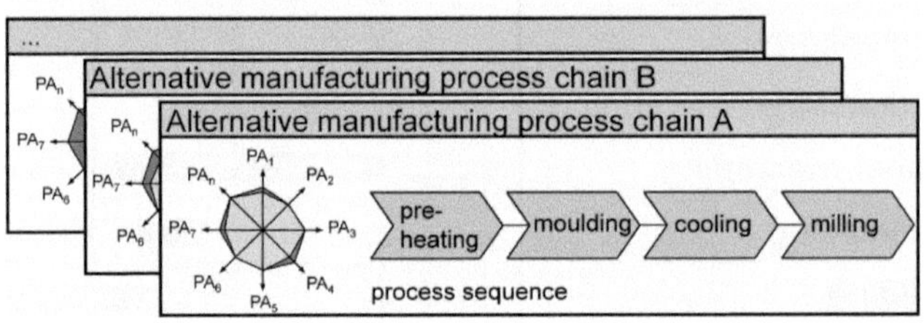

Fig. 6. Set of alternative process chains for the interior door panel given by the Expert System

4 Conclusions and Outlook

Functional graded components offer an innovative and sustainable approach for customisable smart products. Thus a comprehensive framework for the computer-aided planning and model-based optimisation of components with functional graded properties has been presented and demonstrated with an application example.

Future work includes the enhancement of the knowledge base with additional manufacturing process steps, materials and interdependencies as well as the adjustment of the ontology. Furthermore the inference rules of the expert system have to be expanded to realise the synthetisation of more complex manufacturing process chains and their pairwise evaluation. The Expert System of the comprehensive planning framework is able to automatically synthesise process chains for the manufacturing of a component with functionally graded properties. However the final selection of the best process chain for the specific production objective must still be conducted manually since it is not always obvious which alternative fulfils all the requirements according to the objective in the best way. The Analytic Hierarchy Process (cf. [12]) may offer an effective approach to handle the highly diverse characteristics of the decision criteria while not overstraining the decision process with data acquisition and examination.

Acknowledgement. The work in this contribution is based upon investigations of the Collaborative Transregional Research Centre (CRC) Transregio 30, which is kindly supported by the German Research Foundation (DFG).

References

1. Biermann, D., Gausemeier, J., Heim, H.-P., Hess, S., Petersen, M., Ries, A., Wagner, T.: Computer-Aided Planning and Optimisation of Manufacturing Processes for Functional Graded Components. In: Proceedings of the 1st International Conference on Thermo-Mechanically Graded Materials, Kassel, Germany, October 29-30 (2012)
2. Bledzki, A.K., Ries, A., Paßmann, D.: Functional graded self-reinforced polypropylene sheets. Polimery 56 (2011)
3. Paßmann, D.: Prozessinduzierte Gradierung eigenverstärkter Polypropylen-Faserverbunde beim Heißkompaktieren und Umformen, Dissertation, University of Kassel (2010)
4. Gausemeier, J., Bauer, F., Dettmer, D., Reyes Perez, M.: Planning of Manufacturing Processes for Graded Components. In: Proc. International Conference Product Property Prediction – P3, Dortmund, Germany, April 12-13 (2010)
5. Bauer, F., Dettmer, D., Gausemeier, J.: Feature-based component description for functional graded parts. In: 4th International Conference on Changeable, Agile, Reconfigurable and Virtual Production (CARV 2011), Montreal, Canada, October 02-05 (2011)
6. Wagner, T., Paßmann, D., Weinert, K., Biermann, D., Bledzki, A.K.: Efficient modeling and optimisation of the property gradation of self-reinforced polypropylene sheets within a thermo-mechanical compaction process. In: Proceedings of the 6th CIRP International Conference on Intelligent Computation in Manufacturing Engineering (ICME 2008), Naples, Italy, July 23-25 (2008)
7. Sieben, B., Wagner, T., Biermann, D.: Empirical Modeling of Hard Turning of AISI 6150 Steel Using Design and Analysis of Computer Experiments. Production Engineering – Research and Development 4, 115 (2010)
8. Wagner, T., Bröcker, C., Saba, N., Biermann, D., Matzenmiller, A., Steinhoff, K.: Analysis of a Thermomechanically Coupled Forming Process Using Enhanced Design and Analysis of Computer Experiments. In: Digital Proceedings of the International Conference on Design and Analysis of Computer Experiments (ENBIS EMSE), St-Etienne, France, July 01-03 (2009)

9. Gausemeier, J., Dumitrescu, R., Kahl, S., Nordsiek, D.: Integrative Development of Product and Production System for Mechatronic Products. Robotics and Computer-Integrated Manufacturing 27(4) (2011)
10. Fallböhmer, M.: Generieren alternativer Technologieketten in frühen Phasen der Produktentwicklung. Dissertation, RWTH Aachen University, Shaker Verlag (2000)
11. Ashby, M.F.: Materials Selection in Mechanical Design – Das Original mit Übersetzungshilfen. Elsevier, Munich (2007)
12. Saaty, T.L.: The analytic hierarchy process – planning, priority setting, resource allocation. McGraw-Hill, New York (1980)

As-Is Analysis and Efficient 3D Layout Planning during the Conversion of Old and Design of New Production Facilities by Means of 3D-CAD

Björn Schuster

Business Development, N+P Informationssysteme GmbH
schuster@nupis.de

Abstract. Time is money in the area of production, especially when components or assemblies are to edit, assemble or to pack in large numbers. In many cases, an advanced automation is needed to provide everything necessary and to achieve throughput. New planning, simulation and visualization technologies improve the process of factory planning.

Keywords: Digital factory, Factory Design, 3D layout planning.

1 Introduction

The quality of factory planning today is an important factor to ensure the competitive position of factories. Global competition calls for the best possible planning and a permanent optimisation of the production processes, which is why the constant improvement of models, methods and tools for factory planning and operation is very important. The challenge facing many factory planners in connection with this development is to plan and implement the conversion of old and design of new production facilities as efficiently as possible.

2 Digital Factory Planning Using 3D-CAD Software

The majority of CAD data that is relevant for a factory today is available in 2D drawings and can only be transferred to a 3D format in stages and for specific projects on account of the effort involved and the generally existing departure from reality. There are a large number of software tools that allow virtual reality to be used to visualise the manufacturing planning of a factory. Virtual reality has become established in product development under the catchword CAD (Computer Aided Drafting) and is used in factory planning under the concept of a 'digital factory'. 'Digital factory' therefore means a virtual two- or three-dimensional representation of the factory to carry out real planning and a test-driven simulation, in particular to visualise a material flow analysis.

The use of 3D-CAD software accelerates the planning process since the planner can develop and communicate an efficient factory layout using a digital model. The

G.L. Kovács and D. Kochan (Eds.): NEW PROLAMAT 2013, IFIP AICT 411, pp. 423–425, 2013.

subsequent simulation of a factory is understood as meaning the emulation of a factory system, with all of the associated dynamic processes, in an experimental model. The goal is to get as close as possible to reality so that know-how obtained from the simulation can be transferred to reality, in other words the real factory (see Bracht, Geckler, Wenzel 2011, p. 318).

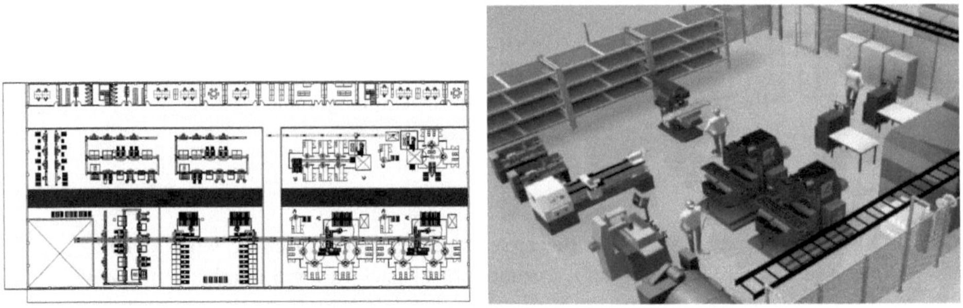

(a) (b)

Fig. 1. From 2D layout planning to 3D layout planning with the Autodesk® Factory Design Suite (Autodesk 2012)

Autodesk® Factory Design Suite is a comprehensive software package that has been well received in the industry and which was written especially for the virtual development of layouts for production lines and complete production facilities (see Figure). This allows the complex process from planning the building, through the actual installation of the production facilities, right down to planning the progress of the building work and transport routes, to be simulated. The suite extends the operational procedures in AutoCAD® and Autodesk® Inventor® with interoperable procedures that are only available in the suite as well as factory-specific components and content. It therefore improves efficiency, accuracy and communication in factory planning projects as well as flexible reactions to altered business requirements.

The Autodesk® Factory Design Suite accelerates factory planning through extensive special commands and a specific environment for the rapid development of factory layouts. Any model can be dimensioned and edited as necessary with the aid of intelligent plant objects such as walls, doors and prop grids. By integrating scatter plot scans, the as-is condition of a plant (hall planning) can be recorded in 3D and then converted or adapted to the virtual environment. Complex conversion work or retrofits can hence be carried out on a 3D layout with the optimum efficiency, quality and flexibility. Analyses of material flows, machinery utilisation and energy consumption are possible in the early engineering and design phases. Non-value-creating costs that are incurred through superfluous transport routes within the factory can be determined on the basis of several simulation scenarios and optimised directly in the 3D layout.

What makes the suite special is that a variety of plant, building and machinery manufacturer data can be managed and compiled in one model. The functions for a digital check and evaluation of the equipment as well as for collision detection and distance analysis thus allow a custom-fit construction of the factory geometry before any extensive investments have been made (see Autodesk® 2012).

3 Outlook

The example of suppliers for the automotive industry shows just how urgent it has become to address the topic of a 'digital factory'. The closer a manufacturer and supplier work together, the greater the pressure to resort to a comprehensive IT solution such as the Autodesk® Factory Design suite. Digital planning, 3D-CAD models, simulation, the compilation of data in a factory model and standardised data exchange are important requirements for this development. Tomorrow's 'digital factory' will also call for an even greater exchange of process data, a topic that is still new for many companies outside the core zone of automobile production, but which is already possible today with the Autodesk® Factory Design suite (see Autodesk® 2011).

The topic of the 'digital factory' as well as the associated AS-IS analysis and efficient 3D layout planning during the conversion of old and design of new production facilities by means of 3D-CAD will also have to be dealt with in different forms by small and medium-sized enterprises. N+P links individual software products and offers an overall, customer-optimised IT solution for factory development and planning, thus creating added value, efficiency and a cutting edge for the customer.

4 Conclusions

Optimize your factory layout before it becomes real. Autodesk ® Factory Design Suite helps you to design and communicate the most efficient manufacturing layout by creating a digital factory model. Evaluate quickly multiple "what-if" scenarios to determine the best solution before any equipment is installed. Factory-specific visualization and analysis tools are helpful by increasing innovations, design efficiency and communication.

References

1. Autodesk®, Autodesk Factory Design Suite 2013: new orders through innovation, p. 14 (2012)
2. Autodesk®. Digital factory – a topics for SMEs? (2011), `http://usa.autodesk.com/adsk/servlet/pc/item?siteID=403786&id=15963402` (August 28, 2013)
3. Bracht, U., Geckler, D., Wenzel, S.: Digital factory: methods and practical examples, pp. 318–320. Springer, Heidelberg (2011)

Author Index

Albert, David 185
Anacker, H. 296

Barth, Andreas 24
Bauer, Frank 391
Bauer, Joachim 364
Bauernhansl, Thomas 167
Bielawny, Dirk 296
Bondar, Sergej 97
Borzykh, Mikhail 320

Chirici, Lapo 61

Damerow, U. 320
Dani, Ines 181
Dusalina, Elmira Kh. 343

Emmer, Christian 286
Erčević, Bojan M. 190
Erčević, Marko M. 190
Erler, Martin 401
Eßers, Marius 401

Filos, Erastos 1
Fischer, Andreas 112, 167
Frank, U. 296
Fritz, Joe 254
Fröhlich, Arnulf 286

Gausemeier, Jürgen 391, 413
Gerth, Nancy 124
Greifzu, Moritz 181
Gruber, Franz E. 36
Gyurecz, György 230

Handschuh, Sebastian 41
Hardwick, Martin 254
Hayka, Haygazun 73
Hedlind, Mikael 254
Henke, Christian 320
Hoeren, Karlheinz P.J. 138
Holtzhausen, Stefan 221
Homberg, W. 320
Hribernik, Karl 85
Hund, Ralf 124

zur Jacobsmühlen, Joschka 153
Jurke, Benjamin 391

Kabitzsch, Klaus 376
Kager, Robert 185
Katzenbach, Alfred 41
Kimura, Fumihiko 11, 330
Kleszczynski, Stefan 153
Kovács, George L. 239
Kretschmer, Ralf 310
Kubisch, Frank 185

Leyens, Christoph 181, 185
Loffredo, David 254
von Lukowicz, Marian 181

Majstorović, Vidosav D. 190
Malek, Roland 124
Matsuda, Michiko 330
Melnikov, Andrey 354
Müller, Bernhard 124
Müller-Lohmeier, Klaus 108

Nestler, Andreas 401
Nowotny, Steffen 185

Paniti, Imre 239
Petersen, Marcus 413
Potjewijd, Leo 97
Pruschek, Peter 391

Rehage, Gerald 391
Renner, Gábor 230
Richter, Arndt 53
Roch, Aljoscha 181
Rommel, Steve 112, 167
Rude, Stefan 31
Rulhoff, Stefan 310

Schichtel, Marcus 273
Schöne, Christine 221
Schreiber, Andreas 31
Schuster, Björn 423
Schwenke, Clemens 376
Sehrt, Jan T. 153
Sembdner, Philipp 221
Shakhmametova, Gyuzel R. 343

Šibalija, Tatjana V. 190
Stark, Rainer 73
Stelzer, Ralph 221
Stepien, Lukas 181
Stjepandic, Josip 97, 286, 310

Thieme, Sebastian 185
Thoben, Klaus-Dieter 85
Trächtler, Ansgar 320
Trippner, Dietmar 31

Vettermann, Steven 41
Vokhmintsev, Alexander 354

Wagner, Thomas 376
Wang, Kesheng 61, 205
Wilson, Heike 369
Witt, Gerd 138, 153
Wolter, Lars 73
Wuest, Thorsten 85

Yu, Quan 205
Yusupova, Nafisa I. 343

Zaepernick, Nikolai 176
Zettl, Reik 369

MIX
Papier aus verantwortungsvollen Quellen
Paper from responsible sources
FSC® C105338

If you have any concerns about our products,
you can contact us on
ProductSafety@springernature.com

In case Publisher is established outside the EU,
the EU authorized representative is:
Springer Nature Customer Service Center GmbH
Europaplatz 3, 69115 Heidelberg, Germany

Printed by Libri Plureos GmbH
in Hamburg, Germany